Pharmaceutical production

An engineering guide

Pharmaceutical production

An engineering guide

Edited by Bill Bennett and Graham Cole

INSTITUTION OF CHEMICAL ENGINEERS

Published by
Institution of Chemical Engineers (IChemE)
Davis Building
165–189 Railway Terrace
Rugby, Warwickshire CV21 3HQ, UK

IChemE is a Registered Charity

Offices in Rugby (UK), London (UK) and Melbourne (Australia)

© 2003 Institution of Chemical Engineers

ISBN 978 0 85295 519 2

Typeset by Techset Composition Limited, Salisbury, UK

Printed and bound by CPI Antony Rowe, Eastbourne

Preface

The pharmaceutical industry aims to produce safe and effective medicines with efficiency and profitability. In order to achieve these aims, qualified personnel from many scientific and commercial disciplines are needed. The industry needs specialists with qualifications in biological, chemical, engineering and pharmaceutical sciences, but there is also a requirement for a wider knowledge of the integral parts of an innovative manufacturing company including research, development, manufacturing, distribution, marketing and sales. Chapter 1 sets the scene by introducing the essential stages, from the synthesis of a new chemical entity through to its development into a licensed medicine.

Further education and advanced training for staff in the industry is needed through in-house or external courses. However, there is a distinct lack of detailed texts written by industrial experts. This book overcomes this deficiency in the area of pharmaceutical engineering and provides detailed information in all principal areas relevant to the manufacture of medicines. It will be a useful reference book for information on topics selected from the vast range of material covered in Chapters 2 to 11. Comprehensive coverage of each major topic, written by experts, provides valuable information for both newcomers and experienced personnel working in the pharmaceutical industry.

Abbreviations and acronyms proliferate throughout the modern world and the pharmaceutical industry has its share. Fortunately, the editors have provided a list of acronyms and a glossary of terms most commonly used in the industry.

The book is divided into ten main chapters, each covering specialist areas with their principal sub-sections clearly set out in the comprehensive list of contents at the beginning of the book. This feature will be very useful for those who need rapid access to detailed information in a specific area.

Chapters 2 to 10 cover all the important aspects of the production of licensed medicines, as indicated in the following précis.

Chapters 5 and 6 cover in detail primary and secondary production from the preparation of bulk bioactive substance by chemical synthesis, biotechnology and extraction from natural products, through to modern packaging technologies required for the finished medicine. Chapter 8 deals with the design of utilities and services, as well as the associated areas of cleaning and maintenance. The design of facilities is continued in Chapter 9 which covers the planning, furnishing and provision of services in laboratories, whereas the special requirements for process development and pilot plant are presented in Chapter 10.

Having provided an outline of the chapters dealing with production, we can turn towards the beginning of the book for coverage of regulatory matters and quality assurance. Chapter 2 is an outline of the main stages in the approval process, post-marketing evaluation and the European and US perspectives.

The concepts and practices embodied in Good Manufacturing Practice are covered concisely in Chapter 3 with special reference to engineering aspects of pharmaceutical production, whereas validation and safety issues are presented in great detail in Chapters 4 and 7.

Finally, in Chapter 11, the special requirements for the development and manufacture of modern bio-pharmaceutical products are dealt with in great detail with reference to small scale and pilot facilities.

After six years working in research and development in the pharmaceutical industry, the rest of my career has been in academic pharmacy. Close contact with the industry has been maintained through education, training, research, consultancy and involvement with the design, delivery, assessment and external examinership of postgraduate diploma and MSc courses for advanced training of personnel in the industry. Such courses by universities or independent consultants provide course material of a high standard, but this should be supplemented by texts written by experts working in the industry. The Engineering Guide to Pharmaceutical Production provides an authoritative and detailed treatment of all major aspects related to the manufacture of medicines.

Geoff Rowley

Professor of Pharmaceutics,
Institute of Pharmacy and Chemistry,
University of Sunderland

List of acronyms

The following is a list of acronyms used in this book. It is followed by a glossary of the more important validation terms.

ADR	Adverse Drug Reaction
AGMP	Automated Good Manufacturing Practice
AGV	Automated Guided Vehicles
AHU	Air Handling Unit
ALARP	As Low As Reasonably Practicable
ANDA	Abbreviated New Drug Application
ANSI	American National Standards Institute
API	Active Pharmaceutical Ingredient
ASME	American Society of Mechanical Engineers
BATNEEC	Best Available Techniques Not Entailing Excessive Costs
BL1	Biosafety Level 1
BL2	Biosafety Level 2
BL3	Biosafety Level 3
BL4	Biosafety Level 4
BMR	Batch Manufacturing Record
BMS	Building Management System
BOD	Biological Oxygen Demand
BP	British Pharmacopeia
BPC	Bulk Pharmaceutical Chemical
BPEO	Best Practicable Environmental Option
BS	British Standard
BSI	British Standards Institution
cAGMP	Current Automated Good Manufacturing Practice
CAMMS	Computer Aided Maintenance Management System
CCTV	Closed Circuit Television
CDER	Centre for Drug Evaluation and Research
CDM	Construction (Design and Management) regulations

CFC	Chlorofluorocarbons
CFR	Code of Federal Regulations
CFU	Colony Forming Unit
cGCP	Current Good Clinical Practice
cGLP	Current Good Laboratory Practice
cGMP	Current Good Manufacturing Practice
CHAZOP	Computer HAZOP
CHIP	Chemical Hazard Information and Packaging regulations
CIMAH	Control of Industrial Major Accident Hazards regulations
CIP	Clean In Place
CMH	Continuous Motion Horizontal
COD	Chemical Oxygen Demand
COMAH	Control Of Major Accident Hazards regulations
COSHH	Control Of Substances Hazardous to Health
CPMP	Committee on Proprietary Medicinal Products
CPU	Central Processing Unit
CSS	Continuous Sterilization System
CV	Curriculum Vitae
DAF	Dissolved Air Flotation
DIN	Deutsches Institüt für Normung
DMF	Drug Master File
DNA	Deoxyribonucleic Acid
DOP	Dioctyl Phthalate
DQ	Design Qualification
EC	European Community
EEC	European Economic Community
EMEA	European Agency for the Evaluation of Medical Products
EPA	Environmental Protection Agency
EPDM	Ethyl Propylene Diene Terapolymer
ERP	Enterprise Resource Planning
EU	European Union
FAT	Facility Acceptance Testing
FBD	Fluidized Bed Dryer
FDA	Food and Drug Administration
FMEA	Failure Mode Effects Analysis
FS	Functional Specification
GAMP	Good Automated Manufacturing Practice
GC	Gas Chromatograph
GCP	Good Clinical Practice
GLP	Good Laboratory Practice

GLSP	Good Large Scale Practice
GMP	Good Manufacturing Practice
GRP	Glass Reinforced Plastic
GSL	General Sales List
HAZOP	Hazard and Operability Study
HEPA	High Efficiency Particulate Arrestor
HFC	Hydrofluorocarbons
HIC	Hydrophobic Interaction Chromatography
HMAIP	Her Majesty's Inspectorate of Air Pollution (now defunct)
HMSO	Her Majesty's Stationery Office
HPLC	High Pressure Liquid Chromatograph
HS	Hazard Study
HSE	Health and Safety Executive
HSL	HAZOP Study Leader
HVAC	Heating Ventilation and Air Conditioning
IBC	Intermediate Bulk Container
ICH	International Conference on Harmonization
IDF	International Diary Foundation
IEC	Ion Exchange Chromatography
IEEE	Institute of Electrical and Electronics Engineers
IMV	Intermittent Motion Vehicle
IND	Investigational New Drug Application
I/O	Inputs and Outputs
IPA	Iso Propyl Alcohol
IPC	Integrated Pollution Control
IQ	Installation Qualification
ISO	International Standards Organization
ISPE	International Society for Pharmaceutical Engineering
LAAPC	Local Authority Air Pollution Control
LAF	Laminar Air Flow
LIMS	Laboratory Information Management System
LTHW	Low Temperature Hot Water
mAb	Monoclonal Antibody
MCA	Medicines Control Agency
MCB	Master Cell Bank
MCC	Motor Control Centre
MEL	Maximum Exposure Limit
MRA	Mutual Recognition Agreement
MRP	Manufacturing Resource Planning
MSDS	Material Safety Data Sheet

NCE	New Chemical Entity
NDA	New Drug Application
NDT	Non-Destructive Testing
NICE	National Institute for Chemical Excellence
NMR	Nuclear Magnetic Resonance
OEL	Occupational Exposure Limits
OES	Occupational Exposure Standards
OQ	Operational Qualification
OSHA	Occupational Safety & Health Administration
OTC	Over The Counter
P	Pharmacy only
PBTB	Polybutylene Teraphthalate
PC	Programmable Controller
PCB	Printed Circuit Board
PDA	Personal Digital Assistants
PEG	Polyethylene Glycol
PFD	Process Flow Diagram
PHA	Preliminary Hazard Assessment
Ph.Eur	European Pharmacopeia
PHS	Puck Handling Station
P&ID	Piping and Instrumentation Diagram
PLA	Product Licence Application
PMI	Positive Material Identification
POM	Prescription Only Medicines
PP	Polypropylene
PPE	Personal Protective Equipment
PQ	Performance Qualification
PSF	Performance Shaping Factors
PTFE	Polytetrafluoroethylene
PV	Process Validation
PVC	Polyvinyl Chloride
PVDF	Polyvinylidene Fluoride
PW	Purified Water
QA	Quality Assurance
QC	Quality Control
QRA	Quantitative Risk Assessment
R&D	Research and Development
RF	Radio Frequency
RH	Relative Humidity
RHS	Rolled Hollow Section

RIDDOR   Reporting of Injuries, Disease and Dangerous Occurrences
            Regulations
RP-HPLC   Reverse Phase High Performance Liquid Chromatography
SCADA   Supervisory Control And Data Acquisition system
SEC   Size Exclusion Chromatography
SHE   Safety, Health and Environment
SIP   Sterilize In Place/Steam In Place
SOP   Standard Operating Procedure
SS   Suspended Solids
THERP   Technique for Human Error Rate Prediction
TOC   Total Organic Carbon
TWA   Time-Weighted Average
UK   United Kingdom
UPVC   Unplasticized Polyvinyl Chloride
URS   User Requirement Specification
USA   United States of America
USP   United States Pharmacopeia
UV   Ultra Violet
VDU   Visual Display Unit
VMP   Validation Master Plan
VOC   Volatile Organic Compound
WCB   Working Cell Bank
WFI   Water for Injection

Glossary

Acceptance criteria   The product specifications and acceptance/rejection criteria, such as acceptable quality level and unacceptable quality level, with an associated sampling plan, that are necessary for making a decision to accept or reject a lot or batch (or any other convenient subgroups of manufactured units).

Action levels   Levels or ranges that may be detrimental to end product quality, signalling a drift from normal operating conditions.

Alert levels   Levels or ranges that signify a drift from normal operating conditions. These ranges are not perceived as being detrimental to end product quality, but corrective action should be taken to ensure that action levels are not obtained.

Audit   An audit is a formal review of a product, manufacturing process, equipment, facility or system for conformance with regulations and quality standards.

Bulk drug substance   Any substance that is represented for use in a drug and that, when used in the manufacturing, processing or packaging of a drug, becomes an active ingredient or a finished dosage form of the drug. The term does not include intermediates used in the synthesis of such substances.

Bulk pharmaceutical chemical   Any substance that is intended for use as a component in a 'Drug Product', or a substance that is repackaged or relabelled for drug use. Such chemicals are usually made by chemical synthesis, by processes involving fermentation, or by recovery from natural (animal, mineral or plant) materials.

Calibration
Comparison of a measurement standard or instrument of known accuracy with another standard or instrument to detect, correlate, report or eliminate by adjustment any variation in the accuracy of the item being compared.

Certification
Documented statement by qualified authorities that a validation event has been done appropriately and that the results are acceptable. Certification is also used to denote the acceptance of the entire manufacturing facility as validated.

Change control
A formal monitoring system by which qualified representatives of appropriate disciplines review proposed or actual changes that might affect validated status and take preventive or corrective action to ensure that the system retains its validated state of control.

Computer validation
The validation of computers has been given a particular focus by the US FDA.

Three documents have been published for agency and industry guidance. In February 1983, the agency published the Guide to Inspection of Computerized Systems in Drug Processing; in April 1987, the Technical Reference in Software Development Activities was published; on 16 April, 1987, the agency published Compliance Policy Guide 7132 in Computerized Drug Processing: Source Codes for Process Control Application Programmes.

In the inspection guide, attention is called to both hardware and software; some key points being the quality of the location of the hardware unit as to extremes of environment, distances between CPU and peripheral devices, and proximity of input devices to the process being controlled; quality of signal conversion, for example, a signal converter may be sending inappropriate signals to a CPU; the need to systematically calibrate and check for accuracy of I/O devices; the inappropriateness and compatibility within the distributed system of command overrides, for example, can an override in one computer controlled process inadvertently alter the cycle of another process within the distributed system? Maintenance procedures are another matter of interest to the agency during an inspection. Other matters of concern are methods by which unauthorized programme changes are prevented, as inadvertent erasures, as well as methods of physical security.

Hardware validation should include verification that the programme matches the assigned operational function. For example, the recording of multiple lot numbers of each component may not be within the programme, thus second or third lot numbers of one component may not be recorded. The hardware validation should also include worse case conditions; for example, the maximum number of alphanumeric code spaces should be long enough to accommodate the longest lot numbering system to be encountered. Software validation must be thoroughly documented — they should include the testing protocol, results, and persons responsible for reviewing and approving the validation. The FDA regards source code, i.e., the human readable form of the programme written in its original programming language, and its supporting documentation for application programmes used in any drug process control, to be part of the master production and control records within the meaning of 21CFR parts 210, 211 (Current Good Manufacturing Practice Regulations).

As part of all validation efforts, conditions for revalidations are a requirement.

Concurrent validation Establishing documented evidence that the process being implemented can consistently produce a product meeting its predetermined specifications and quality attributes. This phase of validation activities typically involves careful monitoring/recording of the process parameters and extensive sampling/testing of the in-process and finished product during the initial implementation of the process.

Construction qualification

The documented evaluation of the construction or assembly of a piece of equipment, process or system to assure that construction or assembly agrees with the approved specifications, applicable codes and regulations, and good engineering practices. The conclusion of the evaluation should decidedly state that the equipment, process or system was or was not constructed in conformance with the specifications.

Critical process variables

Those process variables that are deemed important to the quality of the product being produced.

Design review

A 'design review' is performed by a group of specialists (such as an Architect, a Quality Assurance Scientist, a HVAC Engineer, a Process Engineer, a Validation Specialist, a Civil Engineer and a Regulatory Affairs Specialist) to review engineering documents to ensure that the engineering design complies with the cGMPs for the facility. The thoroughness of the design review depends upon whether the engineering project is a feasibility study, a conceptual design, preliminary engineering, or detailed engineering. Minutes of all meetings for design review will be sent to team members and the client to show the compliance of the design to cGMPs.

Drug

Substances recognized in the official USP; substances intended for use in the diagnosis, cure, mitigation or prevention of disease in man or other animals; substances (other than food) intended to affect the structure or any function of the body of man or other animals; substances intended for use as a component of any substances specified above but does not include devices or their components, parts or accessories.

Dynamic attributes

Dynamic attributes are classified into functional, operational and quality attributes, which are identified, monitored, inspected and controlled during actual operation of the system.

Edge of failure A control or operating parameter value that, if exceeded, may have adverse effects on the state of control of the process and/or on the quality of the product.

Facilities Facilities are areas, rooms, spaces, such as receiving/ shipping, quarantine, rejected materials, approved materials warehouse, staging areas, process areas, etc.

Functional attributes Functional attributes are such criteria as controls, instruments, interlocks, indicators, monitors, etc., that operate properly, are pointing in the correct direction, and valves that allow flow in the correct sequence.

Good manufacturing practice (GMP) The minimum requirements by law for the manufacture, processing, packaging, holding or distribution of a material as established in Title 21 of the Code of Federal Regulations.

Installation qualification protocol An installation qualification protocol (IQ) contains the documented plans and details of procedures that are intended to verify specific static attributes of a facility, utility/system, or process equipment. Installation qualification (IQ), when executed, is also a documented verification that all key aspects of the installation adhere to the approved design intentions and that the manufacturer's recommendations are suitably considered.

Intermediate (drug/ chemical) Any substance, whether isolated or not, which is produced by chemical, physical, or biological action at some stage in the production of a bulk pharmaceutical chemical and subsequently used at another stage in the production of that chemical.

Life-cycle The time-frame from early stages of development until commercial use of the product or process is discontinued.

Master plan
The purpose of a master plan is to demonstrate a company's intent to comply with cGMPs and itemizes the elements that will be completed between the design of engineering and plant start-up. A typical master plan may contain, but is not limited to, the following elements: approvals, introduction, scope, glossary of terms, preliminary drawings/facility design, process description, list of utilities, process equipment list, list of protocols, list of SOPs, equipment matrices, validation schedule, protocol summaries, recommended tests, calibration, training, manpower estimate, key personnel (organization chart and resumes), protocol examples, SOP examples.

Medical devices
A medical device is defined in the Federal Food Drug and Cosmetic Act Section 201(h) as:
An instrument, apparatus, implement or contrivance intended for use in diagnosis, cure, mitigation, prevention or other treatment of disease in man or other animals, or intended to alter a bodily function or structure of man or other animal.
This is the definition used in the code of Federal Regulations 21 parts 800 to 1299. Medical Devices.

Operational attributes
Operational attributes are such criteria as a utility/system's capability to operate at rated ranges, capacities, intensities, such as: revolutions per minute, kg per square cm, temperature range, kg of steam per second, etc.

Operation qualification protocol
An operation qualification (OQ) contains the plan and details of procedures to verify specific dynamic attributes of a utility/system or process equipment throughout its operated range, including worse case conditions. Operation qualification (OQ) when executed is documented verification that the system or subsystem performs as intended throughout all anticipated operating ranges.

Operating range
A range of values for a given process parameter that lie at or below a specified maximum operating value and/or at or above a specified minimum operating

value, and are specified on the production worksheet or the standard operating instruction.

Overkill sterilization process   A process which is sufficient to provide at least a 12 log reduction of microorganisms having a minimum D-Value of 1 minute.

Process parameters   Process parameters are the properties or features that can be assigned values that are used as control levels or operating limits. Process parameters assure the product meets the desired specifications and quality. Examples might be: pressure at 5.2 psig, temperature at $37°C \pm 0.5°C$, flow rate at $10 \pm 1.0\,1\,min^{-1}$, pH at 7.0 ± 0.2.

Process variables   Process variables are the properties or features of a process which are not controlled or which change in time or by demand; process variables do not change product specifications or quality.

Process validation   Establishing documented evidence that provides a high degree of assurance that a specific process will consistently produce a product meeting its pre-determined specifications and quality attributes.

Process validation protocol   Process validation protocol (PV) is a documented plan, and detailed procedures to verify specific capabilities of a process equipment/system through the use of simulation material, such as the use of a nutrient broth in the validation of an aseptic filling process.

Product validation   A product is considered validated after completion of three successive successful lot size attempts. These validation lots are saleable.

Prospective validation   Validation conducted prior to the distribution of either a new product or a product made under a revised manufacturing process, where the revisions may have affected the product's characteristics, to ensure that the finished product meets all release requirements for functionality and safety.

Protocol   A protocol is defined in this book as a written plan stating how validation will be conducted.

Quality assurance   The activity of providing evidence that all the information necessary to determine that the product is fit for the intended use is gathered, evaluated and approved.

Quality attributes Quality attributes refer to those measurable properties of a utility, system, device, process or product such as resistivity, impurities, particulate matter, microbial and endotoxin limits, chemical constituents and moisture content.

Quality control   The activity of measuring process and product parameters for comparison with specified standards to assure that they are within predetermined limits and, therefore, the product is acceptable for use.

Retrospective validation   Validation of a process for a product already in distribution based upon establishing documented evidence through review/analysis of historical manufacturing and product testing data, to verify that a specific process can consistently produce a product meeting its predetermined specifications and quality attributes. In some cases a product may have been on the market without sufficient pre-market process validation.

Retrospective validation can also be useful to augment initial pre-market prospective validation for new products or changed processes.

Revalidation   Repetition of the validation process or a specific portion of it.

Specifications   Document that defines what something is by quantitatively measured values. Specifications are used to define raw materials, in-process materials, products, equipment and systems.

Standard operating procedure (SOP)   Written procedures followed by trained operators to perform a step, operation, process, compounding or other discrete function in the manufacture or production of a bulk pharmaceutical chemical, biologic, drug or drug product.

State of control
A condition in which all process parameters that can affect performance remain within such ranges that the process performs consistently and as intended.

Static attributes
Static attributes may include conformance to a concept, design, code, practice, material/finish/installation specifications and absence of unauthorized modifications.

Utilities/systems
Utilities/systems are building mechanical equipment and include such things as heating, ventilation and air conditioning (HVAC) systems, process water, product water (purified water, water for injection), clean steam, process air, vacuum, gases, etc. Utilities/systems include electro-mechanical or computer-assisted instruments, controls, monitors, recorders, alarms, displays, interlocks, etc., which are associated with them.

Validation
Establishing documented evidence to provide a high degree of assurance that a specific process will consistently produce a product meeting its predetermined specifications and quality.

Validation programme
The collective activities related to validation.

Validation protocols
Validation protocols are written plans stating how validation will be conducted, including test parameters, product characteristics, production equipment, and decision points on what constitutes acceptable test results. There are protocols for installation qualification, operation qualification, process validation and product validation. When the protocols have been executed it is intended to produce documented evidence that the system has been validated.

Validation scope
The scope identifies what is to be validated. In the instance of the manufacturing plant, this would include the elements that impact critically on the

quality of the product. The elements requiring validation are facilities, utilities/systems, process equipment, process and product.

Worst case   A set of conditions (encompassing upper and lower
processing limits and circumstances including those
within standard operating procedures), which pose the
greatest chance of process or product failure when
compared to ideal conditions. Such conditions do not
necessarily induce product or process failure.

Contents

Introduction

1

Everyone is aware of the potential benefits of medicines and the patient takes them on trust expecting them to be fit for the purpose prescribed by the doctor or agrees with the claims of the manufacturer on the packaging or on advertisements. This book is a general introduction for all those involved in the engineering stages required for the manufacture of the active ingredient (primary manufacture) and its dosage forms (secondary manufacture).

All staff working in or for the pharmaceutical industry have a great responsibility to ensure that the patient's trust is justified. Medicines made wrongly can have a great potential for harm.

Most of the significant developments of medicines, as we know them, have occurred in the last 70 years.

From ancient times, by a process of trial and error, man has used plants and other substances to produce certain pharmacological effects. The best example is probably alcohol, which has been developed by every culture.

Alcohol has a number of well-known effects depending on the dosage used. In small amounts it causes flushing of the skin (vasodilatation), larger quantities produce a feeling of well being, and if the dose is further increased, loss of inhibition occurs leading to signs of aggression. Beyond aggression, somnolence occurs and indeed coma can supervene as the central nervous system becomes progressively depressed. This well-known continuum of effects illustrates very neatly the effect of increasing dosage over a period of time with a substance that is metabolized simply at a fairly constant rate. It further illustrates that where small quantities of a drug are useful, larger quantities are not necessarily better — in fact they are usually harmful.

Using the trial and error technique, the good or harmful properties of various other materials were also discovered, for example, coca leaves — cocaine, or poppy juice — opium, which contains morphine.

Today the pharmaceutical industry is faced with escalating research costs to develop new products. Once an active product has been discovered and proven

1

to be medically effective the manufacturer has to produce the active ingredient and process it into the most suitable dosage form.

Speed to market is essential so that the manufacturer can maximize profits whilst the product has patent protection. Companies are now concentrating products at specific sites to reduce the time-scale from discovery to use, to give economics of scale and longer campaign runs.

The manufacture of the active ingredient is known as primary production (see Chapter 5). Well-known examples of synthetic processes are shown in Figures 1.1 and 1.2 (see pages 3 and 4). The manufacturing process for methylprednisolone (a steroid) is complex (see Figure 1.1), but it is relatively simple for phenylbutazone (see Figure 1.2). The processing to the final dosage form such as tablet, capsule (see Figure 1.3 on page 4), or injection, is known as secondary production (see Chapter 6).

Bringing a mainstream drug to market can cost in excess of £200 million (300m US dollars). This involves research, development, manufacturing, distribution, marketing and sales. The time cycle from discovery to launch takes many years and will probably not be less than four years for a New Chemical Entity (NCE). Any reduction in this time-frame improves the company's profitability and generates income.

Many companies conduct the early studies on NCE's for safety, toxicity and blood levels using capsules. This is due to a very small amount of NCE being available and the ease of preparing the dosage form without loss of material. Only when larger quantities become available is a dosage form formulated as a tablet or other form. The product design process must take into account the demands of regulatory approval (manufacturing licences, validation), and variation in demand requiring flexibility of operation. The treatment of hay fever is a good example of a product only being in peak demand in spring and early summer.

All companies will attempt to formulate oral solid dosage forms, such as a tablet or capsule, as this is the most convenient form for the patient to take and the easiest product to manufacture. An estimated 80–85 percent of the world's medicines are produced in this form. Not all products are effective from the oral route and other dosage forms such as injections, inhalation products, transdermals or suppositories are required.

The discovery and isolation of a new drug substance and its development into a pharmaceutical dosage form is a costly and highly complex task involving many scientific disciplines. Figures 1.4 and 1.5 illustrate many of the steps involved.

Figure 1.5 illustrates the various departments and disciplines that need to co-operate once it has been decided that the product will be marketed. This

2

figure assumes that facilities are available for manufacturing the active ingredient (primary manufacture).

Failures by manufacturers led to the establishment of regulatory authorities initially in the USA, then in the UK and more recently in Europe.

In 1938 in America sulphonamide elixir was contaminated by diethylene glycol resulting in a large number of deaths. This led to the Food, Drug and

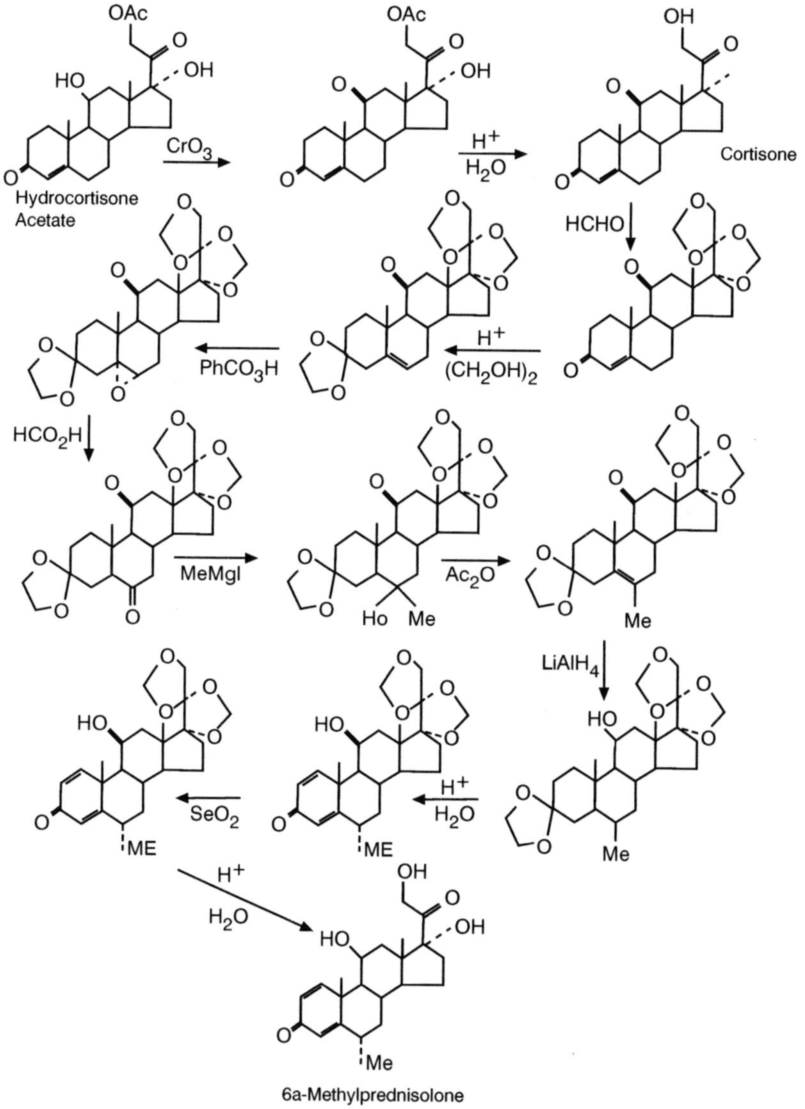

Figure 1.1 Synthetic route for 6a methylprednisolone

3

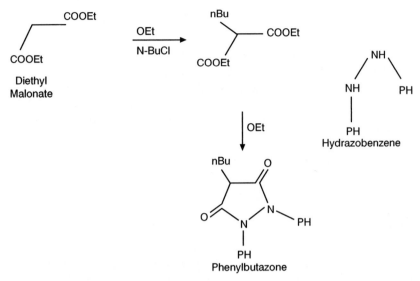

Figure 1.2 Synthetic route for phenylbutazone

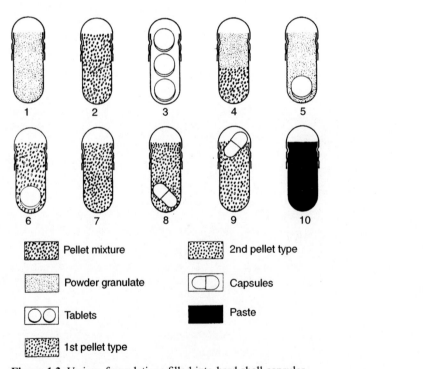

Figure 1.3 Various formulations filled into hard shell capsules

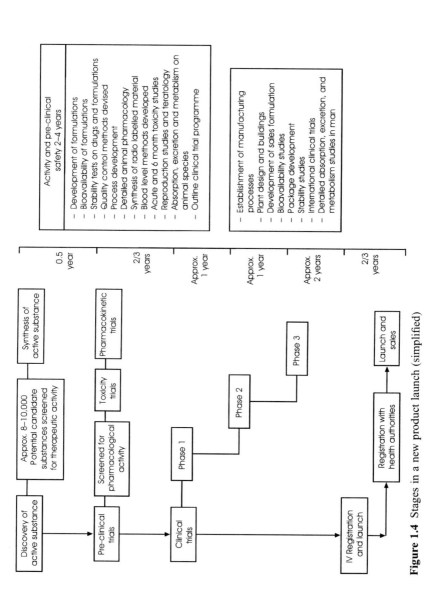

Figure 1.4 Stages in a new product launch (simplified)

Cosmetic Act coming into force in the USA, followed by the establishment of the Food and Drug Administration (FDA).

In 1962, there was the much publicized Thalidomide tragedy leading to the tightening up of the testing of drugs prior to marketing, and eventually to the Medicines Act 1968 in the UK. The Medicines Control Agency (MCA) was established to police the industry and there is now also the European Medicines Evaluation Agency (EMEA) and the National Institute for Chemical Excellence (NICE).

Such legislation (see Chapter 2) has had a considerable impact on the design, construction, operation and on-going maintenance of pharmaceutical production facilities.

The FDA, the MCA and European Regulatory Authorities have all issued codes of Good Manufacturing Practice, providing basic ground rules to ensure adequate patient protection from hazards associated with the poor design of manufacturing processes. Chapter 3 provides background knowledge on the regulatory framework and constraints on the manufacturer.

Validation has been introduced in recent years. This was defined by the FDA as the act of establishing documentary evidence to provide a high degree of assurance that a specific process will consistently produce a product meeting its pre-determined specifications and quality attributes. Chapter 4 provides details of the documentation required including concepts such as the User Requirement Specification (URS), Validation Master Plan (VMP), Design Qualification (DQ), Installation Qualification (IQ), Operational Qualification (OQ) and Performance Qualification (PQ).

It is important that the designer understands these requirements because it is far easier to collect validation documentation throughout the design process rather than to attempt to do so post-design, often known as retrospective validation.

Chapter 5 deals with primary production, or manufacture of the active ingredient. For many years designers considered this to be no different to the manufacture of any other chemical, but codes of good manufacturing practice and validation now apply. Reactions and other key unit operations are discussed with ideas for layouts to satisfy good manufacturing practice and other regulator requirements.

Chapter 6 is a comprehensive review of secondary production, turning the active ingredient into the dosage form.

Chapter 7 covering safety, health and environment explains how risks to these are managed in the pharmaceutical industry and how effective process design can eliminate or control them.

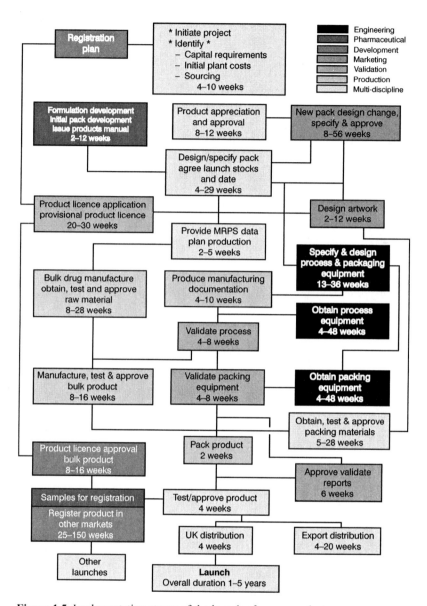

Figure 1.5 Implementation stages of the launch of a new product

The reader may ask why Chapter 8 has been included as process utilities and services are common throughout all industry. This chapter concentrates on aspects that are particularly relevant to the pharmaceutical industry. Regulatory authority inspectors, when inspecting plants, spend a lot of time looking at water supplies, compressed air systems, air conditioning and cleaning systems which are all in the designer's control.

Much of the book is about the production of the active ingredient and dosage forms. However, Quality Assurance departments have an important part to play in ensuring medicines are of an appropriate quality. In fact, regulatory authorities demand that a Qualified Person (usually from the QA department) is legally responsible for the release for sale of the manufactured product. Chapter 9 focuses on the design of quality control laboratories which form an important part of the quality assurance process.

In a similar way, process development facilities and pilot plants are an integral part of the development of the manufacturing process for the active ingredient and its dosage form, particularly in the preparation of clinical trials. Chapter 10 gives ideas on the design, construction, commissioning and validation of these facilities.

Chapter 11 is a review of the special requirements of Bio-pharmaceutical products particularly for pilot-scale manufacture of these products.

Regulatory aspects

2

JOHN WELBOURN

2.1 Introduction

The pharmaceutical industry is distinctive from many other industries in the amount of attention paid to it by regulatory authorities. In all industries there are regulations relating to safety and the environment, rules and directions for services and recommendations from a wide range of authorities about design and maintenance of facilities. Engineers in the pharmaceutical industry also have to cope with a myriad of medicines regulations throughout the design and engineering process. Whilst it is not essential to have a detailed knowledge of all aspects of the regulations of medicinal products, facilities and processes, engineers should at least recognize that many of these regulations are restrictive or impose additional requirements. When products and processes have been registered with the regulatory authorities, it can be difficult and time-consuming to alter these specifications. This makes it important to be aware of the registered processes and quality control systems throughout the design.

In the UK, medicines are regulated by the Medicines Control Agency (MCA). The MCA was launched as an Executive Agency of the UK Department of Health in July 1991. The MCA's primary objective is to safeguard public health by ensuring that all medicines on the UK market meet appropriate standards of safety, quality and efficacy. Safety aspects cover potential or actual harmful effects; quality relates to development and manufacture; and efficacy is a measure of the beneficial effect of the medicine on patients. The MCA operates a system of licensing before the marketing of medicines, monitoring of medicines and acting on safety concerns after they have been placed on the market, and checking standards of pharmaceutical manufacture and wholesaling. The MCA is responsible for enforcing these requirements. It represents UK pharmaceutical regulatory interests internationally; publishing quality standards for drug substances through the British Pharmacopoeia.

A medicinal product (also known as a drug product) is any substance or article that is administered for a medicinal purpose. This includes treating or preventing disease, diagnosing disease, contraception, anaesthesia and preventing or interfering with a normal physiological function.

In all cases, the product must be fit for the purpose for which it is intended. From the consumer's point of view this could be a single tablet, but each tablet cannot be tested to ensure it is of the correct quality as many of the tests needed to demonstrate this are destructive. Manufacturers have to assure quality by ensuring all aspects of the process are consistent every time.

As a result of well-publicized failures, resulting in patients deaths, regulations have become more and more stringent. Regulation is now achieved through the licensing of both the product and the facilities in which it is manufactured and the monitoring of medicines after a licence has been granted. The way medicinal products are supplied depends upon the nature and the historical experience of the product. Products may be Prescription Only Medicines (POM), Pharmacy only (P) or General Sales List (GSL). This categorization provides an important element in the control of medicinal products.

In the UK, the Medicines Act 1968 and the Poisons Act 1972, together with the Misuse of Drugs Act 1971, regulate all retail and wholesale dealings in medicines and poisons. Certain non-medicinal poisons and chemicals are also subject to the labelling requirements of the Chemicals Hazard Information and Packaging Regulations (CHIP).

It is important to appreciate at the outset that the Medicines Act 1968 applies only to substances where they are used as medicinal products or as ingredients in medicinal products.

2.2 Key stages in drug approval process

To obtain the evidence needed to show whether a drug is safe and effective, a pharmaceutical company will normally embark on a relatively lengthy process of drug evaluation and testing. Typically this will begin with studies of the drug in animals (preclinical studies) and then in humans (clinical studies). The purpose of preclinical testing is two-fold. Firstly, it is used as an aid to assessing whether initial human studies will be acceptably safe, and secondly, such studies are conducted to predict the therapeutic activity of the drug. If the drug looks promising, human clinical studies are proposed. In the USA, for example, this requires the submission of an Investigational New Drug Application (IND) to the regulatory authority, which in this case would be the Food and Drug Administration (FDA).

The IND must contain sufficient information about the investigational drug to show it is reasonably safe to begin human testing. An IND for a drug not previously tested in human subjects will normally include the results of preclinical studies, the protocols for the planned human tests, and information on the composition, source and method of manufacture of the drug.

Provided the IND application is successful, drug testing in humans then proceeds progressively through three phases (called Phase 1, 2 and 3).

- Phase 1 includes the initial introduction of an investigational drug into humans and consists of short-term studies in a small number of healthy subjects, or patients with the target disease, to determine the metabolism and basic pharmacological and toxicological properties of the drug, and if possible, to obtain preliminary evidence of effectiveness.
- Phase 2 consists of larger, more detailed studies; usually including the first controlled clinical studies intended to assess the effectiveness of the drug and to determine the common short-term side effects and risks of the drug.
- Phase 3 studies are expanded controlled and uncontrolled trials. They are performed after preliminary evidence of effectiveness has been established and are designed to gather the additional information necessary to evaluate the overall benefit-risk relationship of the drug and to provide an adequate basis for professional labelling.

If the results appear to be favourable at the end of clinical trials and the company decides to market the new product, they must first submit an application to do this. In the USA the company must submit the results of the investigational studies to the FDA in the form of a New Drug Application (NDA). The NDA must contain:

- full reports of the studies (both preclinical and clinical) to demonstrate the safety and effectiveness of the drug;
- a description of the components, chemical formulation, and manufacturing controls;
- samples of the drug itself and of the proposed labelling.

Many companies choose to prepare a Drug Master File (DMF) to support the NDA. A DMF is submitted to the FDA to provide detailed information about facilities, processes or articles used in the manufacturing, processing, packaging and storage of one or more human drugs. In exceptional cases, a DMF may also be used to provide animal or clinical data. A DMF is submitted solely at the discretion of the holder, the information being used in support of the NDA.

The application is reviewed. Typically this includes reviews of product chemistry, labelling, bio-equivalency, clinical data and toxicity. In addition, and of particular relevance to pharmaceutical engineers, the review will also include a pre-approval inspection of the facilities in which the drug is manufactured.

The pre-approval inspection will generally consist of a review of the facilities, procedures, validation (discussed in Chapter 4) and controls associated with formulation development, analytical method development, clinical trial manufacturing, manufacturing (if applicable), quality control laboratories, bulk chemical sources and contract operations. If the application is successful the pharmaceutical company will receive approval to market the product.

A similar (although not identical) situation exists in Europe. For example, in the UK regulation is achieved through a Clinical Trial Certificate, Animal Test Certificate and Product Licence (also in certain circumstances Product Licence of Right and Reviewed Product Licence) for the product and a Manufacturer's Licence, Assembly Only Licence, Special Manufacturer's Licence, Wholesale Licence and Wholesale Import Licence for the Manufacturer/Supplier.

2.3 Example of requirements

An example of the 'regulatory environment' in the UK is summarized in Figure 2.1:

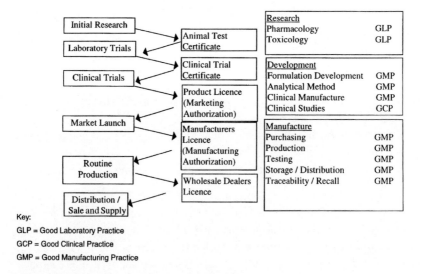

Figure 2.1: The UK regulatory environment

GLP is concerned with the organizational processes and the conditions under which laboratory studies are planned, performed, monitored, reported and recorded. The UK GLP regulations (Statutory Instruments No. 654) came into force in April 1997 and are monitored by the UK GLP Monitoring

Authority, which is part of the MCA. Currently about 150 test facilities are registered under the scheme and are inspected on a two-year cycle.

GCP is 'a standard for the design, conduct, performance, monitoring, auditing, recording, analysis and reporting of clinical trials that provide assurance that the data and reported results are credible and accurate, and that the rights, integrity, and confidentiality of trial subjects are protected' (Definition from the International Conference on Harmonization (ICH) Note for Guidance on Good Clinical Practice (CPMP/ICH/135/95)). In the UK, the GCP Compliance Unit was established within the Inspection and Enforcement Division of the MCA in 1996. GCP inspectors assess compliance with the requirements of GCP guidelines and regulations, which involves conducting on-site inspections at pharmaceutical sponsor companies, contract research organizations' investigational sites and other facilities involved in clinical research.

GMP is 'the part of Quality Assurance (QA) which ensures that products are consistently produced and controlled to the quality standards appropriate to their intended use and as required by marketing authorization or product specification.' (Definition from the EU Guide To Good Manufacturing Practice and Good Distribution Practice). GMP is discussed in more detail in Chapter 3.

2.4 Post-marketing evaluation

2.4.1 Pharmacovigilance

No matter how extensive the pre-clinical work in animals and clinical trials in patients, certain adverse effects may not be detected until a very large number of people have received the new drug product. The conditions under which patients are studied pre-marketing do not necessarily reflect the way the new drug product will be used in hospitals or in general practice. Pharmacovigilance is the process of monitoring medicines as used in everyday practice to:

- identify previously unrecognized (or changes in) patterns of adverse effects;
- assess the risks and benefits of medicines in order to determine what action, if any, is necessary to improve their safe use;
- provide information to users to optimize safe and effective use of medicines;
- monitor the impact of any action taken.

Information from many different sources is used for pharmacovigilance including spontaneous adverse drug reaction (ADR) reporting schemes, clinical and epidemiological studies, world literature, morbidity and mortality databases. In the UK the MCA runs the spontaneous adverse drug reaction reporting scheme (called the Yellow Card Reporting Scheme) which receives reports of suspected drug reactions from doctors, dentists, hospital pharmacists and coroners. The scheme provides an early warning of adverse effects of medicines.

2.4.2 Variations and renewal of marketing authorizations

Drug products may undergo changes over time in relation to production, distribution and use. These will require authorization by the licensing agency. Also, authorizations are normally renewed on a regular period — marketing authorizations are valid for five years in the UK.

2.5 Procedures for authorizing medicinal products in the European Union

In 1995 a new European system for the authorization of medicinal products came into effect, and a new agency was established — the European Medicines Evaluation Agency (EMEA) based in London, UK. Two new registration procedures for human and veterinarian medicinal products have become available. The first system, known as the centralized procedure, is compulsory for medicinal products derived from biotechnology and is available at the request of companies for other innovative new products. Applications are submitted directly to the EMEA who undertake the evaluation and submit their opinion to the European Commission. The European Commission then issue a single market authorization.

The second system, known as the decentralized procedure, applies to the majority of conventional medicinal products and is based upon the principal of mutual recognition of national authorizations. It provides for the extension of the marketing authorization granted by one Member State to one or more other Member States identified by the applicant.

2.6 European and US regulatory perspectives

On the 18 May 1998, the European Union and the USA signed a 'Joint Declaration to the agreement on Mutual Recognition between the EU and the USA'. This agreement lays down the framework for mutual recognition of GMP regulations under the principal of 'equivalence' and the mutual recognition of pre-approval and post-approval inspections.

The agreement covers human medicinal products (prescription and non-prescription drugs, biologicals including vaccines and immunologicals); veterinary pharmaceuticals (prescription and non-prescription drugs premixes and preparations for medicated feeds); active pharmaceutical ingredients and intermediate product, starting materials, bulk pharmaceuticals. The agreement excludes human blood, human plasma, human tissues and organs, veterinary immunologicals, human plasma derivatives, investigational medicinal products, human radiopharmaceuticals and medicinal gases.

Reading list

1. Rules and guidance for pharmaceutical manufacturers and distributors 1997. London. The Stationery Office, 1997. ISBN 0 11 321995 4. (Also known as the 'Orange Guide'). (Incorporating EC Guides to Good Manufacturing Practice and Good Distribution Practice; EC GMP Directives (91/356/EEC & 91/412/EEC); Code of Practice for Qualified Persons and Guidance for Responsible Persons; Standard provisions for manufacturer's licences; Standard provisions for wholesale dealers licences; Guidance on reporting defective medicines).

2. Good Laboratory Practice Regulations 1999 (GLP Regulations); Statutory Instrument 1999/3106; Department of Health, The United Kingdom Good Laboratory Practice Monitoring Authority.

3. Guide to UK GLP Regulations 1999, Feb 2000, Department of Health, The United Kingdom Good Laboratory Practice Monitoring Authority.

4. International Conference on Harmonization (ICH) Note for Guidance on Good Clinical Practice (CPMP/ICH/135/95)

5. Research Governance in the NHS, Guidance on Good Clinical Practice and Clinical Trials in the NHS, Department of Health

6. Royal Pharmaceutical Society of Great Britain; Medicines, Ethics and Practices, A guide for Pharmacists, 18 Edition, July 1997.

7. US Food and Drug Administration, Centre for Drug Evaluation and Research (CDER), Department of Health and Human Services; Code of Federal Regulations 21 CFR (in particular, but not limited to, Parts 10b, 11, 210, 211, 600, 820). Guidance for Industry, including: Guideline For Drug Master Files September 1989; Content and format of Investigational New Drug Applications (INDs) for Phase 1 Studies of Drugs, Including Well-Characterized, Therapeutic Biotechnology-derived Products; Guideline for the Format and Content of the Microbiological Section of An Application (Docket No. 85D-0245); February 1987; Guideline for the Format and Content of the Chemistry, Manufacturing and Controls Section of An Application; Preparing Data for Electronic Submission in ANDAs [HTML] or [PDF], Sep 1999; Regulatory Submissions in Electronic Format; General Considerations Jan 1999; Regulatory Submissions in Electronic Format; New Drug Applications Jan 1999.

8. Agreement on Mutual Recognition between the European Community and the United States; US – EC MRA Pharmaceutical Good Manufacturing Practice Annexe; Sectorial Annex For Pharmaceutical Good Manufacturing Practice; Signed 18 May 1998; Exchange of Letters 30 October 1998; Published in Official Journal L 31, 4 February 1999

Web Sites

www.fda.gov/cder/guidance/index.htm
www.emea.eu.int/
www.mca.gov.uk
www.rpsgb.org.uk/

Good manufacturing practice

3

JOHN WELBOURN

3.1 Introduction

This chapter explains what is meant by current Good Manufacturing Practice (cGMP) and, in particular, how it applies to the engineering aspects of pharmaceutical production. The chapter also shows how it is possible to develop the GMP requirements to allow the facility to be engineered, and looks at the GMP design review process.

3.1.1 Definition

A key part of the control of medicinal products and facilities relates to GMP.

The EU Guide To Good Manufacturing Practice and Good Distribution Practice defines GMP as 'the part of Quality Assurance (QA) which ensures that products are consistently produced and controlled to the quality standards appropriate to their intended use and as required by marketing authorization or product specification.'

'Engineering for cGMP' may be defined as those activities performed throughout the project life-cycle, which ensure that it will be easy and natural to operate the completed facility in accordance with current Good Manufacturing Practice.

The 'Project Life-Cycle' means from project inception through feasibility studies/conceptual design, engineering, construction, installation, start-up, operation, maintenance to final plant decommissioning or modification.

GMP is controlled by the US Code of Federal Regulation (CFR) 21 in the USA. European pharmaceutical companies wishing to supply this market must also comply with these regulations.

The various regulatory authorities produce different types of applicable documentation, which broadly fall into two categories:

- directives, rules, regulations, including for example:
  - US Code of Federal Regulations CFR 21 Parts 210 and 211 (Drug products) and CFR 21 Parts 600 to 680 (Biological products);

- o EU GMP Directive 91/356/EEC, Commission Directive Laying Down The Principles and Guidelines of Good Manufacturing Practice;
- o Rules Governing Medicinal Products For Human Use in the European Community, Volume IV; Guide to Good Manufacturing Practice for Manufacture of Medicinal Products.
- guides, guidelines, points to consider, including for example:
  - o FDA Guide to Inspection of Bulk Pharmaceutical Chemical Manufacturing;
  - o FDA Guide to Inspection of Validation of Cleaning;
  - o FDA Guide to Inspection of Computerized Systems in Drug Processing;
  - o FDA Guidelines on General Principles of Process Validation.

Although not necessarily in a strict legal sense, the first category is mandatory and must be complied with. The second category, although classed as guides or guidelines, is also very important and generally must be complied with.

The US Food and Drug Administration prepares guidelines under 10.90 (b) of the regulations (21 CFR Part 10) to help with compliance. A comprehensive listing of potentially relevant guidelines, guidance and points to consider is provided by Center for Drug Evaluation and Research, 'Guidelines for Regulations that are applicable to the Center for Drug Evaluation'.

As well as the formal documents outlined above, there are other ways that cGMPs have evolved. These include the interpretation of the various rules and regulations and what is generally considered to be good practice by the industry. For example, the US Food and Drug Administration, through the freedom of information service, produces reports on inspections and inspection failures. These reports are in effect 'legal rulings' or interpretations of the regulations, e.g. Form 483. It is important to keep up to date on these requirements through publications such as GMP Trends or QC Gold Sheet. As a rule of thumb in terms of good practice, if more than 50% of the industry is moving over to something then it becomes cGMP.

In addition to the codes laid down by the various regulatory authorities, there are parallel industrial quality standards that are deemed to apply to all industries. In Europe these tend to be grouped around ISO 9000, and the US equivalent are ANSI standards grouped around Q90. It is obvious that common standards should be applied and to this end the International Committees for Harmonization of Standards have published relevant recommendations as ICH guidelines.

3.1.2 General GMP requirements

When first embarking on a new pharmaceutical facility, consideration will need to be made as to what cGMP requirements will apply to the project and how they will impact on the project life-cycle. These may vary. Although the words differ, there are common general requirements that run through virtually all the cGMPs worldwide. Common elements are:

- the establishment and maintenance of an effective quality assurance system;
- control of the process;
- personnel that are suitably qualified, trained and supervised;
- premises and equipment that have been located, designed, installed, operated and maintained to suit intended operations;
- maintenance of adequate records of all aspects of the process so that in the event of a problem being identified, an investigation can trace the complete history of the process, including how, when, and where it was produced, under what conditions and by whom (i.e. an audit trail);
- the prevention of contamination from any source, in particular from components, environment, premises and equipment by the use of suitable premises and equipment and through standard operating procedures.

3.1.3 Project assessment to determine applicable standards

Whilst the objectives of most cGMPs are generally the same (i.e. to safeguard consumers), the nature of pharmaceuticals dictate that different sets of specific requirements have evolved depending upon the type of product, its stage of development or manufacture, and where it will be manufactured and sold. In addition the different regulatory authorities have prepared slightly different sets of standards, and apply them in different ways. One of the first steps when preparing to undertake a new project is to establish under what cGMP regulations the plant will operate. An assessment should be made to determine the:

- stage of product development;
- stage of production;
- category of the product and production processes employed;
- facility location and location of the markets that the facility will serve.

Based on these factors a judgment can be made as to applicable standards that need be applied.

19

Stage of product development

For the purposes of this book, the stage of product development may be divided into three parts:

- laboratory trials (pre-clinical animal trials);
- clinical trials;
- routine production.

Generally speaking, cGMPs regulations do not apply during laboratory trials, 'Basic cGMPs' apply during clinical trials, and 'full cGMPs' apply during routine production. cGLPs (Current Good Laboratory Practice) may apply during laboratory trials and cGCPs (current Good Clinical Practice) may apply during clinical trials.

Essentially GLP is concerned with the organizational processes and the conditions under which laboratory studies are planned, performed, monitored, reported and recorded. The UK GLP regulations (Statutory Instruments No. 654) came into force in April 1997 and are monitored by the UK GLP Monitoring Authority, which is part of the Medicines Control Agency (MCA). Currently about 150 test facilities are registered under the scheme and are inspected on a two-year cycle.

GCP is 'a standard for the design, conduct, performance, monitoring, auditing, recording, analysis and reporting of clinical trials to provide assurance that the data and reported results are credible and accurate, and that the rights, integrity, and confidentiality of trial subjects are protected' (Definition from the ICH Note for Guidance on Good Clinical Practice (CPMP/ICH/135/95)). In the UK the GCP Compliance Unit was established within the Inspection and Enforcement Division of the MCA in Autumn 1996. GCP inspectors assess compliance with the requirements of GCP guidelines and regulations, which involves conducting on-site inspections at pharmaceutical sponsor companies, contract research organizations investigational site and other facilities involved in clinical research.

Stage of production

The stage of production means what the facility is used for. The stages of production can be divided into the following four parts for the purposes of this book:

- Bulk Pharmaceutical Chemicals (BPCs) manufacturing;
- finished product manufacturing;
- packaging;
- warehousing/holding.

Different regulatory authorities apply certain specific cGMPs to different stages of production. In some cases facilities may be used for more than one stage of production, and in such cases more than one set of cGMPs may apply.

Category of the product and production processes employed
Broadly speaking most active ingredients are manufactured by one of the following routes:

- chemical synthesis;
- biotechnology;
- blood derived;
- animal or plant extraction.

By the nature of these routes certain methods of production to produce dosage forms have evolved, and in each case specific GMP requirements have been developed. From a GMP point of view regulatory authorities categorize products as following:

- sterile medical products:
  - o terminally sterilized products;
  - o aseptic preparations;
- biological medical products:
  - o microbial cultures, excluding those resulting from r-DNA techniques;
  - o microbial cultures, including those resulting from r-DNA or hybridoma techniques;
  - o extraction from biological tissues;
  - o propagation of live agents in embryos or animals;
- radiopharmaceuticals;
- veterinary medicinal products;
- medical gases;
- herbal medicinal products;
- liquids, creams and ointments;
- metered dose aerosols;
- products derived from blood;
- tablets and hard gel capsules;
- soft gel capsules;
- transdermals;
- implants.

Clearly some pharmaceuticals represent a combination of these types.

Facility location and market location
cGMPs regulations are produced by a number of different countries or groups of countries world-wide, in addition to the World Health Organization. The key

21

regulations are from the USA, the EC and, to some extent, Japan. However no assumption must be made that these are suitable standards to apply. Clarification should be sought from the pharmaceutical manufacturer before the design commences.

3.2 GMP design requirements

Based on an assessment of the regulatory requirements (as described above) we can begin to define the GMP requirement for the project. Generally, issues and areas to be considered during the conceptual design phase will include:

- **process** issues:
  - o closed or open (Is it to be completely contained with piping and equipment at all times or will it be exposed to the surrounding environment? In which case, what measures are to be taken to prevent/minimize contamination?);
  - o level of batch to batch integrity required (Is simultaneous filling and emptying of vessels with different batches in known proportions or limits to be permitted? Do systems need to be engineered to be self-emptying? Will process systems need to be subject to cleaning, drying or sterilization between batches?);
  - o level of segregation or containment required (Is it acceptable to manufacture product A in the same facilities as product B? Will processes be campaigned?);
  - o level of production required.
- **layout** issues:
  - o site location and layout (including existing site, brown field, green field, overall site layout and its suitability in terms of space, general layout);
  - o facility layout (including cored versus linear layout; use of transfer corridors, segregation of areas, environment, containment strategy, modularization/expansion, security and access control).
- **automation strategy** issues:
  - o level of technology, use of design tools and models, number of layers — hierarchy;
  - o availability/redundancy/maintainability, modularization/expansion;
  - o instrumentation/cabling/field devices;
  - o paperless batch records, electronic signatures.
- **flow** issues:
  - o people (security, access, occupancy level, shift patterns);

- equipment (mobile or fixed, use of hard piping, flexible piping or disposable transfer bags, cross-contamination/mix-ups);
- components/materials (materials handling systems, cross-contamination/mix-ups).
- **regulatory** issues:
  - stage of product development, stage of production, category of the product and production processes employed, facility location, and location of the markets that the facility will serve.
- **validation strategy** issues:
  - validation required, validation team(s), validation plan(s).

These basic requirements can then be refined for the various aspects of the project to allow the facility to be engineered. The following categories are suggested for guidance:

- facilities and environment;
- services and utilities;
- personnel flows;
- material flows;
- equipment flows;
- equipment design;
- computerized systems;
- maintenance and services;
- waste management;
- procedure and documentation.

The following sections provide guidance to the type of criteria that will need to be considered. It may be appropriate to formulate these (and other applicable criteria), into a checklist for use during the development of the design and the design review.

3.2.1 Facilities and environment

These are the buildings, rooms and environment containing the production processes. They are of prime concern wherever the product or product components may be exposed. Typical criteria include the following:

General considerations for the entire facility:

- local environmental considerations (including pollution and security);
- suitability/acceptability of physical segregation of processes for manufacturing and holding products, (such as segregation of production stages of the same, similar and different products and the use of dedicated or shared facilities);

23

- overall layout of the facility (including use of cored environmental layout, position of technical and other non-production areas with respect to processing areas);
- general layout of production processes (logical flow through the facility with no/minimal cross-over of processing streams);
- pedestrian and vehicular access;
- pest control.

Specific considerations for each area:

- available space and ergonomics for operators, equipment, materials and products;
- cavities/penetrations and how they are sealed;
- surfaces of walls, floors and ceilings (they should be easily cleanable, low particle shedding, minimal dust traps);
- materials of construction of the walls, floors and ceilings and their suitability for the intended operations;
- types of doors, windows, light fittings and void closures (for example, flush fitting, methods of sealing);
- provision and location of support utilities (both for production and maintenance/housekeeping purposes);
- provision of suitable electrical outlets and communications systems (electrical sockets, telephones, speak through panels, network termination points, intercoms);
- furniture (quantity, suitability for the operators, surfaces, cleanability).

Environment:
This is of prime concern wherever the product or product components may be exposed. Typical criteria include:

- assessment of the environmental classification of the various areas against the level of quality required by the product (including non-viable particulate and microbiological contamination in both the unmanned and manned conditions);
- airflow regime and types of processing operations (turbulent or laminar, horizontal or vertical);
- air pressure differentials between areas;
- air change rates per hour;
- location of ventilation ducts relative to processing points and other equipment;
- emissions within the area (water vapour, compressed air, toxic fumes);
- humidity (comfort level, static hazards, growth promoting);
- environmental control at point of access to area;

- illumination levels (relative to operations performed);
- adverse operating conditions (start-up/shutdown, dirty filters/blockages, power failure, redundancy);
- methods of monitoring, recording and controlling the environment (including temperature, pressure, humidity, air flow/velocity, particulate and microbiological);
- maintenance and cleaning of environmental systems (such as routine maintenance, safe change systems, redundancy).

GMP requirements may generally be limited for external areas such as administration buildings, canteens, plant rooms. Staff will be able to access these areas in street clothing or working garments unrestricted by GMP, but there may be other reasons why specific garments are required. Personnel access should be controlled to all areas within a pharmaceutical facility (by access cards or pass-codes, for example). Pest control measures must be employed to prevent insect and rodent infestation.

For areas such as packaging, warehousing, technical areas or where the product is fully contained in pipework, typical GMP requirements would include that:

- clothing consists of general factory overalls or lab coats and hats, with personnel to enter these areas via cloakroom facilities (primary change);
- environment air should be filtered to Eu 3 or above. Air pressure should normally be ambient;
- surfaces should be easily cleanable, finished flush and sealed. Equipment should be readily accessible for cleaning;
- measures should be taken to minimize the risk of cross-contamination.

For areas where specific environmental control is required such as in secondary pharmaceutical manufacturing where products or ingredients are exposed, or for the preparation of solutions and components for terminally sterilized products, and in BPC plant areas handling exposed products or critical step intermediates, GMP requirements may include the following in addition to the above:

- personnel must enter the area via a secondary change and the area must not contain toilets or eating areas;
- process materials and components should enter via an airlock;
- filtration and air circulation should achieve EU GMP Guide Grade D or equivalent;
- drains should be sealed during normal operation with air breaks provided between sink or equipment outlets and floor drains;

- compressed air exhausts should be vented outside the area;
- the preferred material of construction for process equipment is generally stainless steel and pipework lagging should be avoided where possible. Operators should be protected by mechanical guarding. Separate belt conveyors should be in different grade rooms, with dead plates at the wall opening.

Where tight microbial control is required, such as areas used for the preparation of solutions to be filtered before aseptic filling, GMP requirements may also include that:

- filtration and air circulation should achieve EU GMP Guide Grade C or equivalent with pressure positive (typically 15 Pa) to adjacent lower grade areas;
- strategically located local environmental protection, such as positive pressure Grade A LAF units, should be in place for exposed operations.

For areas where specific microbial control is to be exercised continually, such as for aseptic preparation and filling operations, additional GMP requirements will need to be applied such as:

- all operations should be performed aseptically with filtration;
- air circulation should achieve Grade B (EU GMP Guide, Annex 1) at positive pressure to lower grade areas;
- any process or equipment drains should be sealed and fitted with a sterilizable trap;
- strategic Grade A protection should be provided at all points of product exposure.

3.2.2 Services and utilities

Services and utilities that come into direct product contact (or form part of the product) are of particular concern. Some typical criteria for commonly used critical utilities include:

High purity water systems such as WFI systems:

- assessment of the proposed water quality against the level of quality required by the product (in terms of chemical quality, microbiological, pyrogenic, and physical particulate contamination);
- materials of construction (including piping, gaskets, valve diaphragms);
- internal surface finishes (Ra ratings, use of electropolishing, passivation);
- water pre-treatment and control (adequacy);
- system sizing (minimum and maximum demand);

- key design considerations such as minimum flow rates, minimum deadlegs with no cavities, vents and how they are sealed/filtered, drainage air gaps and backflow prevention devices;
- use of security devices, such as 0.2 micron sterilizing grade filters, UV sterilizers, ozone injection;
- instrumentation and control of critical process parameters (for example, temperature, velocity, flow, conductivity control limits and alarms, use of dump valves and recirculation of bad quality water, monitoring, recording and controlling systems);
- storage (such as storage temperature, maintenance of circulation and wetting of all internal surfaces, vent filter integrity and sterilization);
- methods and adequacy of cleaning and sanitization;
- adverse operating conditions (start-up/shutdown, power failure, redundancy, etc.);
- proposed method of construction (including procedures, control and inspection of material stock, fabrication, welding, field installation, passivation, preservation).

Clean steam systems:

- similar considerations to those described for high purity water systems can generally be applied to clean steam systems.

Gases (such as compressed air, nitrogen, hydrogen and oxygen):

- assessment of the proposed gas quality against the level of quality required by the product (in terms of chemical quality, microbiological, pyrogenic, and physical particulate contamination);
- materials of construction (including piping, gaskets, valve diaphragms);
- internal surface finishes (Ra ratings, use of electropolishing, passivation);
- system sizing (minimum and maximum demand);
- use of security devices, such as 0.2 micron sterilizing grade;
- instrumentation and control of critical process parameters (for example, temperature, pressure and dew point, monitoring, recording and controlling systems);
- methods and adequacy of cleaning and sanitization;
- adverse operating conditions (start-up/shutdown, power failure, redundancy, etc.);
- proposed method of construction (including procedures, control and inspection of material stock, fabrication, welding, field installation, passivation, preservation).

Typical GMP criteria for Water for Injection (WFI):

- quality to conform to compendia requirements (such as USP and/or Ph.Eur Monographs);
- production to be by distillation (also reverse osmosis allowed in some regions) from purified water and to conform to USP and/or Ph.Eur Monographs;
- WFI to be sterile and pyrogen free with an action limit set to less than 10 CFU/100 ml (Colony Forming Units) with a sample size of between 100 and 300 ml and an endotoxin level of less than 0.25 EU/ml (endotoxin units).

Design of WFI systems:

Firstly it is important to ensure that there is adequate pre-treatment and control of feed water, using methods such as deionization, ultrafiltration and reverse osmosis. Pre-treatment by deionization alone may prove to be unsatisfactory.

Key features of the WFI system itself include:

- still to be of multi-effect type, heat exchangers of double tube sheet design and holding tank employing tube type external jacket;
- WFI system to be fitted with a hydrophobic sterilizing grade vent filter to protect system from ingress of non-sterile air;
- vent filter to be jacketed to prevent condensate blocking the filter and to be steam sterilizable and integrity tested in place;
- provision for continuous ring main circulation at temperatures over 70°C at velocities sufficient to achieve a Reynolds number of >25000;
- provision for periodic sterilization of the system;
- provision for sampling at all loop take-offs (the start and end of the loop) with take-offs design to prevent re-contamination of the system by air-drying, steam locking or trace heating;
- WFI to be stored in a nitrogen atmosphere where appropriate to minimize the absorption of oxygen;
- product contact materials be supplied with material certification and PMI (Positive Material Identification) and stainless steel contact surfaces to be <0.5 µm Ra and passivated;
- pipework joints and couplings to be minimized with pipework being orbitally welded where possible. Detailed weld records to be supplied with weld logs and NDT reports on specified minimum proportions of all welds. Couplings and equipment to be crevice free — clamp fittings IDF couplings or similar are preferred. Deadlegs in vessels and pipework be minimized, by for

example use of zero deadleg diaphragm valves. System to be designed to allow for periodic complete flushing or draining such that all lines will slope to low drain points at a slope of greater than 1 in 100.

3.2.3 Personnel flows

This includes the influence personnel have on the quality of the product that might be caused by their contact with the product. Typical criteria include:

- clothing requirements (suitability of proposed plant clothing against the types of operations being performed within that area);
- changing regimes (stages of changing);
- changing facilities (adequacy of changing and washing facilities, doors, step over barriers, provision of adequate space for clothing, use of vision panels and their position relative to/from production areas);
- security and access control including potential short cuts and back doors;
- types of movements within the area (including passing through, local operations, supervisory support);
- occupancy levels;
- shift patterns (what supervisory and maintenance support is available);
- potential points of cross-contamination between personnel (such as transfer hatches, changing rooms — gowning/ungowning, finger streak stations);
- activity levels (i.e. sedentary or active and how this compares to the required room environment, occupancy level and clothing regime).

3.2.4 Material flows

This includes all the movement of materials. Typical criteria include:

- general flow of materials through the area (for example, linear flow through with no cross-over of production streams);
- methods of handling and prevention of cross-contamination;
- frequency of movements and available space;
- possible points of cross-contamination between materials (for example, temporary storage points, processed and non-process materials, bulk containers);
- identification and segregation of materials;
- storage conditions (refrigerated, toxic, hazardous, filtered).

3.2.5 Equipment flows

It is important to consider that not all equipment may be fixed in one position; it may either be moved routinely as part of the production process, or at least be capable of relocation for plant maintenance or reconfiguration. Typical criteria include:

- methods of handling and prevention of cross-contamination;
- frequency of movements and available space;
- physical size and weight of equipment against room construction (heavy equipment may damage welded sheet vinyl floors or fracture gyprock walls — trowelled on epoxy cement or blockwork may be more appropriate);
- possible points of cross-contamination between equipment (such as temporary storage points, washing machines and bays);
- identification and segregation of mobile equipment;
- storage conditions (refrigerated, toxic, hazardous, filtered);
- provision of non-routine access, such as removable wall or ceiling panels.

3.2.6 Equipment design

The examination of the GMP issues within a machine or system is a 'micro' version of those for a facility, and includes many of the same questions such as surfaces, flow of materials and personnel issues. The amount of detail will vary with the complexity of the equipment and its effect or potential effect on product quality. Typical criteria include:

- pedigree of the machine (established for pharmaceutical use, 'off the shelf' or specially developed prototype);
- pedigree of the manufacturer (specialist supplier to the pharmaceutical industry who manufactures more than 50 identical units per year or first development machine by a new manufacturer);
- materials of construction and surface finishes of primary and secondary contact parts (i.e. primary — direct product contact, secondary — contact with local environment);
- equipment sizing (minimum and maximum demand);
- key design considerations (minimum deadlegs with no cavities, all critical surfaces accessible and cleanable, drainage air gaps and backflow prevention devices);
- instrumentation and control of critical process parameters (temperature, pressure, speed control limits and alarms, monitoring, recording and controlling systems);
- methods and adequacy of cleaning and sanitization;
- adverse operating conditions (start-up/shutdown, power failure, redundancy);
- proposed method of construction (including procedures, control and inspection of material stock, fabrication, field installation);

- maintenance (access for maintenance during and outside production, use of maintenance free items, requirements for special tools/no tools).

For equipment and pipework that does not come into contact with the product or product components, there are no specific GMP requirements.

For process pipework and equipment there is no need for sophisticated Clean in Place (CIP) or Steam in Place (SIP) but plant washing and flushing with water or chemicals may be used. Typical requirements include the following:

- dismantling and inspection should be easy and involve minimal use of tools;
- all pipework should slope towards the drain points;
- product contact materials should be supplied with material certification and stainless steel surface finishes in contact with the product should be $<1.0\,\mu$m Ra and passivated. Pipework couplings and equipment should be crevice free. Clamp fittings, IDF couplings or similar are preferred and deadlegs in vessels and pipework should be minimized.

For areas where CIP and SIP effectiveness is critical, GMP requirements may include, in addition to the above, that joints and couplings are minimized with pipework being orbitally welded where possible and that stainless steel product contact surfaces are $<0.5\,\mu$m Ra electropolished.

For certain types of equipment, specific GMP requirements have been issued — one example of this is for sterilization equipment. Typical criteria for porous load moist heat sterilizers include:

- the complete chamber space should achieve a uniform temperature distribution of less than $\pm 1°$C at the sterilization temperature for the complete sterilization period, and the equilibrium time to achieve this distribution should be less than 30 seconds;
- the chamber should be resistant to corrosion and the leak rate of the chamber should be less than 1.3 mbar per minute;
- monitoring instrumentation and recording charts should be independent of control instrumentation and utilize an independent time/temperature and pressure chart or equivalent of a suitably large scale to record the sterilization process;
- an air detector should be fitted such that a difference in temperature of greater than 2°C between the centre of a standard test pack and chamber temperature at commencement of equilibrium time is detected;
- drains should be trapped and vented and not connected to other drains which could cause a backpressure or obstruction to flow — an air break is necessary;

- steam used for the sterilization process should have a dryness fraction of not less than 0.95 and the superheat measured on expansion of the steam to atmospheric pressure should not exceed 25°C with the fraction of non-condensable gases not exceeding 3.5% by volume. The steam generator should be designed to prevent water droplets being carried over into the steam and should operate so as to prevent priming. The steam delivery system should be fitted with a water separator and traps to virtually eliminate condensate build up, and be resistant to corrosion with minimum deadlegs to reduce the risk of water collection and biofilm formation.

3.2.7 Computerized systems

The amount of detail will vary with the complexity of the computerized system and its effect or potential effect on product quality. In particular the pedigree of the manufacturer, type of hardware and type or category of software to be used need to be carefully considered. The systems manufacturer is generally responsible for providing the validation documentation and ensuring that the system complies with GMP. Typical criteria include:

General:

- up to date specifications, including principles, objectives, security measures and scope of the system and the main features of the way the system will be used and how it interacts with other systems and procedures;
- the development of software in accordance with a system of quality assurance;
- system testing including a demonstration that it is capable of achieving the intended results;
- procedures for operation and maintenance, calibration, system failure (for example, disaster recovery, restarting), recording, authorizing and carrying out changes, analysis of errors, performance monitoring;
- pedigree of the machine and manufacturer;
- type of hardware (for example, standard 'off the shelf' components from reputable suppliers operating a recognized quality system, installed in a standard system such as a PC or fully bespoke hardware developed specifically for the system);
- type/category of software (operating system, can be configured, bespoke software);
- adequacy of system capacity (in terms of memory, I/O, etc.).

Control/access/security:

- built in checks of the correct entry and processing of data;
- suitable methods of determining unauthorized entry of data such as the use of keys, pass cards, passwords and restricted access to computer terminals;
- control of data and amendments to data, including passwords. Records of attempts to access by unauthorized persons;
- additional checks of manually entered critical data (such as weight and batch number of an ingredient during dispensing);
- entering of data only by persons authorized to do so;
- data storage by physical and electronic means. The accessibility, durability and accuracy of stored data. Security of stored data;
- data archiving, remote storage of data;
- recording the identity of operators entering or confirming critical data. Amendments to critical data by nominated persons. Recording of such changes;
- audit trail for system;
- change control system;
- obtaining clear printed copies of electronically stored data;
- alternative arrangements in the event of system breakdown, including the time required to recover critical data;
- positioning of the equipment in suitable conditions where extraneous factors cannot interfere with the system;
- form of agreement with suppliers of computerized systems including statement of responsibilities, access to information and support;
- release of batches, including records of person releasing batches.

Personnel/training:

- personnel training in management and use;
- expertise available and used in the design, validation, installation and operation of computerized systems.

Replacement of a manual system:

- replacement of manual systems should result in no decrease in product quality or quality assurance;
- during the process of replacement of the manual systems, the two systems should be able to operate in parallel;
- reducing the involvement of operators could increase the risk of losing aspects of the previous system.

3.2.8 Maintenance and servicing

This applies to all the facility and everything within it. It is important to consider that not all equipment may be fixed in one position, it may either be moved routinely as part of the production process or at least be capable of relocation for plant maintenance or reconfiguration. Typical criteria include:

- methods of handling and prevention of cross-contamination;
- frequency of movements and available space;
- physical size and weight of equipment against room construction (heavy equipment may damage welded sheet vinyl floors or fracture gyprock walls; trowelled on epoxy cement or blockwork may be more appropriate).

3.2.9 Procedures and documentation

In order to support the facility, adequate procedure and documentation are required. During the design stage many of the documents required for normal operation of the facility may not yet be available. At this stage, it is probably too early to consider exactly what documentation will be required, but it is possible to begin to consider how documentation will be accommodated and organized. Typical criteria include:

- adequate workspace, storage capacity and personnel to control stored documentation;
- security of documentation (including access control, fire protection, additional remote storage capacity);
- adequate, rapid access to stored data, including suitable provisions for the local retrieval of data stored electronically.

3.3 GMP reviews of design

To ensure that the project remains in compliance with cGMP as it progresses through its life-cycle, periodic GMP design reviews must be undertaken.

3.3.1 Organizing the GMP design review team

Reviewing a design for compliance to cGMP requirements can often be a daunting prospect. It requires a range of knowledge that no single person is likely to possess. For this reason it is often more effective if the review is performed by a small team that has an understanding of the basic requirements and works methodically. The team should consist of persons selected for both their depth of knowledge in a particular area and for general knowledge of cGMP principles applicable to the project. A good mix for a suitable team would be:

- cGMP compliance/validation specialist (knowledge of regulatory, QA, validation, etc.);
- architect (knowledge of finishes, layout, personnel/materials flows, etc.);
- process engineer (knowledge of process, equipment, utilities, etc.).

Depending upon the nature of the facility the architect or process engineer could be substituted for more suitable disciplines. For example, the design review of an automated high bay warehouse may be better performed using a materials handling specialist and an automation specialist. The team would normally be lead by the cGMP compliance/validation specialist who would organize the team, co-ordinate the review and prepare the report(s). It is recommended that the team be kept as small as practicable, since it will be able to operate more efficiently and flexibly and be easier to co-ordinate. If issues arise that are beyond the combined knowledge of the team then they can be referred for further investigation by specialists in the particular subject.

3.3.2 Information required to perform the review
Two basic types of information are required to perform an effective review:

- specification of the pharmaceutical product and manufacturing process;
- specification of the equipment and facility.

Note that some facilities are used for a variety of products that may utilize different processes. In this case a separate review of each process may be performed. However, often it is possible to base a review on a 'typical' product that runs through the entire process.

As part of the cGMP review all information sources used must be documented. Regulatory authorities always demand to see original information. It is, therefore, essential that a good record keeping system be established — for example, original design calculations must be retained. All engineering drawings must be authorized and signed off.

Specification of the pharmaceutical product and manufacturing process
General details of the process are required rather than exact details of, say, a particular chemical reaction involved. Sources of information may include:

- regulatory documents such as:
  - New Drug Application (NDA), Product Licence Application (PLA), Investigational New Drug Application (IND);

- o manufacturer's licences such as Product Licence, Wholesale Dealers Licence;
- o Drug Master File (DMF);
- technology transfer documents;
- batch manufacturing documentation prepared for similar facilities;
- process description;
- process flow diagrams (PFD).
   The type of information required will typically include:
- description of processing operations including:
   - o manual operations such as loading, sampling testing, adjustments;
   - o automatic operations such as process unit operations, cleaning cycles and materials handling;
- quantities and throughputs;
- components and processing chemicals;
- critical parameters such as temperature, pressure, time and volume;
- batch size and frequency;
- regulatory requirements in original product licence/regulations;
- technical requirements identified during laboratory/pilot scale production.

Specification of the equipment and facility
Clearly the review will utilize the GMP design philosophy as a key document, but this should be compared with what has actually been specified. General details of the equipment and facilities are required. Sources of information may include:

- architects/facility engineers;
- process engineers;
- engineers from the various technologies as appropriate — for example, mechanical, electrical, civil, control, instruments;
- R&D;
- QC/QA.

   The type of information required will typically include:

- process description, materials and personnel flow diagrams;
- general arrangement drawings, axiometric drawings and room layouts;
- process and instrumentation diagrams (P&IDs);
- HVAC basic layouts, specifications and area classification drawings;
- main equipment items list with specifications;
- utilities list with specifications;

- user requirement specifications;
- control system functional design specifications.

3.3.3 Divide up the facility into manageable sized areas

The best way to divide up the facility for the review largely depends on the type of facility and nature of the process. The following approach is suggested for guidance.

Bulk pharmaceutical chemical manufacturing

Typically for BPC manufacturing the process is contained within closed vessels and pipework arranged as an integrated/interconnected process. In this case it is probably easiest to break the cGMP review up into a series of reviews of each main P&ID. Each P&ID is then considered by the review team along with any associated equipment and utility specifications, control system descriptions etc., as a package.

Secondary manufacturing

Typically for secondary manufacturing, the process is carried out in a series of discrete stages in separate areas such as:

Goods in.	Weighing.	Services and utilities.
Warehousing.	Mixing/blending.	Goods out.
Amenities.	Filling.	
Changing rooms.	Sterilizing.	
Equipment preparation.	Labelling.	
Dispensing.	Packing.	
QC testing laboratory.	Administration area.	

The best method here may be to perform the review on each area of the facility. The review will centre on the room layout drawings along with associated environmental classification drawings, equipment and utility specifications and control system descriptions, as a package. It may also be possible to identify specific areas that have no cGMP implications — these can be considered to be 'outside the GMP area' and need not form part of the review although any decisions made to include or exclude particular areas should be documented.

In some cases, a combination of both the above methods may be the most appropriate. The key point is to break the task down into logical, manageable-sized portions, which can then be reviewed.

Validation

JOHN WELBOURN

4.1 Introduction

Validation first started in the 1970s on sterilization processes, when it became clear that end product testing alone could not show that every container within every batch of product was sterile and the time and cost associated with testing each individual container was too great, or the testing was too destructive to the product. Validation offered a way of providing evidence that the process was capable of consistently producing a product with defined specifications.

This type of work spread gradually through from sterile and aseptic processes to non-aseptic processes (tablet manufacture, for example) by the mid 1980s. By the late 1980s, the concept of validation was reasonably well established. Regulatory authorities and the pharmaceutical industry have co-operated to define validation requirements and agree upon the definition. The principle is the same for whichever process is being investigated — that is, to provide documented proof of GMP compliance. Validation and GMP go hand in hand.

4.1.1 Definition

Even before the current definitions of validation, industry was operating to the concept in the first edition in 1971 of the British Guide to Good Pharmaceutical Manufacturing Practice (the 'Orange Guide'), which suggested that procedures should undergo a regular critical approach to ensure that they are, and remain capable of, achieving the results they are intended to achieve.

Although the US Federal Register does not contain an official definition, US CFR Part 211 section 211.100 states that:

'There should be written procedures for the production and process control designed to assure that the drug product has the strength, quality and purity they purport or are represented to possess.'

The FDA has issued a 'Guideline on General Principles of Process Validation' which defines process validation as:

'Establishing documented evidence which provides a high degree of assurance that a specific process will consistently produce a product meeting its predetermined specifications and quality attributes.'

The EU 'Rules Governing Medicinal Products in the European Community' Vol IV define validation as:

'Action of proving, in accordance with the principles of Good Manufacturing Practice, that any procedure, process, equipment, material, activity or systems actually leads to the expected results.'

The EU Rules also define the term 'Qualification', which arises many times within validation work, as:

'Action of proving that the equipment works correctly and actually leads to expected results. The word validation is sometimes widened to incorporate the concept of qualification.'

Validation for the engineer is the act of proving with the necessary formal documentation that something works. It is advisable to create the documentation throughout the design process since it is often expensive and time-consuming to produce retrospective documents.

4.1.2 The need for validation

There are three reasons why the pharmaceutical industry is concerned about validation:

- government regulation;
- assurance of quality;
- cost reduction.

Government regulation

The requirements for validation are now explicitly stated in both the US and European regulations (US Code of Federal Regulations US CFR Part 211, subpart L, 211.220 and 211.222 and within the EU 'Rules Governing Medicinal Products in The European Community' Vol IV, Part 5.21, 5.22, 5.23, 5.24).

In CFR 211.220 it says:

'The manufacturer shall validate all drug product manufacturing processes ...'

39

and:

'... validation protocols that identify the product and product specifications and specify the procedure and acceptance criteria for the tests to be conducted and the data to be collected during process validation shall be developed and approved...'

and:

'... the manufacturer shall design or select equipment and processes to ensure that product specifications are consistently achieved. The manufacturer's determination of equipment suitability shall include testing to verify that the equipment is operating satisfactorily....'

Similar requirements are stated in the EU Rules.

Assurance of quality

Without process validation, confidence in the quality of products manufactured is difficult to prove. The concepts of GMP and validation are essential to quality assurance. Frequently, the validation of a process will lead to quality improvement, as well as better consistency. It may also reduce the dependence upon intensive in-process and finished product testing. It should be noted that in almost all cases end-product testing plays a major role in assuring that quality assurance goals are met, i.e. validation and end-product testing are not mutually exclusive.

Cost reduction

Experience and common sense indicate that a validated process is a more efficient process that produces less reworks, rejects, wastage, etc. Process validation is fundamentally good business practice.

In summary, validation should be applied to all aspects of the process, including the equipment, computer systems, facilities, utilities/services and in-process testing (analytical methods). From the above discussion, the following key points have developed:

- documented evidence must be written down (if it's not documented it's not done);
- formal documentation — all design documents should be signed off. Signatures, page numbering, control copies, storage/retrieval, etc., should be installed;
- acceptance criteria — decide what is acceptable before testing;
- repeatable — one-off results are not acceptable;

- validation and qualification — processes are validated whereas the equipment used within the process is qualified.

4.2 Preliminary activities

Prior to embarking on a validation project, it is necessary to establish an organizational framework in which validation resides. This must start with the commitment and sponsorship of the senior management within the company, for without this commitment to validation any validation project is likely to fail.

4.2.1 Establishing policies and procedures

One of the first steps is to establish the policies and procedures that will govern the validation project — for example, the development of policies to define general concepts involved such as:

- how validation 'fits' within the overall QA structure and its relationship with cGMP;
- commitment to cGMP and its reinforcement through validation (i.e. the pharmaceutical company's commitment);
- definition of key terms such as critical process step, critical equipment and instrumentation, the various qualification activities including DQ, IQ, OQ, PQ (more about this later);
- how validation is structured and applied with respect to plant, processes, computer systems, analytical methods, etc. (how is it organized, what steps are performed in each case and how does it all fit together).

More specific procedures will need to be generated later for:

- validation documentation preparation (including house style, standard document sections, document numbering);
- validation documentation review and approval process;
- validation document change control system;
- validation master plans and final validation reports (preparation, content and structure);
- pre-qualification activities;
- cGMP reviews of design;
- vendor assessment and auditing (especially computer systems);
- equipment/computer system protocols and reports (i.e. DQ, IQ, OQ, PQ) preparation, content and structure;
- instrumentation and calibration;
- execution of field work;

- set-up and operation of validation test equipment;
- cleaning validation;
- process optimization and experimental work;
- process validation protocols and reports;
- analytical methods validation;
- documentation filing and management systems.

Note that it is particularly important at an early stage in the project to agree aspects such as document format, structure, content and numbering. This agreement needs to be recorded in the project quality plan.

At this early stage it is a good idea to establish the key validation team members and prepare an overall organizational chart.

Some of the first activities for the validation team to address will include:

- process evaluation to determine validation requirements;
- identification of systems and system boundaries;
- preparation of user requirement specifications;
- development of the validation master plan.

4.2.2 Process evaluation to determine validation requirements

Process evaluation involves a review of the process to identify the process steps and process variables, to determine how they are controlled/monitored and to identify what processing, equipment, utilities, instrumentation and control systems are associated with these steps. This should identify which systems need to be qualified and which parameters and instrumentation are important to the process and will need to be evaluated in the validation study or will become 'critical instruments.' As part of the development work done on the process, much of this should already have been defined, however, the documents where this is recorded need to be collated and reviewed.

The specification and procedures required for the process such as equipment operation and maintenance, calibration, set-up, cleaning and in-process testing should be identified, since these will need to be prepared for the new facility.

The various components used to manufacture the product should be reviewed to establish that all items have been specified and are under control. This may then point to requirements for analytical methods, validation or supplier audits, for example.

Based on an evaluation of the process a decision can be made as to what does and does not require validation. To perform such an evaluation requires a thorough understanding of the process and may include process components, process chemistry, plant (equipment, automation systems, etc.), specifications and procedures, in process controls and analytical testing methods.

User requirement specifications (URS)

These should be prepared by the user to formally document the requirements for each system to be qualified in terms of the final process requirements. A URS should typically include specific, but non-detailed information relating to, for example, quantity, quality, compatibility, performance, environment and finishes, in terms of:

- materials of construction;
- cleanability requirements;
- maintenance requirements;
- operator interface requirements;
- performance criteria;
- critical parameters;
- essential design criteria;
- requirements of computerized/automation system;
- training and documentation requirements.

It should make reference to relevant in-house standards and regulatory documents. It is essential that input to the URS includes persons with 'hands on' knowledge of the system and persons with a wider knowledge of the overall project.

4.2.3 Identification of systems and system boundaries

In parallel with process evaluation, systems and system boundaries need to be defined. The objective is to break the facility down into logical, manageable-sized packages of qualification work, and concentrate the validation effort in the most important areas to allow structured qualification.

A system may be an area of the facility (group of rooms), a group of functionally related process items, a utility or part of a utility, a HVAC, a computerized/automation system or any combination of these.

Determination of system boundaries involves the evaluation of the proposed facility design to establish the boundaries and break points for each package of qualification work. It is important that at the earliest stage practicable any 'grey' areas are removed, such as overlaps between areas of responsibility, missing areas, break points, IT systems interfaces.

Systems may then be categorized as 'Primary' or 'Secondary', (it may be appropriate to develop several more intermediate categories, such as in the case of IT systems). For example, primary systems could be defined as large, complex, purpose built or configured, generally fixed in place units. Examples include an aseptic filling suite, low temperature hot water system, water for injection system, electrical power distribution system, a piece of automated manufacturing equipment or a plant supervisory control and data acquisition system (SCADA).

Secondary systems could be defined as smaller, simple, 'off the shelf', generally portable items with no or minimal unique features or configuration, such as a bench top balance, filter integrity tester, a pallet-bailing machine and a 10-litre standard holding tank. Typically these systems may be bought direct from a supplier's catalogue.

Systems may be further categorized as 'critical' or 'non-critical.' Typically the following criteria are used to evaluate if a system is critical:

- stage of the process — is it used before, during or after a critical process step;
- effect on product quality;
- contact with product or product components;
- monitoring or controlling elements related to product quality.

Examples of primary critical systems are an aseptic filling suite, a water for injection system, a piece of automated manufacturing equipment, or a plant supervisory control and data acquisition system (SCADA).

Examples of primary non-critical systems are a low temperature hot water distribution system or an electrical power distribution system.

Examples of secondary critical systems are a bench top balance, filter integrity tester, and a 10-litre standard holding tank. An example of a secondary non-critical system is a pallet-bailing machine.

All critical systems should be validated. For primary critical systems this may involve the development of detailed plans, protocols, reports, certificates; for secondary critical systems, however, the use of simple, standard, check-sheet type documents may be more appropriate.

Non-critical systems do not require qualification — standard, well-structured project documentation is adequate.

4.3 Validation master planning

The initial activities described above can be formalized and consolidated into a validation master plan (VMP). This is a formal, approved document that describes in clear and concise wording the general philosophy, expectations, intentions and methods to be adopted for the validation study. Everyone involved in a project will have their own interpretation as to what validation is and what should be done. The VMP is an agreed document acting as a road map or guide for all team members to follow.

Once complete, it becomes a useful tool to show regulatory bodies that compliance with regulations is being sought and that there is a plan describing in detail the steps and programmes to be implemented to assure a validated and compliant facility.

To prevent the VMP becoming too unwieldy, it is common practice to develop separate validation plans for various parts of the overall project such as process equipment, utilities, computer systems, process and analytical methods. On large projects it may be necessary to have several levels of plans.

In terms of when to begin to develop the VMP, this will vary from project to project but it should normally be in place by the early part of detailed design. The VMP will then be a living document, updated regularly and amended during the course of the project. At the end of the project the VMP should define how the validation was actually performed.

The VMP, as with all formal validation documents, should be prepared, reviewed, approved and controlled under pre-defined company policies and procedures with final approval by QA. It must have a document number and a document revision history and page numbering must pass the 'drop test' (i.e. it is possible to reassemble the document from the page numbering and know that all sheets have been accounted for). The number of copies should be controlled.

4.3.1 Contents of the VMP

This will differ slightly from project to project and company to company, but the following items should usually be included:

(1) approval page;
(2) introduction;
(3) the aim;
(4) descriptions of:
- facility;
- services/utilities;
- equipment;
- products;
- computer systems;
(5) validation approach:
- overall;
- detail (matrix of validation documents);
(6) other documentation.

Approval Page

The approval page is the title page to the entire document and should contain the name of the company, the title and a space for approval signatures. Usually the author and three approvers sign the approval page. The approvals should come from the people affected by the validation project, such as production, QA and engineering functions related to the facility. A development signature may be necessary if the project relates to the manufacture of a new product.

As a general rule it is not a good idea to have too many approvers as there is a danger that scrutiny and understanding starts to suffer because each approver will be expecting others to have checked certain items. It is important that the approvers know what they are signing for. As with all validation documentation, the continuity of the dates from the signatures is important. The author should sign first, followed by the others, with QA input last.

Introduction

The introduction should explain why the project is being undertaken, where it is going to be located and the broad timetable.

Aim

The aim should explain that this is to be a formal validation study on a specific project and show that the approach conforms to cGMP. The aim may point to the various company policies and procedures under which the VMP is to be prepared and controlled.

Description

This section should describe the main features of the project in concise terms, picking out particularly critical features or acceptance criteria.

Facility

This section of the VMP should outline the facility's intended use, briefly discuss how it is to be built and state whether it is an entirely new facility or an expansion of an existing one.

For example, it could describe the size of the facility, the number of floors the facility occupies, the processing areas and, if necessary, the segregation for contamination; how many HVAC systems there are, and what the classifications are; any special gowning procedures or other procedures to be followed. Some simple outline drawings will generally be included with the description — typical drawings to insert are:

- facility location in relation to site;
- cross section of the facility (if relevant);
- floor plan (one for each floor) with equipment locations;
- HVAC zone identifications;
- personnel flow;

- component flow;
- raw material flow;
- product flow.

Services/Utilities

This section may consist of a list of plant utilities and services, such as cold potable water, purified water, water for injection, plant air, instrument air, nitrogen, chilled water.

In addition to this listing, there should be a brief description with simple line diagrams for each system, which should include any key performance criteria such as minimum flow rate or pressure, and quality. However, detailed requirements of the systems can be written into individual protocols — this helps keep the VMP to a sensible size and makes it easier to control.

Equipment

As with the previous section, this could start with a list of all the major items of equipment that are going to be installed into the facility, for example, porous load steam sterilizer, bench top balance, or powder mixer. It is a good idea to divide up the list by facility area or stage in the process. The list that is generated should include a unique plant item number for each major piece of equipment for reference purposes. For the most important items it is a good idea to include a brief description with a simple line diagram with any key performance.

Products

In this section, information should be provided about the products that are going to be manufactured in the facility in question. For each product this may include:

- batch size;
- ingredients:
  - quantities per unit dose;
  - quantities per batch;
- the steps by which the product is manufactured:
  - process flow diagrams;
  - summary of manufacturing method.

Computer systems

This section lists all the computer systems associated with the facility, process equipment and utilities as well as IT systems to operate the plant such as LIMS, SCADA and MRP systems, and provides descriptions of each system picking out any important performance.

Validation approach – overall

This section of the VMP is used to describe how the validation work is to be performed and documented (see Figure 4.1 on page 49).

It gives the design engineer's viewpoint of the Validation Master Plan. Note that it starts with the User Requirement Specifications (URSs), which is usually prepared by the user in discussion with the design engineer. This document forms the basis for the design.

This flow chart forms an excellent checklist for the validation process and underlines the importance of preparing validation documentation right from the issue of the URS to the performance qualification of the plant built to the final design. The main aspects of this flow chart, which provide the design engineer with a good background to the validation process, are detailed.

Process evaluation and validation systems

This section should explain how the facility has been divided up into separate systems and how the process has been evaluated to determine what aspects are critical to product quality. It should introduce concepts such as 'critical para-meters' and 'critical instrumentation' and relate these to the validation require-ments, in line with the method described in Section 4.2.2 and 4.2.3.

Validation team

This section defines the role and responsibilities of key personnel involved. It is often a good idea to use job titles rather than names since individual personnel may change, and to include a project organization chart. In particular, it is important to explain the role of QA in the approval processes.

Validation methodology

The validation methodology should describe what types of documents will be generated within the project (protocols and reports — Design Qualification (DQ), Installation Qualification (IQ), Operational Qualification (OQ), Perfor-mance Qualification (PQ), and Process Validation (PV)) and how they will be prepared, reviewed, approved and controlled. This section should draw on company policies and procedures, which should define each part in more detail. In addition, as appropriate, the methodology should discuss cleaning valida-tion, analytical methods validation and computer systems validation (there will be more about the various validation activities later in this chapter).

The section should then describe the execution strategy for the protocols including, for example, how results are recorded and how any problems encountered are dealt with, and the role of equipment vendors in validation

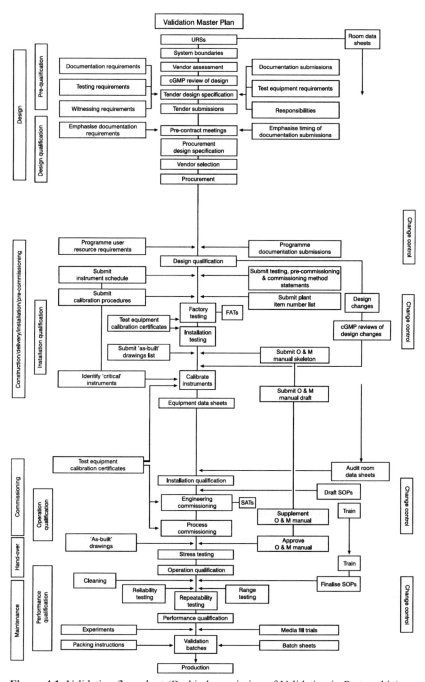

Figure 4.1 Validation flow chart (By kind permission of Validation in Partnership).

(i.e. utilize vendors as much as possible in the preparation and execution of validation work or do as much of the work as possible 'in-house').

This section can also be used to describe the organization and management of project documentation, including document flow and filing (for example, documentation filing structure, use of document management systems, IT).

Validation schedule

It is often useful (although not obligatory) to include a time schedule in the plan. It is probably best to keep this relatively simple, as schedules tend to change frequently during a project. The VMP is not intended as a document to convey this type of information.

Validation approach – detail

This section includes details of which types of documents are going to be produced for each system to be qualified and which processes are to be validated. This is often done by a validation matrix (see Table 4.1).

Other documentation required

This section should establish links to other types of documents that could be required at regulatory authority inspections. The type of documents which come under this heading include:

- batch production records;
- packing instructions;
- training;

Table 4.1 Example of a validation matrix

Item	Item no.	Document type				
		DQ	IQ	OQ	PQ	PV
Utilities						
HVAC	ABC123	✓	✓	✓	–	–
WFI	ABC456	✓	✓	✓	✓	–
Equipment						
Tablet Press	XYZ789	✓	✓	✓	–	–
Autoclave	XYZ123	✓	✓	✓	✓	–
Product						
Tablet A		–	–	–	–	✓
Tablet A, cleaning		–	–	–	✓	–

- SOPs;
- maintenance and calibration records;
- organizational charts and CVs;
- change control procedure;
- drawings.

4.4 Development of qualification protocols and reports

The VMP defines which systems are to be qualified and how the work is to be organized and controlled. The next step involves the preparation of qualification protocols and the generation of associated reports.

4.4.1 Qualification protocols

There are various different approaches to the format and content of qualification protocols — for example, protocols can be developed as stand-alone documents or can cross-reference other project engineering documentation. They can be designed so that results are recorded within the body of the protocol or that all the detail is left for recording in the reports. The former results in bulky protocols but brief reports, whereas the latter results in slim protocols and bulky reports. As with all validation work the protocols should be developed in accordance with company policies and procedures. There should be SOPs for protocol preparation, execution and reporting.

Whatever approach is taken, there are certain key features that the protocol must have. These can be summarized as follows:

- **formal documents:** The protocol must go through a review and approval process with final approval by QA; this must be numbered, the number of copies must be controlled and have a document revision history, page numbering must pass the 'drop test' (see Section 4.3);
- **defined scope:** The protocol must define what area, equipment, etc., it addresses. This may be achieved by, for example, a system description, diagram or list of items;
- **objective:** The protocol should describe the purpose and how this relates to the overall validation activity and scope of the protocol;
- **test structure:** Each test must describe the objective and purpose of the test, the test procedure and the method of recording results. This should be in sufficient detail so that it could be understood by a third party, and repeated if necessary;

- **acceptance criteria:** Each test must have acceptance criteria as to what constitutes a pass or a fail. The acceptance criteria must be approved before execution of the protocol.

A typical table of contents for a qualification protocol would consist of the following:

- title page;
- revision history;
- table of contents;
- introduction/background;
- purpose;
- scope;
- reference documents;
- system description;
- prerequisites;
- personnel performing the qualification;
- test equipment details;
- method;
- acceptance criteria;
- list of attachments.

4.4.2 Qualification reports

Once the protocol has been executed the results should be documented in a qualification report. At least one report should be written for each protocol. A typical table of contents for a qualification report would consist of the following:

- title page;
- revision history;
- table of contents;
- purpose;
- scope;
- executive summary;
- results;
- deficiencies and corrective actions;
- assumptions, exclusions and limitations;
- conclusions;
- appendices (depending on the protocol style adopted, one of the appendices may be the complete protocol).

The reports are also formal documents and should follow a similar preparation, review and approval process as protocols.

Deficiencies

As a general rule the report should be prepared by exclusion; that is, if a test was successful with no problems then only a brief mention is required in the report. The report should concentrate on the tests that failed and describe what remedial action was necessary and what retesting or further work was/is required. Examples of deficiencies include:

- conflicts with specifications — for example, the pump seal material was viton rubber not EPDM rubber as specified;
- information which is unavailable or incomplete;
- documentation discrepancies (incorrect reference number, issue number).

Each deficiency should be given a unique identification number and a complete list of deficiencies encountered during the execution of the protocol should be included in the report. An audit trail should be established to show how the deficiency was resolved.

4.5 Design qualification (DQ)

The purpose of design qualification is to ensure that the final design:

- accords with all relevant specifications and design intentions;
- meets the requirements of the process, product and user;
- adequately specifies all necessary supporting documentation;
- complies with the requirements and principles of GMP.

DQ is providing documented evidence that quality is built into the design. DQ is an auditing function to provide formal documentation that the facility has been designed to meet the requirements of the user and the GMP guidelines. DQ activities may include:

- GMP reviews of overall facility design;
- establishing the suitability of vendors and vendor deliverables through vendor assessment and auditing where appropriate;
- review and approval of equipment specifications and design documentation to ensure user requirement specifications (URS) have been adequately interpreted in the design process and that the design is in compliance with GMP.

DQ comes down to carrying out a formal comparison of what is required against the proposed design. There should be DQ documentation for:

- the overall facility;
- each system within the facility.

4.5.1 GMP reviews of overall facility design

The GMP review of the overall facility/project design can be defined in the same terms as an audit, that is a formal documented review of the design of a plant (including facilities, equipment, utilities, computerized/automation systems and procedures) to give assurance that:

- it complies with the applicable statutes and associated published current Good Manufacturing Practices;
- it complies with applicable regulatory licence(s) and registrations submitted for the particular process(es) or product(s) to be manufactured, held or stored.

Note that because of the confidential nature of the process, including licensing application details, the second point may be considered separately from the first.

Typically, topics to be dealt with include:

- facility (construction, finishes of walls, floors and ceilings, corners and crevices, cleanability, durability, access control, pest control, etc.);
- environment (area classification, temperatures, humidity, air pressures, air change rates, viable and non-viable particle levels, etc.);
- personnel flows (access authorization, change regimes, gowning requirements, occupancy levels, cross-contamination, etc.);
- materials flows (solids, liquids, gases, toxicity, hazard risk, containers, transportation, storage, cross-contamination, etc.);
- equipment flows (size, weight, mobility, cleaning, method of handling, cross-contamination, etc.);
- general equipment design (proprietary, purpose built, materials of construction, finishes, cleaning, change parts, control systems, etc.);
- automation philosophy (monitoring or controlling, level, protection, environment, access control, archive storage and retrieval, electronic signatures, disaster recovery, etc.);
- maintenance/servicing (access, space, tools, diagnostic equipment, materials, power, lighting, authorization, training, etc.);
- documentation (SOP's, permits, history records, training, log books, etc.);
- waste management (liquids, solids, gases, packaging materials, cleaning, etc).

4.5.2 DQ of each system

Vendor assessment

Vendor assessment is the documented evaluation of the suitability and capability of the vendor to provide the 'system' to be procured to the quality required to fulfil user and cGMP requirements, including all necessary supporting documentation. Where appropriate this may include vendor auditing.

Vendor assessment may stretch over several stages including assessment of the vendor's suitability to tender, assessment of preferred vendor and follow up vendor audit(s). Vendor assessment would generally involve, for each primary critical system including primary critical computer system, sending out self-assessment questionnaires and then, where appropriate, auditing vendors prior to placement of orders. Subsequent audits may be required throughout the design and construction/implementation process depending upon the nature of the system and the findings of the assessments and audit.

DQ of system plant

Design Qualification (DQ) of system plant (in other words, equipment, piping, valves and in-line fittings, field instrumentation, ductwork, insulation etc., or combinations of these) is the documented evidence that quality is built into the design of the system. It should include verification that the 'system' design incorporates the requirements of the user and of cGMP. Typically the DQ activities will include.

- cGMP review of design;
- specification review (URS/design specification(s) review);
- compilation of design documents;
- QA/QC review;
- facility acceptance testing (FAT).

4.6 Installation qualification (IQ)

Installation qualification is the documented evaluation of the equipment or system to establish that it has been installed in accordance with design specifications, cGMP requirements and manufacturers recommendations. Typically it will consist of various static checks, which may include for example:

- **system completion:** Check that the system is mechanically complete and all critical punch list items have been cleared. Check that all work which should have been completed and documented during the construction and installation of the system has been performed. This will involve checking through the various construction check sheets and certificates;

- **security/utility connections:** Check that the correct connection of utilities has been made and that, where appropriate, utilities have been IQed;
- **documentation inventory:** Check that all necessary supporting documentation such as specifications, operation and maintenance manuals are available and have been reviewed and approved;
- **equipment inventory:** Check that installed equipment name plate data complies with specification and record equipment serial numbers;
- **materials qualification:** Check that, where appropriate, contact part materials, surface finishes and lubricants are in accordance with the specification. This may involve a review of material certificates, chemical data sheets etc., or performing physical inspection and testing of materials;
- **drawing validation:** Perform a P&ID walk-down to check that all main components are as shown and in the sequence indicated. Where appropriate check pipework slopes (is it free draining?), measure pipework dead legs and drainage air gaps, check accessibility of manually operated devices;
- **main equipment features:** Check that each main component is in accordance with the construction drawing, check critical specifications such as filter grade, perform any static checks required prior to start up, such as checking lubricant levels, drive belt tension and torque settings;
- **instrument calibration:** Check that all critical instruments have been calibrated and that the calibration is traceable to national standards;
- **spares and maintenance:** Check that adequate spares provision has been made and maintenance requirements have been considered. This may involve, for example, getting a copy of the spares list reviewed and approved by the maintenance department and then checking that all spares have been supplied, and checking that the maintenance and calibration programme for the system is in place and that equipment log book(s) have been prepared.

4.7 Operational qualification (OQ)

Operational qualification is the documented evaluation of the system to show that it operates as intended throughout the anticipated operating ranges. Typically it will consist of various functional checks on the equipment, generally performed using inert materials such as water or compressed air and in the absence of real product.

Tests should be designed to show that the equipment would perform as intended and to specification. The tests should encompass upper and lower processing limits and circumstances, including those within normal operating conditions, which pose the greatest chance of process or product failure compared to ideal conditions. These conditions are widely known as 'worst case' or 'most appropriate challenge' conditions.

For utilities it is important to show that the utility can be delivered within the requisite parameters (such as flow rate, temperature, quality, etc.) under conditions of maximum diversity (i.e. with the greatest or least preserved normal operating demand on the system from the most or least users of the system).

It is difficult to provide typical examples of tests conducted during OQ because they will be dependent upon, and specific to, the system under test, but for example the tests on a dispensary area downflow booth could consist of:

- air supply system:
  o downflow and bleed air velocity (check that when correct velocity is achieved inside the booth the volumetric flow rate is within range);
  o green zone velocity test (to ensure that the green zone of safe airflow is set to correspond to an average filter face velocity of between 0.45 and 0.55 $msec^{-1}$);
  o filter pressure differential test (to ensure that the pressure drop across each filter is within the correct operating range and to provide a baseline clean filter reading);
  o dirty filter simulation test (to ensure that the airflow rate is controlled to maintain correct downflow velocity with dirty filters);
- control and indication system:
  o temperature control and indication system (to demonstrate the functionality of the temperature control and indication system and show that booth temperature can be maintained with specified limits with maximum heat load generated in the booth);
  o dehumidification control and indication system (to demonstrate the functionality of the dehumidification control and indication system and show that booth humidity can be maintained with specified limits with maximum moisture load generated in the booth);
- containment systems:
  o HEPA filter integrity testing (check that all HEPA filters are integral and pass the DOP test);
  o smoke containment test (to demonstrate using smoke that the booth contains emissions generated within the safe working zone at both the minimum and maximum safe airflow setting, and that fresh make-up air

drawn in from outside the booth is drawn in and maintained below bench top height through to the back of the booth);

- light and sound levels:
  - light levels (to confirm that the lighting levels are within range for an industrial working environment);
  - sound levels (to confirm that the sound levels are within range for an industrial working environment);
- safety systems:
  - air flow alarm (to demonstrate the functionality of the unsafe flow alarm system);
  - emergency stop (to demonstrate the functionality of the emergency stop system and check that all devices move to fail safe condition).

OQ and commissioning

OQs demonstrate the functionality of the installed system and are often carried out as part of commissioning. Engineering commissioning is normally undertaken by a 'system' vendor and is geared to starting up the 'system.' OQ work is more concerned with the operating parameters of the 'system' and with the identification and independent measurement of operating variables over their normal operating ranges.

However, depending on how contracts are let and the responsibilities for the 'system' testing are specified, the vendor or installer may be requested to carry out certain OQ activities as part of commissioning work. For instance, in the case of the commissioning of a HVAC system, it may fall within the scope of the engineering activities to stimulate certain 'worst case' conditions such as the effects on the air pressure regime of a power dip.

The OQ protocol should require verification of the satisfactory completion of all such commissioning activities.

4.8 Handover and process optimization

Most projects undergo a period of plant handover following completion of OQ. This is normally the time that 'ownership' of the facility is transferred from the engineering function to the user function. If a main process contractor is running the project then this is often the point that completes their contractual responsibilities.

Generally, before the next stage of the validation can begin, a period of time is spent optimizing the process. Process optimization can take various forms depending upon the nature of the process and facilities. For example in BPC plants this may encompass 'solvent trials', where solvents to be used in the facility are first introduced. This may require re-tuning of control loops that have only previously operated with water. The nitrogen system may now switch from running on compressed air over to running with nitrogen. Plant safety is clearly of primary concern during this phase.

Typically during this period operator training will be underway and the SOP's required to operate the facility, run the process, and maintain the equipment will be finalized.

4.9 Performance qualification (PQ)

Prior to commencement of PQ all operators involved must be trained and the procedures that will be required during production must be available, since they should be used during the PQ.

Performance qualification is the documented evaluation of the system to show that the system operates as intended throughout the anticipated operating ranges, under conditions as close as possible to normal production. Typically it will consist of various functional checks on the equipment, generally performed using actual product.

PQ work should be performed on systems whose performance or process parameters are critical and could affect the quality of the product. Examples of the systems requiring PQ work are pieces of process equipment such as a production sterilizer and critical utilities such as a WFI system.

As with an OQ, the critical parameters and acceptance criteria of the system under consideration should be defined. Once these have been defined, the test that is required to show the parameters are met can be designed. To successfully complete PQ work it is necessary to examine a number of consecutive batches or runs. One should also consider the variability to be expected to show that it does not affect product quality — i.e. 'worst case' conditions.

Normally any samples taken during PQ testing work will be taken by the user's personnel, not by vendors or outside contractors responsible for installing and commissioning of the system.

The contents of a PQ protocol may include for example:

- approval page;
- system description;
- purpose;
- sampling regime;
- testing regime;
- acceptance criteria;
- deviation and corrective action.

4.10 Process validation (PV)

Process validation is defined as:

'Establishing documented evidence which provides a high degree of assurance that a specific process will consistently produce a product meeting its predetermined specifications and quality attributes.'

In essence, a PV is a PQ of the manufacturing process. As with a PQ, the critical parameters and acceptance criteria of the process steps should be defined. The parameters can be associated with the raw materials used in the process, with the equipment used, or with process variables (time, pressure, temperature, etc.). Identifying the critical parameters and understanding how each of them can adversely affect the finished product is the first step in the validation cycle.

The second step is to examine the effect of each of the critical parameters on the process to ensure that the variability in the parameter anticipated during routine production does not adversely affect the quality of the product. This procedure of examining the practical limits of the critical parameters is often referred to as 'worst case' validation or 'most appropriate challenge' conditions. It is essentially examining the robustness of the process.

The third step to successfully complete PV work is to examine a number of consecutive batches (usually three). The sampling and testing of these batches should be designed around the critical parameters. This step is what many companies have traditionally undertaken to validate their process. It is essentially examining the reproducibility of the process, and is acceptable if the process being validated is robust; but this is often not the case — hence the need for the first two steps.

The process should be considered as a series of functional steps. Each step should have a recognizable end point, or deliver a significant change to the material such as an increase in bulk, change of identity, change of physical or chemical form, change of container.

Process validation is associated with the process and not with the product. It is the list of instructions that is being qualified. An alternative process that produces the same product will be subjected to a separate process validation. Each functional step must be examined three times. In many instances a batch will comprise a number of sub-lots — it is not necessary to examine every functional step in all sub-lots of the three subject batches.

The protocol is often based on demonstration batches or manufacturing batch records. The contents of a typical PV protocol should include:

- approval page;
- system description;

- purpose;
- sampling regime;
- testing regime;
- acceptance criteria;
- deviation and corrective action.

Process validation data is presented as a report. It is important to note that it is the review of all the batches involved together, not a series of separate individual reviews.

4.10.1 Retrospective process validation

When a product has already been manufactured successfully for at least three years (and at least twenty batches have been made), a statistical review of all the data pertaining to at least the last twenty batches can be carried out.

No batches may be omitted from this review unless documented reasons are included to explain each individual case (examples would include equipment failure, or contamination not associated with the process). If more than 20% of past batches are omitted, the retrospective process validation should be abandoned, as it is likely that influencing systems are not under control. Only when these are identified and addressed should the validation project recommence.

4.10.2 Sterile products

Process validation for sterile products can be considered in two parts:

- validate the process to gain assurance that the system can deliver a sterile product. This would include, for example, themal mapping, thermal commissioning, filter integrity testing and control systems testing;
- validate the manufacturing process of the actual product including process technology and biological testing.

4.10.3 Bulk pharmaceutical chemicals (BPC)

For BPCs process validation starts at the point where the drug substance is chemically formed or where other impurities will not be readily removed.

4.11 Cleaning validation

The creation and implementation of effective cleaning processes is an essential part of any pharmaceutical production process. The two main reasons for this are:

- to ensure that the appropriate level of general cleanliness is maintained in order to prevent the accumulation of dirt and microbial contamination which could affect the quality of the product;

- to minimize the risk of cross-contamination from one active product into the subsequent product, which could lead to serious adverse effects on patients. Cross-contamination could also result in degradation of the main product and loss of potency.

4.11.1 Choice of cleaning method

Various approaches can be taken to ensure that cross-contamination levels are minimized between two different products.

The simplest approach is to dedicate a complete facility, its building, services and equipment, to a single product. Obviously this is a very expensive approach, unless the product is required in sufficient quantity to justify a dedicated facility. For very active products such as penicillin, cephalosporin and hormones, where cross-contamination at very low levels is not acceptable, this is the safest option and is a regulatory requirement.

In dedicated facilities effective cleaning procedures still need to be developed and validated, although the stringent cross-contamination levels that are usually applied to multi-product facilities can be relaxed somewhat and the emphasis placed on general levels of cleanliness in accordance with GMP.

In most circumstances though, facilities are multi-product and effective cleaning processes must be developed and validated by means of sampling and measuring the levels of cross-contamination.

The most common type of cleaning process involves the full or partial dismantling of equipment, followed by solvent washing and subsequent drying of the separate parts. Water/steam (with or without added detergent) is the most common cleaning solvent, but organic solvents can also be utilized.

Manual cleaning is still used extensively in the pharmaceutical industry but 'clean-in-place' (CIP) systems are rapidly expanding and 'sterilization-in-place' (SIP) is also being introduced.

It is quite common and also highly desirable to dedicate specific parts of the equipment which are difficult to clean, thereby reducing the overall time and cost of the cleaning process. Examples of this are the woven fibre filter bags used in fluid bed dryers or the rubber/plastic o-rings found in pipework.

These examples illustrate the importance of designing an effective cleaning process using a variety of techniques before embarking on any validation work. Remember, successful validation will only confirm that the cleaning process is effective, it will not make an ineffective one effective!

4.11.2 Measuring the level of cleanliness

As part of the overall validation programme the actual level of cleanliness that has been achieved by the cleaning process must be measured. This involves a three-stage process:

- a sampling method to detect and pick up the remaining contaminants;
- a method of analysis to quantify the amount of contaminant remaining;
- a calculation to extrapolate the results.

The usual sampling methods are:

- swabbing;
- aqueous/solvent rinses;
- non-active product follow through.

(a) Swabbing

Swab testing involves the use of dry or solvent impregnated swabs, which are wiped over a known area of the processing equipment. The contamination picked up is extracted in the laboratory by soaking the swab in a suitable solvent, and the solvent is then analysed to give a quantitative result. The total quantity of the contamination is calculated by multiplying the total area of the equipment by the swabbed area. In practice, the swab is unable to pick up 100% of the contamination, but it is possible to run a laboratory test beforehand to estimate the percentage pick up. This is done by deliberately contaminating the stainless steel plates (or sample of whichever material is in contact with the product) with a known quantity of contaminant, usually letting a solution evaporate on the plate. The plate can then be swabbed and the swab analysed to demonstrate the percentage of the contaminant that has been picked up. The analytical method must also be checked to ensure that the swab itself does not interfere with the result by running blank swab tests.

(b) Aqueous/solvent rinses

Aqueous/solvent rinses are commonly used in areas where it is difficult to swab (such as pipework or a sealed reactor in a bulk chemical plant). The method involves rinsing with a known volume of water/solvent and then analysing a small quantity of the rinse. The total amount of contaminant is simply:

$$\frac{\text{Quantity in sample} \times \text{Total volume of rinse}}{\text{Volume sample}}$$

The solvent used must provide sufficient solubility to pick up the contamination effectively but must not degrade the contaminant. The contact time must be controlled.

The main drawback of this method is that only material dissolved in the rinse water/solvent would be analysed and it would not be possible to find out how much was left inside the pipework, vessel, etc. The solubility of the contaminant, contact time and physical force of the rinse will all affect the final results, and it may not be possible to ensure all the areas have been adequately rinsed.

(c) Non-active product follow-through

The non-active product follow-through is sometimes used, and involves processing a non-active substance through the whole process and then analysing samples for the contaminant. The calculation is analogous to that used for the rinse method, but this method has the advantage that it mimics the real situation of a subsequent batch being processed, and that it covers all the equipment involved. However, as with the rinse method, only the contaminant that has been picked up can be measured, and not the contaminant left behind. Also, in the case of solid dosage forms, the contaminant may not be uniformly mixed throughout the non-active substance.

The swabbing method is generally preferred because it permits the areas likely to be most heavily contaminated to be targeted more thoroughly and also makes allowance for contamination not recovered, provided the laboratory tests are undertaken. Despite all this, it is still prone to variability since no two samplers will swab in exactly the same manner. The inherent variability in any of the sampling methods is one of the reasons for the use of a 'Safety Factor' when calculating the acceptable contamination limit.

4.11.3 Setting limits

When a cleaning process is used only between batches of the same product (or different lots of the same intermediate in a bulk process), it is normally only necessary to meet a criteria of 'visibly clean' for the equipment. Such between-batch cleaning processes do not normally require validation.

Chemical cross-contamination limits

One of the basic concepts of validation is that a process is proven to be capable of performing to a *pre-defined limit*. There is no exception with cleaning validation and although agreeing a pre-defined limit can be difficult, it is essential to establish one prior to commencing the validation work itself.

As there are often no obligatory legal or regulatory limits, manufacturers have come up with their own viable methods for setting limits.

The simplest of these methods is to set a blank limit to all products. A typical limit would be 1 to 10 ppm. This approach has been used in the bulk pharmaceutical chemical production and product development areas where a large number of compounds are processed and for many of them relatively little is known about their properties. The scientific rationale for limits in the region of 1 to 10 ppm is that this is somewhere near the limit of detection for suitable analytical methods for many compounds, and pharmacopoeia limits for heavy metals and other adulterants tend to lie in this region. The problem with this approach is it makes no allowance for the different pharmacological effects of different compounds. This will lead to excessive cleaning and wasted time and resources in some cases, whilst in other cases it may leave patients exposed to potentially hazardous levels of contamination.

Several companies have adopted a limit where the maximum amount of contaminant (A) that can be ingested by a patient taking the product B, manufactured immediately after product A, is one thousandth of the minimum normal therapeutic daily dose. The figure 1000 is used as a safety factor, which not only reduces the daily dose below pharmacological activity level but also allows for the errors inherent in the sampling and testing methods used.

Finally, the limit of detection for the assay method must be considered. Setting a limit of 0.001 mg per swab when the assay limit is 0.01 mg is pointless. Either the assay method needs developing, or the limit of assay will have to be the acceptance criteria.

Microbiological cross-contamination limits
Most cleaning validation protocols do not include sampling and testing procedures for microbial contamination. This is because the sterilization itself is validated for processes where minimization of microbial contamination is important (sterile and aseptic).

It is important that the cleaning procedure does not actually increase the level of microbial contamination. This requires the cleaning agents to have a low level of microbial contamination, and the drying procedures to adequately remove all traces of water. Storage of equipment is also important — it should be kept clean and dry and well covered or wrapped. There should be a maximum storage time defined, after which the equipment is cleaned again.

Where it is felt necessary to confirm that a particular level of microbial contamination has been achieved, swabs can be impregnated with a suitable growth media. The use of media impregnated swabs or media solutions will itself contaminate the equipment, which must be cleaned thoroughly before routine use.

4.11.4 Validation of CIP systems
For CIP systems there are several steps to be undertaken before any actual sampling and testing is carried out.

CIP validation cycle
- Assess design of CIP system including analytical method development;
- Experimental work to optimize cycle and cleaning agents and including analytical method validation;

- Change control system;
- Operational qualification;
- Cleaning validation protocol;
- Cleaning validation report for three successive cleaning cycles.

CIP systems are usually fitted to large immovable pieces of equipment, such as dryers and coaters. Often the CIP system will adequately clean the large flat surfaces of the equipment, but will leave excessive amounts of material in the corners, crevices, inlet/outlet ports, and around and behind seals and flaps. Therefore, before starting with validation protocols, the design of the CIP system should be assessed to eliminate (or at least minimize) any obvious weak areas. For example, one simple test often performed to determine coverage involves coating the item to be cleaned with an appropriate dye, then operating the cycle to determine if all the dye can be removed. If alterations to the CIP system itself are impractical, then it may be possible to remove part of the equipment for separate manual washing.

The main advantage of a CIP system is that it should provide a reproducible cleaning process. This process needs to be effective and optimized to provide the best chance of successfully validating the cleaning process. Experimental work can be performed using different wash cycles, rinse cycles, detergent types, drying conditions, etc. to establish the most effective conditions. If a range of products is to be cleaned then experiments should be performed on the most difficult to clean product.

Having established the most effective conditions, the CIP system and cleaning cycle should form part of the formal OQ for the equipment, to demonstrate that the critical parameters used in the cleaning cycle can be satisfactorily achieved and reproduced.

In parallel to the experimental work and OQ activities, analytical methods will have to be established and validated.

Finally, the cleaning validation/PQ protocol can be written and executed. This protocol can be either a stand-alone document or part of the general PQ protocol. Either way, the cleaning validation protocol is specific to a particular changeover between two products on a specific set of equipment.

The protocol should include the following sections:

- definition of equipment being used;
- definition of the product(s) being cleaned from the equipment, and the product that will subsequently occupy the equipment;
- explanation of the parameters being used in the cleaning process (temperature, times, pressures, detergent types and concentrations, etc.);
- sampling regime (sampling method(s), number and location of samples);

- testing procedures (description of tests to be performed on samples);
- acceptance criteria (acceptable maximum levels of contamination in each of the samples).

The validation protocol should be performed on at least three successive occasions to demonstrate reproducibility.

When the analysis of the samples is complete, the data should be collated, summarized and presented in a validation/PQ report. Comparison of the data to the pre-determined acceptance criteria will form the basis of the conclusions. Any missing data or data that is outside the acceptance criteria should be accompanied by an explanation. If the validation has failed then the cleaning process will have to be altered and the work repeated.

On completion of a successful cleaning programme, the validated cleaning procedure must become subject to the plant's change control system.

4.11.5 Validation of manual cleaning

Manual cleaning validation cycle:

- Experimental work (optimize cleaning method, drying cycle, etc.);
- Change control system;
- Prepare standard operating procedure (SOP);
- Operator training including retraining/re-evaluation;
- Evaluation of training;
- Cleaning validation protocol;
- Cleaning validation report.

Most equipment is relatively small, easily dismantled and portable to facilitate frequent and rapid cleaning. Operators often dismantle, clean and reassemble the equipment.

Operators are people and are therefore variable. Whilst it is virtually impossible to totally eliminate this variability, it can be minimized to an acceptable level by the use of clear and concise instructions (SOPs) together with regular training and assessment of the operators. Part of the validation of any manual cleaning method should involve the evaluation of the process to determine the level of variability — a high variability (even if within acceptable limits) suggests a process that is poorly controlled.

The actual validation protocol will be very similar to that used for the CIP system validation, but it must refer to any SOPs associated with the cleaning procedures.

4.12 Computer system validation

Automated or computerized systems are validated using the same general validation approach identified for equipment and utilities. However the nature of computer systems means that certain activities become particularly critical. A software programme is not a tangible thing and cannot be tested exhaustively (i.e. with large programmes it is impractical to prove the code) since to test every possible path through the code under every possible set of circumstances would take an inordinate length of time. For this reason the quality and confidence must be 'built in'. Software development must be carefully planned and controlled under a quality assurance system following a life-cycle approach. It should be noted that the term 'computer system' refers to the computer hardware and software as well as the interface between the computer and the machine/plant/environment.

Various models have been developed for the validation of computerized systems such as that proposed by IEEE (IEEE Standard for Software Verification and Validation Plans); the PDA report on the validation of computer-related systems or the GAMP (Good Automated Manufacturing Practice) Supplier Guide for Validation of Automated Systems in Pharmaceutical Manufacture. All these models are fairly similar. This section will not cover in detail the 'engineering' associated with the design, development and testing of computer systems but will concentrate on the validation activities associated with each stage.

4.12.1 Assessment of computer systems to determine validation requirements

The necessity for computer system validation is based on several criteria. The first of which is that the element in question is to be classified as a computer system (for example, some instruments may be programmable and may or may not be treated as a computer). The following criteria should help determine whether the element is a computer system:

- **inputs and outputs (I/O):** The presence of physical channels (digital, analogue, pulse, serial, etc.) for importing or exporting data that is used or has been calculated by the element;
- **memory:** A means of storing executable code is used;
- **Central Processing Unit (CPU):** Use of a device for interpreting executable code using data accessed from inputs, and presenting the result via outputs.

If all the above criteria are present then the element can be assumed to be a computer system and should be treated as such from a validation point of view.

The next step is to determine if validation is required. This involves a process of evaluating the role that the computer system plays. Assessment criteria include:

- **GMP implication:** Generally any computer system with GMP implications should be validated. This includes for example critical operations such as controlling or monitoring operations that can affect product quality;
- **system functionality:** If the computer system is only used for supervisory tasks, with no computer-generated information being used by or forming part of the batch record information then generally the computer system does not require validation;
- **safety critical systems:** Although GMP does not cover safety critical systems, there is a good argument for them being treated in the same way;
- **system configuration:** Although a computer system may be involved with critical operations, it might be that another independent system provides a full check of the operation of the computer system. In this case the computer system does not generally require validation;
- **system operability:** Although the system may be computerized, the corresponding operating procedures may introduce so many manual operations and checks that all computer controlled operations are duplicated by the way the system is operated. In this case the computer system does not generally require validation.

Once it has been determined that computer system validation is required, the detailed validation activities will need to be determined. The extent of computer system validation depends upon two main factors — the level of standardization and the complexity of the system. A standard system has been largely validated by its wide use, so most of the validation effort should go into validating the system with respect to the user's particular circumstances. The issue of system security (prevention of modification or reconfiguration) must also be addressed. Generally the simpler the system, the less validation effort is required. There is a risk that because simple systems are easier to understand they tend to be more 'fully' validated. Instead increased emphasis should be placed onto more complex systems.

These two criteria should be applied to both the computer hardware and software.

Hardware
The hardware can be classified as either standard hardware (produced in large quantities over an extended period) or application specific (mainly produced for the applicable project only). Both will require validation but the approach to

standard hardware is simpler, mainly being concerned with the configuration, installation and functional testing aspects. The design and design process must also be considered for application specific hardware. This may involve assessing the methods employed, critical components, compatibility between units, standards used for design and testing, type testing carried out, etc.

Software

There are generally three types of software that can be identified for computer systems:

- **system software:** This is the software required to run the computer system itself. It includes all the operating systems (the software controlling the CPU, memory, I/O, operator interfaces, etc.) as they are configured for a particular computer system. Normally this software does not require validation because it is classified as 'standard software' (see below).
- **configurable software:** As the name implies, this type of software would normally be standard software, which can easily be adapted to an applicable project, such as Lotus 1-2-3 for example. The software purchased from Lotus is classed as standard software, which does not require validation (because of the wide use of this software), but its use with formulae applicable to a specific project must be validated. Configurable software is also sometimes referred to as 'canned software'.
- **application software:** This software is produced or configured specifically for the applicable project and must be validated.

The term 'standard software' is often used as a reason for not performing validation. The following criteria may be used to determine if a piece of software is standard:

- **the supplier's QA system:** Ideally this should be a recognized system such as ISO9000 or similar and it should demonstrate that development and testing of the software is controlled and documented;
- **the product being widely used:** This is generally interpreted as meaning more than 100 similar units. It is of further advantage if the software has been applied to a wide range of applications, and thus more thoroughly exercised and tested;
- **product age:** Product history and experience including knowledge of 'bugs' will increase with age. Standard software is usually expected to have been in wide use for a minimum of twelve months.

- **version control:** Software is usually developed and corrected during its life-cycle. The number of software versions can be great, so a system of version control must be in place to be able to take all versions into account with respect to product age and usage;
- **user feedback:** The vendor must be able to demonstrate that feedback from users is handled and acted upon;
- **not application specific:** The software cannot be classed as standard if parts of it are specific to the particular application.

If all the above factors are fulfilled then the software can be classed as standard and does not require validation. However the computer system may still require validation including functional testing.

The results of the above assessment should be documented and included in the Validation Master Plan.

4.13 Analytical methods validation

Analytical methods can be validated in a number of ways. Compendial methods such as methods appearing in the USP are generally considered validated, but it is important to demonstrate that the method works under the actual conditions of use. If a compendial method exists but a company elects not to use it, they must demonstrate that the in-house method is equivalent or superior to the official procedure.

Validation data from repetitive testing should be consistent, and varying concentrations of test solutions should provide linear results.

4.14 Change control and revalidation

4.14.1 Change control

All process and plant subject to validation should be covered by a change control system that enables formal reporting and recording of changes, reviews the impact of a change on the validation status and permits revalidation requirements to be identified.

Change control standard operating procedures should define which changes do and do not require change control. Generally, items subject to change control include:

- procedures that contain validated activities or processes (for example, cleaning, equipment operation, sterilization);

- process equipment and plant;
- facilities;
- utilities;
- production processes;
- commodities (primary packaging components, filters, sterile clothing, disinfectants, cleaning agents);
- raw materials;
- computer systems;
- test methods and specifications.

Standard operating procedures and change control forms should allow all proposed changes to be considered, commented upon and approved or rejected by relevant experts. These experts generally represent Quality Assurance (whose authorization is always required), Production, R&D, and Engineering, though other experts may be consulted as necessary. Reviewers should identify whether the change needs to be validated and, if so, outline the nature and extent of validation required.

It is recommended that change control forms reference qualification protocols in those cases when revalidation is necessary. The date of re-introducing the process or plant subject to change into operation should be recorded so that it is clear that revalidation, when required, has been completed before use.

On occasions, where an emergency situation occurs, an unplanned change may have to be implemented without prior formal consultation. In such cases details of the change should be introduced into the change control system as soon as possible.

Where a planned change is not approved, it must not be implemented. Where an unplanned change is not approved, the process or plant must immediately be returned to its original state.

4.14.2 Revalidation

In order to maintain the plant, facilities, systems, procedures, methods and processes, once initially qualified, in a state of validation throughout their life-cycle there should be continuous review of the need for revalidation and implementation of revalidation whenever it is agreed to be necessary.

Revalidation requirements should be defined based on a technical review of the initial qualification(s), change control data and documentation supporting the performance of the item subject to validation. Revalidation will be undertaken if a change is likely to affect the validated status or if the performance of the validated system is seen to have deteriorated. Revalidation exercises should be built into the Validation Master Plan.

The need for revalidation may be identified via several mechanisms:

- through a change control procedure;
- by regular review of the performance of a validated item to a predetermined schedule;
- by the use of a plant certification system;
- through annual product reviews;
- through internal audits.

Critical items of the plant are frequently covered by a routine certification and re-certification programme. Revalidation intervals and the test to be conducted are normally specified at the time of certification.

Summary

The key points from this chapter are as follows:

- validation is required to provide documented proof of GMP compliance. Validation activities should be organized as a scientific study that follows a life-cycle approach;
- validation activities should be conducted in accordance with pre-defined company validation policies and procedures under a validation master plan;
- the validation master plan(s) should define what will be validated, describe the validation approach to be adopted (this will reference the policies and procedures developed) and explain how the validation work will be organized and related documentation will be controlled;
- the validation activities should be lead by a validation team, which should consist of members from relevant disciplines participating within the project including members of the QA/QC function. The team will be responsible for organizing the validation activities and reviewing and approving associated documentation;
- the processes should be evaluated to determine what aspects are critical and require validation. This may include determining critical process steps, critical parameters and critical instrumentation and systems;
- in parallel with process evaluation, systems and system boundaries should be defined. This allows validation work to be broken down into logical, manageable sized packages and concentrates the validation effort in the most important areas;

- cGMP reviews should be performed at key points in the project life-cycle to confirm that the design complies with cGMP requirements and the specification;
- User Requirement Specifications (URS) should be prepared by the user for each system to be validated to formally document the final process requirements. These will form a key part of the basis for subsequent validation activities;
- validation activities should be documented and controlled through the use of qualification protocols and reports, typically these will fall into categories including DQ, IQ, OQ, PQ and PV.

Primary pharmaceutical production

5

ROGER SHILLITOE, PHIL MASON and FRED SMITH

This chapter considers the production of the bulk active ingredient or bulk pharmaceutical chemical (BPC) that is subsequently converted by physical means into the final drug's presentation form.

This area of the pharmaceutical industry has much in common with fine chemical manufacture. The unit operations carried out are similar and many fine chemical and speciality chemical manufacturers also manufacture pharmaceutical intermediates.

Traditionally, the bulk production was carried out on a different site to the R&D and secondary processing. The style of operation, attention to cGMP and culture of a primary site, was more associated with the type of chemistry or operation carried out.

Three main influences are changing the face of the BPC industry:

- regulators, particularly the FDA, are putting greater emphasis on reviewing BPC production, and recognize the effect that failure in quality can have on the finished dosage form;
- major pharmaceutical companies are focusing on 'Research and Development' and 'Marketing and Selling of the finished product'. Secondary manufacture to a limited extent, and primary or BPC manufacture to a greater extent, is being sub-contracted out to third parties;
- BPCs are becoming more active and tonnage requirements are dropping as a result. Linked with this, the size of the equipment used in the manufacture is reducing. The increased activity also brings increased handling considerations and limits for exposure, which in turn drives towards closed processing operations, which is also consistent with improvements to cGMP.

5.1 Reaction

The production of the BPC is by three main methods:

- **chemical synthesis**: Examples of synthetic conversions include aspirin, diazepam, ibuprofen. This method produces the largest tonnage;

75

- **biotechnology or microbial action**: Examples include antibiotics, vaccine production, blood plasma products. This method produces the high value products;
- **extraction**: This can be by extraction of natural materials from animal or plant material such as the opium alkaloids, dioxin, heparin, insulin (pigs pancreas), thyroxine (animal thyroid gland).

This chapter will concentrate on the first two methods. The extraction method for naturally occurring materials was the main source of drugs up to the 1930s but was being gradually replaced with synthetic routes to products. There is resurgence now in extraction techniques linked to the biotechnology area, where specifically developed or altered organisms are allowed to grow and produce a desired product that is harvested and extracted. This is discussed in Section 5.1.2.

5.1.1 Synthetic chemistry based processes

Various general synthetic chemical reactions are utilized in the synthesis of BPCs. These include simple liquid/liquid reactions, complex liquid reactions with catalysis such as Grinards, Freidel Craft, reaction with strong reagents such as phosphorous oxychloride, thionyl chloride or elemental halogens such as bromine or chlorine. Gas reactions with liquids are common for example with hydrogen, hydrogen chloride or phosgene.

Most reactions in the pharmaceutical industry are carried out on a batch basis, in non steady state operation. Continuous processing is occasionally used for a few generic tonnage commodity BPCs or where safety can be improved by the benefits continuous processing can bring by inventory minimization.

Conventional batch reactor systems

The batch reactor is the workhorse of the synthetic BPC industry. Typically made from stainless steel or glass lined mild steel, capacities ranges from 500 litres at the small scale to 16 m^3 at the large scale. Some processes employ reactors of even greater capacity but this is becoming unusual as the activity of new drug substances increases.

The reactor is typically fitted with an external jacket or half pipe coils so that the temperature of the contents can be adjusted. Occasionally if a high heat duty is required, further coils can be placed inside the reactor.

Typical operating conditions are from $-25°C$ to $+160°C$, and full vacuum to 6 bar g. Generally, reactions at elevated pressures above 1 bar g are uncommon, with the exception of specific gas reactions such as hydrogenation. However, more processes are now being developed where working at an elevated pressure brings benefits — for example, it can allow the selection of the ideal solvent for a reaction that could not normally be used at the ideal reaction temperature because this would be above its atmospheric boiling point.

The temperature is normally adjusted by indirect contact with a heating or cooling medium circulating through the coil or jacket, but direct heating with live steam or quench cooling with water or other materials is possible. The medium used for the heating and cooling fall into two main areas:

- **multiple fluids**: typically steam, cooling water, refrigerated fluid such as ethylene glycol or brine. These are applied in sequence to the coil or jacket as required;
- **single fluids**: typically some form of heat transfer oil, heated or cooled by indirect contact with steam, cooling water or refrigerant, and blended to provide the correct fluid to the coil or jacket.

Agitation is provided to the reactor to ensure good heat transfer and good mixing for reaction. Depending on the process requirements, various agitation regimes can be set up using different agitator profiles, speeds and locations.

Connections are made to both the top and bottom of the reactor to allow material to be charged into the reactor, material to be distilled from the reactor, and liquids to be drained out.

Reactors are normally fitted with a manway to allow entry for maintenance purposes. Historically, this was also the way in which solids were added to the reactor and samples were extracted, but this practice is becoming less common.

Alternative reactor systems

Other types of reactor systems exist with each having their own specific advantages for specific processes. These include the loop reactor that specializes in gas-liquid reactions at elevated pressures, such as hydrogenation, and the batch autoclave reactor that specializes in high-pressure reactions of 100 bar g and higher.

Materials of construction

Reaction modules can be constructed from other materials dependant on the chemistry being employed and requirements for heat transfer. These include glass, plastics and exotic metals such as hastelloy or titanium.

5.1.2 Biotechnology based processes

The processes in biotechnology are based on cultivation of micro-organisms, such as bacteria, yeast, fungi or animal and plant cells. During the microbial process the micro-organisms grow the product, which is either contained within the cell or excreted into the surrounding liquor. The micro-organisms need carbon substrate and nutrient medium for growth and the microbial process is normally performed in water.

There are essentially three steps to biotechnology processing, namely:

- fermentation;
- recovery;
- purification.

The equipment in which the microbial process is carried out is called the fermenter and the process in which micro-organisms grow or format product is called fermentation.

Once the product is formed it is recovered from the biomass or the liquor by downstream processing, e.g., centrifugation, homogenization or ultrafiltration.

Purification of the recovered product is then required. Two differing techniques are required depending on whether it is for bulk large-scale or for small-scale genetically manipulated organisms. Large-scale recovery can be likened to bulk chemical organic synthesis operation.

Fermentation

The fermenter is the equipment used to produce the micro-organisms. Biotechnology applications of fermentations divide conveniently between microbial types and mammalian cell culture. Microbial fermentation, which can encompass very large-scale antibiotics as well as smaller scale recombinant products, is characterized by fast growth rates with accompanying heat and mass transfer problems. Mammalian cell culture is characterized by low growth rates and high sensitivity to operating conditions. Both techniques have common design principles.

Several different types of vessel are used for large-scale microbiological processes, and their degree of sophistication in design, construction and operation is determined by the sensitivity of the process to the environment maintained in the vessel.

The following is a brief description of the main types of fermenters:

(a) Open tank

The simplest type of fermenter is an open tank in which the organisms are dispersed into nutrient liquid. These have been used successfully in the brewing industry. In the anaerobic stage of fermentation, a foam blanket of carbon dioxide and yeast develops which effectively prevents access of air to the process. Cooling coils can be fitted for controlling temperature during fermentation.

(b) Stirred tank

Stirred-tank fermenters are agitated mechanically to maintain homogeneity, to attain rapid dispersion and mixing of injected materials, and to enhance heat-transfer in temperature control and mass-transfer in dissolving sparingly soluble gases such as oxygen. The extent to which these are achieved depends mainly on the power dissipated into the medium by the agitator, so that the agitator is essentially a power transmission device. The effectiveness of the power input depends on the configuration of the agitator and other fermenter components.

For aerobic fermentations, air is injected through a sparger, a single nozzle or a perforated tube arrangement, positioned well below the lowest impeller to avoid swamping it with gas. The sparger should have provision for drainage so that no culture medium remains in it after the vessel is discharged.

The rate of air supply must be sufficient to satisfy the oxygen demand of the fermentation after allowing for the efficiency of oxygen dissolution achieved.

Instead of a rotating stirrer, some systems obtain the mechanical power input by using a pump to circulate liquid medium from the fermenter vessel through a gas entrainer and then back into the fermenter. This separates the liquid movement and gas dissolution functions into separate specialized units. Two designs have evolved using this principle — the 'loop' fermenter and the 'deep jet' fermenter. In the loop fermenter, the gas dissolution device is a subsidiary vessel into which gas is injected, and the gas-saturated liquid is recirculated to the main growth stage. In the deep-jet system, gas is entrained into a high-power jet of liquid injected into the liquid in the fermenter, re-entraining gas from the vessel headspace. Exhaust gas is purged partly from the vessel headspace and partly from the specially designed circulation pump, from which the degassed liquid passes through a supplementary cooler before passing to the gas entrainer. This system gives high gas dissolution rate, but has correspondingly high power consumption compared to conventional systems. The liquid and entrained gas can also be introduced into the fermenter through a 'bell', which holds the gas bubbles in contact with the recirculating liquid to enhance gas utilization.

(c) Gas-lift and sparged-tank fermenters

This design has no mechanical stirrer and the power required for mixing, heat-transfer and gas dissolution, is provided by the movement of gas through the liquid medium. The gas is, therefore, the power transmission system from the gas compressors into the vessel. While the relatively low efficiency of gas compression seems to make this design unattractive, it has some important advantages compared to the stirred-tank system. Firstly, the absence of a rotating agitator shaft removes the major contamination risk at its entry point to the vessel. Secondly, for very large vessels, the required power input for agitation is just too large to be transmitted by a single agitator. Thirdly, the evaporation of water vapour into the gas stream makes a small contribution to cooling the fermentation. The fermenter interior does, however, need careful design to ensure that the movement pattern of the gas through the system produces satisfactory agitation.

The various designs of non-mechanically agitated fermenters can be grouped broadly into sparged vessels and gas-lift (including air-lift) fermenters. Sparged-tank fermenters are usually of high aspect ratio, with gas introduced at the bottom through a single nozzle or a perforated or porous distributor plate. The gas bubbles rise through the liquid in the vessel and may be redispersed by a succession of horizontal perforated baffle-plates sited at intervals up the column. In the gas-lift fermenters, internal liquid circulation in the vessel is achieved by sparging only part of the vessel with gas. The sparged volume has a lower effective density than the bubble-free volume, and the difference in hydrostatic pressure between the two sections drives the liquid circulation upwards in the sparged section and, after gas disentrainment, downwards in the bubble-free section. The two sections may be separated by a vertical draught-tube.

Important design considerations for good fermenter operation

The following are important design considerations in fermenter operation:

(a) Aeration and agitation

Animal cells are shear-sensitive (mild agitation is therefore required) and they are often sensitive to air bubbles. These considerations impose significant constraints on oxygen transfer design. One way in which this problem has been addressed is by the use of gas exchange impellers. Another strategy is to circulate medium through the reactor while simultaneously oxygenating it in an external loop. A third method is to use silicon tubing through which air diffuses into the liquid medium.

Cell culture medium often contains serum, which has a tendency to cause foaming. Since defoamants may inhibit growth, agitation and aeration systems must be designed to minimize this potential problem. However, care must be taken in the amount of agitation applied because, although it provides good oxygen and heat transfer characteristics, it can result in mechanical degradation of the cells. Usually systems with gentle agitation also minimize foaming. The type of impeller, baffles, and tank dimensions influences the degree of mixing. Note that mammalian cell cultures are more easily damaged by these mechanical forces than microbial cultures.

(*b*) *pH*

The internal environment of living cells is approximately neutral, yet most microbes are relatively insensitive to the external concentrations of hydrogen and hydroxyl ions. Many organisms grow well between pH 4 and 9, although for any particular organism the required pH range is small and accurate control is essential. Note, however, that there are exceptions where growth outside this range can occur.

(*c*) *Sterile design*

The importance of sterile design cannot be over emphasized; even the presence of a single contaminant will be disastrous. The fermenter must be designed to be easily cleanable (smooth surfaces and no crevices), after which it must be sterilized. The most effective form of sterilization is to utilize clean steam to kill both the live micro-organisms and their spores. This is usually defined as maintaining 121°C for 20 minutes. Shorter times and higher temperatures can be used but not vice versa. The quality of the steam supply is important; clean steam is required for mammalian cell culture, whereas, plant steam with approved additives can be used for large-scale antibiotics.

If the fermentation design calls for sterility, the following special precautions are required:

- air should be provided by an oil free compressor;
- Clean in Place (CIP) and Sterilize in Place (SIP) systems should be incorporated into the design;
- the fermenter and all associated piping and vessels should be designed to allow sterilization initially by 1.5 bar g steam. Branch connections should be minimized. All lines should be free draining and have minimum dead legs with the correct type of valves specified. Selection of internal surfaces, piping design, and valves is critical in ensuring effective removal of unwanted organisms during sterilization and preventing subsequent ingress of contaminants from outside the sterilized system;
- many fermentation media, at the large scale, can be sterilized continuously by heat. Economies can be achieved by incorporating heat recovery exchangers in the system to preheat the feed;
- all seals and instruments must be designed to withstand steam sterilization;
- the equipment should be designed to maintain sterility e.g. to include the use of steam seals on agitator inlets, double O-rings for probe insertion and steam blocks on transfer lines;
- piping should be stainless steel;
- an integrated approach should be taken to the physical layout, the piping and instrumentation (P&ID) flowsheeting and the sequencing to ensure that sterility is an integral part of the design;

(d) Temperature control

The temperature for organism growth ranges from approximately $-5°C$ to 80°C. However, the actual temperature is important, particularly for cell cultures, so temperature control is critical. The lower limit is set by the freezing point of water, which is lowered by the contents of the cell. The upper limit depends on the effect of temperature on the vital constituents of the organisms — for example, protein and nucleic acids are destroyed in the temperature range 50° to 90°C.

(e) Media sterilization

Medium ingredients should be controlled through a careful quality assurance programme. However, sterilization is also required and there are essentially three methods used:

- continuous sterilization for large scale. The time and temperature of the continuous sterilizer should be optimized based on the most heat resistant contaminant. The hold section of the continuous sterilizer should be designed for plug flow to prevent back mixing;
- in-situ batch sterilization by heat for smaller batches;
- sterilization by filtration for heat sensitive products such as cell culture.

Recovery and purification

The product separation and purification section is critical to the design of a fermentation plant; indeed, the bulk of capital and operating costs for a typical plant are often connected with this area. The design of product recovery systems encompasses both intracellular and extracellular products from both microbial and mammalian cell fermentation broths:

(a) Large-scale extracellular products

Technologies for recovering the simpler extracellular products consist of conventional unit operations such as vacuum filtration, crystallization, liquid-to-liquid extraction, multi-effect evaporation, precipitation and distillation. These are similar to the basic organic synthesis processes detailed earlier in this section.

(b) Recombinant products

Recombinant therapeutic products can be intra- or extracellular depending upon the host micro-organism. Recovery facilities for the more complex intracellular protein products involve cell harvesting, debris removal, pellet washing and recovery, product concentration, desalting, purification and sterile product finishing operations.

The recovery and purification of protein products from fermentation broths involves rapidly-evolving, state-of-the-art unit operations. The complexity of these operations is increased due to the heat and shear sensitivity of the proteins being recovered.

The use of recombinant-DNA organisms can also affect the design of the cell recovery area. If the organisms are not killed in the fermentation area, the recovery area handling the live organisms must be designed in accordance with applicable guidelines for containment.

Typical methods for recombinant product isolation and purification include:

(a) Cell disruption

For intracellular products the product of interest is inside the cells. The objective of cell disruption is to release this product for further separation. Cell disruption is usually carried out by mechanical means. This can be by use of homogenizers, grinding by beads or by high pressure liquid jet impacting. Other methods are use of sound, pressure changes or temperature changes and chemical methods. The separation of product from the cell debris after cell disruption is usually done by centrifugation.

(b) Centrifugation

Centrifuges are commonly used for cell harvesting, debris removal, and pellet washing operations. Cells can be separated using disc-stack or scroll decanter centrifuges. The latter allows cell washing prior to subsequent processing. The arrival of steam sterilizable, contained designs have made the use of such machines more suitable.

(c) Ultrafiltration

Ultrafiltration is widely utilized in the recovery and purification of protein products. The main uses of ultrafiltration are as follows: concentrating protein products; desalting product solutions by diafiltration; exchanging product buffer solutions by diafiltration; and depyrogenating of buffer solutions used in the process. Ultrafiltration is also finding increasingly wider use in the cell harvesting operation. It has an advantage over centrifugation in this situation since it subjects the protein to less heat and shear effects. Ultrafiltration is excellent for processes using cell recycle and in particular for mammalian cell applications.

(d) Electrodialysis

Electrodialysis is sometimes used to remove salts, acids and bases from fermentation broths. A unit will consist of compartments separated by alternate anion and cation exchange membranes. A direct electric current is then passed through the stack to effect the separation.

(e) Chromatography

Chromatography is the main technique for final purification of the product protein. Chromatographic separations take various forms depending on the driving force for the separation. There are essentially two basic forms of chromatography; partition chromatography (such as gel filtration) and absorption chromatography (for example, ion exchange and affinity chromatography).

Gel filtration, also called molecular sieving, separates molecules based on size. It is sometimes used to desalt protein solutions. In this method the product and impurities travel at different speeds through the bed effecting the separation. Gel filtration is essentially a low capacity technique and not suited for high volume processes.

Absorption chromatography is where the product binds to the matrix in the bed and is subsequently eluted by a change in the buffer composition. Common forms of separations include ion-exchange chromatography (which separates proteins based an electrostatic charge) and affinity chromatography (which separates a product or removes an impurity by means of a biospecific attraction between the molecule and a liagand attached to the gel or resin).

In order to achieve the required purity it is necessary to run the chromatographic units in series to reach the purity needed.

Automated programmed chromatography controllers are recommended for the reproducibility of their operation and reduced labour requirements. Once initiated, the programmed chromatography controller automatically loads the product on to the column, washes and elutes the product.

Scale up of chromatographic systems is reasonably straightforward and follows well-documented guidelines.

Solutions for purification operations

Solutions required during purification are generally prepared in solution preparation areas. Smaller volumes can be filled into portable containers or mobile vessels whereas larger volumes are generally piped to the user point. An important aspect of buffer preparation is to identify where Water for Injection (WFI) is required. In cell culture systems, where endotoxins are not produced by the culture, WFI is generally recommended for all buffer solutions so as to prevent the introduction of endotoxins, which would then need to be removed in a later chromatographic step. In microbial systems where endotoxins are produced (such as *E. Coli lipopolysaccharides*), WFI may not be needed until a later stage where the pyrogens are reduced to low levels or effectively eliminated. For very large volumes, storage of diluted buffer solutions is impractical. One approach is to make up concentrated solutions and dilute as required — this approach can result in significant space and cost savings.

5.2 Key unit operations

5.2.1 Liquids materials handling

Materials to be added to a reaction system can come in liquid, solid or gas form. However, the easiest to handle are liquids and consequently materials are used in the liquid form where possible. If not the natural state at ambient conditions then the material can be made liquid either by melting or more commonly by making a solution by dissolving in a solvent.

Liquids can fall into three categories when used in a reaction:

- **solvent**: this allows the reactant to mix and react and to create a mobile mixture that can be controlled for temperature by heat transfer with surface contact. Solvent liquids generally form large quantities in a batch make up;
- **reactant**: the active compound used to react with another material to synthesize the desired intermediate or final molecule stage. Use of liquid reactants is generally desirable as they can easily be transferred and added to the reactor system under controlled conditions;
- **catalysts**: these are usually required in small amounts. Handling small quantities can bring difficulties; it is easy to dispense the correct quantity in a laboratory or fume cupboard, but getting it safely into a reactor system needs to be carried out via an air lock or charge flask arrangement.

Liquids are usually handled in drums if the quantities are small or the duration of production is short — this typically applies to reactants, particularly where there is no source for bulk deliveries. If the material is used in larger quantities then bulk delivery in road tankers and storage in a bulk tank system is preferred as it minimizes the manual handling requirements, and hence, reduces the operator inputs.

Liquids delivered in bulk quantities from road tankers must be shown to be suitable for use in the process — that is they are of the correct purity, strength or even the correct chemical composition. This may be by reliance on the supplier's audited quality control/assurance system and certificates of conformity, or by sampling the road tanker and then analyzing the contents before offloading. Alternatively where analysis is lengthy and would incur waiting time charges from the delivery company, special quarantine bulk storage tanks can be used which allow a segregated offload of the material and then the appropriate testing prior to release for use or reject and return.

With the increasing legislation on Volatile Organic Compound (VOC) emissions, it is common to vent the bulk storage tank back to the road tanker during offload to avoid release of VOC.

Liquids are charged to the process either by direct pumping from the bulk tank or drum into the reaction system or to an intermediate addition vessel such as a head tank which allows more accurate determination of quantity and greater control over rate of addition. Alternatives to pumping include closed vacuum or pressure charging, although these methods are not commonly used now because of the safety issues associated with them.

5.2.2 Solids materials handling

Solids are most commonly used in processes as reactants but can also be used as catalysts, purification agents such as activated charcoal, or seed for crystallization process stages.

One of the main sources of solid is as an intermediate stage in a lengthy multi-stage synthesis production operation.

Solid material is most commonly stored in sacks, plastic drums or lined fibreboard kegs. The most important consideration during use is the safe, contained dispensing of the required quantity and the charging of this into the reactor system.

Open manway charging used to be the main transfer method but this is now considered unacceptable because of the risks of exposing the operators to the chemicals inside the reactor. Similarly the risk of exposing the process to cross-contamination from surrounding activities is also unacceptable in many circumstances.

Current methods involve creating a protected area for charging, either directly to the reactor via a weigh hopper or charge lock, or to an intermediate bulk container (IBC). This IBC can then be moved to a docking station to allow enclosed charging to the reactor system. The protected area involves controlled clean air flows to minimize risk to product and operator by reducing contamination and exposure within a purpose designed charge booth.

The use of split butterfly valves or contained transfer coupling systems is now a very popular way of making the connection between the IBC and the process system, as it allows the handling of very active materials with increased safety and ensures minimal contamination of the reaction process.

The use of solids in bulk is not very common unless for large tonnage products where a dedicated plant with silo storage and transfer techniques such as pneumatic transfer, screw feeders or conveyors can be used.

5.2.3 Liquid/liquid separation techniques

As part of either the reaction stage or purification stages of the synthesis, it is often necessary to separate one liquid from another. There are two main types of technique available for this, those involving heat and those using other properties of the liquids to achieve the separation.

Thermal processes

Thermal processes are commonly used for removing materials, such as an inhibiting by-product formed during a reaction, typically water, or operations where evaporation techniques give an effective and efficient method of separation. These can be either single stage such as a flash distillation or involve the use of fractional distillation by utilizing distillation packing materials in a column.

Batch distillation is not an easy process to perform due to the unsteady composition of the still vessel and the fall in efficiency as volumes drop, and therefore, so does contact with the heat transfer surface. A supplementary heat transfer surface can be provided by pumped or thermo-syphon circulation through a heat exchanger.

Another problem with thermal processes is that they can result in the degradation of product if it is sensitive to heat. To minimize this, the pressure at which the distillation is carried out can be reduced by vacuum pump systems to allow evaporation at lower temperatures. In the event of particularly sensitive or labile materials this can be carried out in small continuous units operating at extremely low vacuums known as short path stills.

An alternative extractive technique is azeotropic distillation. Here an additional material is added to create an azeotrope, which will preferentially be distilled out achieving an otherwise impossible thermal separation. The entrainer is then separated from the removed material and recycled if possible.

Non-thermal processes

It is a relatively common process to add a liquid to the process into which impurities or even the product is preferentially soluble. The added liquid is immiscible with the process stream and forms a separate phase, which can then be separated by various techniques. This process is commonly carried out with water or aqueous solutions and is known as washing.

The immiscible phases can be separated by allowing the layers to settle in the reactor vessel and then running the lower layer out until the interface is seen. It is common to run this layer to a receiver; it could be the product layer or if it is the waste layer it could be held prior to discharge.

Interface detection can be difficult. Automatic detection devices have mixed success and generally an illuminated tubular sight glass and trained operator is the most successful technique.

In large production plants, mechanical techniques such as decanter centrifuge, multi-plate disk centrifuge or counter flow liquid–liquid extraction devices can be used to increase the efficiency of the separation.

Techniques that were previously used mainly in the biotechnology field are now becoming more available to achieve difficult separations and purifications in the synthetic process arena. These include chromatography techniques and selective membrane processes, which are becoming more feasible with the developments in membrane technology.

5.2.4 Crystallization

Most synthetic processes involve the isolation of a solid stage. This can be an intermediate stage, a byproduct or most commonly the final active BPC. The formation of the solid form can be carried out in several ways:

- crystallization by cooling;
- crystallization by evaporation/concentration and cooling;
- precipitation by reaction or pH change;
- precipitation or crystallization by solvent change.

This operation can be carried out in the standard or slightly modified batch reactor described earlier. The allocation of a specific or dedicated reactor for crystallization use is becoming more common and provides a way of avoiding contamination of the final product. The need to provide controllable agitation with gentle profiles to avoid crystal damage and good heat transfer are the main areas addressed along with the rate of addition of precipitant or cooling profiles to allow for optimal crystal form and size. In order to promote the desired crystal form, seed materials of the desired crystal type can be added at the correct stage to initiate crystallization of the appropriate form.

The crystallization activity is becoming increasingly sophisticated. Known as crystal engineering, it is of growing importance especially in tailoring the product form of the final BPC to suit the demands of the secondary operations, avoiding comminution or granulation to achieve desired product form.

Most crystallizations are carried out on a batch basis. However, if production quantities demand or specific product form/size distribution profiles are required then continuous crystallization arrangements can be used. New developments involving the use of ultrasound to form a nucleus for crystallization (known as Sonocrystallization) have been developed. They can produce mono-size distributed slurries accurately engineered for the desired property and are of particular interest for sterile production where seed introduction is more difficult.

5.2.5 Solids isolation

Once the solid form has been produced, it needs to be isolated from the liquid or mother liquor.

Separation of solid from liquid generally involves some form of filtration since techniques such as sedimentation are not routinely applied in the pharmaceutical industry. Filtration involves creating a medium through which the liquid can pass but the solid is retained. Once the medium has been formed, a driving force to cause the liquid to flow is needed; the way in which the driving force is generated is the main area where differences in technique or equipment occur and can be created by vacuum, gas pressure, mechanical pressure or centrifugal force.

The other main area which differentiates the filter type is the quantity of solid involved and whether it is a by-product to be removed or a product.

Filters

Solid impurities in small quantities up to 10 kg can be removed using cartridge, bag or multi-plate filters such as the calmic filter.

The single sheet, nutsche filter is a common unit that has developed greatly. The original form was an open box filter that used vacuum in a lower section of the box to draw filtrate through a filter medium or cloth. The disadvantage with this type is that they offer little to protect the general plant area, contain the process to protect the operator or prevent cross-contamination. The other main disadvantage is the level of vacuum that can be generated limits the driving force.

The first development of the nutsche filter was the agitated pressure nutsche filter. This unit has an integral pressure chamber above a filter media, typically a cloth element. The unit is fitted with an agitation arm that can be used to smooth the cake and discharge the damp solid. The driving force for separation is generated by either applying vacuum to the filtrate receiver and sucking the filtrate out of the slurry to leave a damp cake, or by applying pressure above the slurry and forcing the filtrate out to leave a cake.

Occasionally both pressure and vacuum are used to generate the driving force, but it is commonly found that increasing the driving force above 3 bar has little benefit on filtration rate due to compression of the cake and the closing off of the route by which filtrate can flow out. The pressure is most commonly generated by nitrogen and because the materials are typically flammable solvents, nitrogen also provides an inert atmosphere. It can be provided either once-through from a mains supply leaving via the filtrate receiver or recycled taking low pressure nitrogen from the receiver, increasing the pressure, then putting it above the cake to displace more filtrate. This has the advantage of minimizing the amount of nitrogen used and reducing emissions to the atmosphere as the nitrogen entering the receiver is laden with solvent vapour. The recirculated nitrogen can also be heated prior to entering the filter to aid drying of the cake. The nitrogen is then taken directly from below the cloth to the compressor package where it is chilled to remove the solvent, then repressurized and heated before recirculating back above the cake.

The cake can be washed in the filter to remove soluble impurities. This is done in two ways, either a displacement wash or a reslurry wash. In the displacement wash the wash fluid is sprayed onto the cake surface whilst vacuum or pressure is applied to cause the wash fluid to quickly pass down through the cake, taking out the impurities and out to a receiver. This is commonly used where the impurities are very soluble in the wash and can be easily removed or where the product cake itself is soluble in the wash so that residence time is minimized to avoid losing product with the wash. With a reslurry wash, a volume of wash fluid is added to the filter and the agitator is used to mix the cake with the wash fluid to form a slurry. By this process the impurities can then dissolve into the wash fluid. The resultant slurry is then filtered again to remove the wash fluid and the impurities. The wash filtrate is often collected in a separate receiver to allow for recovery of product that may have been dissolved and lost as well. This is known as second crop recovery.

Discharge from the filter can be in one of three ways. Most commonly the product is discharged as a damp cake; here the agitator is lowered to the cake surface and rotated to start to break up the cake. By altering the direction of rotation, the cake can be drawn to the outside edge of the filter where an outlet hatch is opened to allow discharge of the cake out of the filter to the next process unit. As discharge proceeds the agitator is lowered gradually to the bottom of the filter to ensure all the cake is discharged. The nature of this operation results in slugs of damp cake being discharged as the arm goes past the discharge hatch, which may cause problems for the next processing module. An alternative approach is to have a central opening in the middle of the filter element and dig the cake and bring it to the middle. This provides a continuous flow of solid out but reduces the area for filtration and can give problems of sealing the central outlet. The other methods of discharge involve either making a slurry or solution of the cake in a solvent and charge as per the wash fluid. This is then agitated and discharged via a valve and pipe arrangement from the side of the filter above the filter cloth.

The nutsche pressure filter has also been developed into a filter dryer. Here heat can be applied to the cake once filtration has occurred via coils on the side and top of the filter body and via heating passages through the agitator. A single fluid heating medium, often hot water, is circulated through these coils and this provides heat to the product to remove the remaining solvent to give a dry solid.

At the same time as the heat is applied, the space above the cake is subjected to a vacuum pulled on the system normally via an integral dust filter to avoid any losses of product solid with the evaporated filtrate. The filter dryer has proved a very successful item of plant and minimizes the exposure of the product during its transfer from the filter to another dryer. The disadvantage of the unit is that the time taken to filter, wash and dry a batch in the filter dryer is overall rate limiting for batch time cycles.

Other types of filters exist which provide different methods of presenting a filtration element and a driving force of pressure to separate solids and liquids and then discharge the solid. These include rotary vacuum filters, tube filters, disc filters and belt filters, but they are not common in the pharmaceutical industry and are used for specialized applications only.

Centrifuges

These devices generate a centrifugal force to drive the liquid through the separating medium leaving the solid. There are four main types:

- **vertical axis — top discharge by basket lift out:** This is the traditional type and is not commonly used now except in small sizes. The main problem is the exposure of the operator when emptying the basket and the risk to the product of cross-contamination in the open process;
- **vertical axis — bottom plough discharge:** This allows contained discharge of the solid from the basket by a movable knife or plough that cuts the solid out of the basket and down a chute at the bottom of the machine;
- **horizontal axis — peeler discharge:** This unit has advantages over the vertical axis machine in that it can spin at higher speeds, and hence, create a higher G-force or driving force for separating the liquid. Discharge of the solid is carried out in a similar way by a knife or peeler blade, which is used to channel the solid into a chute and away from the machine;
- **horizontal axis — inverting bag discharge:** This is the most current development. It has the benefits of the higher G-force for separation but the cake is removed by inverting the filter cloth. It also has the benefit of being able to remove the entire heel to ensure ease of further separations and minimize batch-to-batch contamination. Most modern centrifuges are automatically controlled. This covers inerting and purging cycles, filling, spinning, washing and discharge.

5.2.6 Drying

The final step for most BPC processes is to dry the intermediate or final product. This removes any residual solvent from the solid. Often this is done to produce a fine free-flowing powder that can easily be handled in the secondary processing. Alternatively if the solid is an intermediate then subsequent processing often involves the use of a different solvent. Drying reduces the moisture level of solvent to an acceptable level, usually to below 1% w/w of the solvent present.

Dryers can be classified into two main types — direct and indirect. With a direct dryer, air or more commonly nitrogen is heated and passed through the solid. An example of this type of dryer is the batch Fluid Bed Dryer (FBD). This unit uses a basket that would be filled either by hand or by gravity from the filtration or centrifugation unit. The basket has a perforated base and when placed in the fluid bed dryer, the heated air or nitrogen flows up through the solid, fluidizing it and evaporating the solvent. The off-gas stream is filtered, usually by a cyclone or a bag filter system to prevent loss of product. The filtered stream can be cooled to remove the evaporated solvent, then reheated and passed back through the basket. Whilst the units are relatively cheap, they are not favoured for the following reasons:

- VOC losses are high without the high additional cost of a nitrogen gas recycle system;
- there is a high risk of static discharge;
- effective filtration of the heated air stream is required to avoid introducing contamination;
- open handling of the cake does not provide a contained system, particularly for very active products.

For these reasons, indirect or enclosed dryers have replaced the direct dryer. Many pharmaceutical products tend to be thermally sensitive and as a result most are dried under vacuum, since this allows for solvent evaporation at lower temperatures. Jacket temperatures of typically 40–100°C are used with hot water or a single fluid system as the heat source. A dust filter is installed on the dryer body or in the vapour line to prevent loss of product with the vapour stream. A vacuum is generated by liquid ring pumps, once-through oil lubricated pumps, dry running vacuum pumps or more rarely ejectors. Solvent condensing is carried out either before or after the vacuum pump depending on the capability of the pump to handle liquids and condensation of the solvent. Often this is not desirable for corrosion reasons and all the condensation is carried out after the pump. The ideal solution is to use a liquid ring pump with the same or compatible solvent, chilled, as the ring fluid, then condensation can occur directly into the ring fluid.

The fundamental principle of the indirect dryer is to provide a heated surface and a means to ensure good heat transfer from that surface to the solid, whilst maintaining a vacuum above the solid to efficiently vaporize the solvent. Various designs for achieving this exist and can be categorized by the means used to achieve the heat transfer, as follows:

(a) No agitation

The vacuum tray dryer is the only example still in routine use under this category. Here, solid is laid in thin layers onto trays and placed onto heated shelves in a vacuum chamber where heat and vacuum are applied to evaporate the solvent. The dryer is not very efficient as it takes a long time to dry the product due to the lack of agitation, and hard dried lumps can form because there is no agitation to break down agglomeration during drying. The biggest failing with the dryer is that it is messy to load and unload the trays, requiring a high degree of containment and equipment to protect both the operator and the product. It is, however, very popular in R&D environments where its flexibility is a benefit, and in instances where mechanical work on the product will damage crystal size or shape or cause safety problems such as detonation of a shock sensitive solid.

(b) Horizontal axis agitated vacuum dryers

This type of dryer, the 'paddle dryer', is most widely used in BPC manufacture. It consists of a horizontal cylindrical chamber, the outside of which is fitted with heating and cooling jacket or coils. Inside, the dryer is fitted with a slow rotating paddle that moves the solid to give good mixing and allows replacement of the solid in contact with the heating surface, aiding drying. Horizontal axis dryers have high jacket surface area to volume ratios and are efficient dryers giving low drying times. Vapour is withdrawn via a dust filter fitted to the top of the body, allowing collected powder to be routinely shaken or blown back into the batch. They also have low headroom and can be fitted into process buildings without adding a full floor whilst utilizing gravity in the isolation train. They can be difficult to clean particularly because both shaft seals are immersed in the solids. Some designs allow for easy and complete removal of the end plate and agitator shaft.

(c) Vertical axis vacuum dryers

There are a number of variations of vertical axis, agitated vacuum dryers; the main difference between them being the ratio between diameter and depth of dryer. Short large-diameter dryers, often referred to as pan dryers, are popular. A variant of this utilizes a specially designed agitator that provides a very efficient mixing regime giving good heat transfer and efficient drying. This type of dryer has been termed a turbo dryer. Some designs allow the lid to be hydraulically lifted for internal inspection and cleaning. High-speed impellers known as lump breakers can be fitted in addition to the main stirrer to break up any agglomeration. The drive can be either top or bottom mounted. The bottom drive has the disadvantage of requiring a seal in the product contact area, whilst the top mounted drive takes up a lot of space on the dryer lid, reducing the opportunity for additional nozzles and restricting the opening of the lid. The top mounted drive allows for the agitator to be raised and lowered through the solid, adding to the range of agitation profiles for drying. A variant of the vertical axis vacuum dryer is the filter dryer, referred to in the previous section, which combines the functions of a pressure nutsche filter with a vacuum pan dryer. The compromise tends to be due to the retention of the filter cloth during the drying process and the design of the agitator.

When the depth of the dryer exceeds the diameter, the dryer is referred to as a cone dryer. Deep cone dryers have a double rotating screw inside, which performs three functions: wall to centre solids movement for heat transfer by horizontal and vertical turning; delumping of solid initially and during drying; assisting bottom valve discharge by reversing the screw direction.

This design is favoured by a number of companies since it offers reasonably efficient heat transfer, delumping, relative ease of cleaning by refluxing with solvent and caters for variable batch sizes. Top and bottom drive mechanisms are available. From a GMP viewpoint, internal drive mechanisms must not shed particles. The one disadvantage of these dryers is that they are relatively tall compared to the other types and can add a floor to the isolation area, although protruding the discharge cone region into the clean pack-off room can compensate this.

5.2.7 Product finishing

Historically, BPC products were simply packed off from the dryer into fibre-board kegs and shipped, via a QC sample and check stage, direct to the secondary plant. Here finishing operations such as mixing, comminution or milling and granulation were generally carried out.

However, with the change in the profile of the BPC manufacturer, the end user for the BPC is often a different business or group within the same pharmaceutical manufacturer, or the BPC manufacturer is a different company to the pharmaceutical secondary company. In these instances there is an increasing need to provide some of the finishing operations to produce a product with specific physical characteristics in addition to the correct chemical composition. The increasing demands of 'speed to market' have also caused a blurring of the activities traditionally seen as 'secondary operations' and have increasingly come to be expected as part of the BPC manufacture.

Milling, sieving and granulation

Milling is an operation to reduce the particle size of a solid down to an acceptable profile or range of sizes typically below a certain maximum size. It is best carried out in-line after the dryer to avoid double handling, particularly since dryer discharge is often a low rate, semi-controlled process. If carried out off-line after quality approval, then a separate milling line in a clean room suite is needed. Intermediate bulk containers (IBCs) are usually used for solid transfers and act as feed hoppers to the mill feed system.

There are various types of mill used in the BPC industry, including pin mills, hammer mills and more commonly jet mills and micronizing mills. Further details are given in Chapter 6 covering secondary processing.

Sieving is an operation to classify the solid into a range of particle sizes. The equipment is often used in-line with the discharge from the dryer. The sieve operation consists of passing the solid through a series of screens. The first screen removes particles that are larger than the specification; these are discharged and recycled to the mill. The second screen then retains particles of the minimum size and above. The solids passing through the screen 'fines' is too small and may be recycled to the crystallization stage. The material is encouraged to pass through the screens by either vibration or by the use of rotating arms. The material that does not pass through the screen is removed from the sieve in either a batch or a continuous method to be packaged. Oversize and fine material can be reworked in some cases, but sometimes has to be destroyed.

There are some cases where more than two screens are used. This provides a series of size fractions that can be used for products that require specific drug related release profiles of for filling directly into hard shell gelatin capsules.

5.2.8 Packaging

The final packaging of a BPC is carried out in a controlled environment to protect the product from contamination by external sources and also to protect the operator from exposure to the active material. Most BPCs are solid powders and are packaged in sacks, drums or IBCs. A small number of products are liquids and these are packaged into the appropriate containers in either a manual or automated filling system.

5.2.9 Solvent recovery

Solvents are widely used in the production of BPCs and, as previously stated, provide several functions including dilution of the reactant concentration and mobility to allow good mass and heat transfer. Solvents are important in obtaining the correct final product form and in washing the product in isolation equipment. When used in a reaction, the solvent generally does not react or break down to other components. In order to maximize the efficiency of the process, solvent remaining after a processing stage can be recovered for use in the same process from which it originated.

Solvent recovery can be either a batch operation or, more commonly if larger volumes are involved, a continuous recovery plant.

The type of recovery used largely depends on the contamination present and the properties of the solvent being recovered. Flash stripping is the simplest operation and is often sufficient. Fractionation, often by the use of random or structured packing, is used where complex mixtures require separating.

Pre-treatment is often used to allow a simpler recovery. This can involve crude solids filtration to more complex precipitations or pre-stripping.

Most solvent recoveries result in a residue, which will then require further treatment or handling — most commonly incineration or landfill.

5.3 Production methods and considerations

5.3.1 Production

Pharmaceutical production is mainly carried out on a batch basis for a number of reasons. The main reasons are normally linked to the trace ability of the product, validation and regulatory issues, but others include the scale of operation, the flexibility of operation required, inventory optimization or even technology development.

Production is arranged into three main types of facility:

- dedicated — the facility is designed and built for one specific process;
- multi-purpose — the facility is designed and built to carry out a number of known and defined processes, potentially with a minor amount of modification to configure the plant to the next process;
- General purpose — the facility is designed to handle a variety of processes, both known and envisaged for the future.

Batch chemical processes with cycle times typically of 16 hours or more are most commonly carried out on a 24-hour a day, seven days a week operation.

5.3.2 Automation and control issues

Any automation system must provide tangible benefits to justify the investment. In general, the benefits of automation will derive from:

- higher levels of safety;
- the ability to apply sophisticated control strategies;

- consistent product quality;
- higher levels of plant utilization for a given manning level;
- more efficient usage of materials and reduction in waste;
- provision of timely and relevant information.

The logic and numerical processing capabilities of modern process control systems enables operating conditions to be tightly regulated to the specified profiles, optimizing processing time, delivering consistent quality of product and providing a higher level of safety.

While the use of properly designed and implemented process automation systems enhances the safety of the plant (by improved control and reporting/notification of potential risks) these systems should not be relied upon to ensure plant safety. The recently published international standard IEC 61508: Functional Safety of Electrical/Electronic/Programmable Electronic Safety Related Systems addresses the requirements of safety related systems.

The key issues to be considered when embarking on automation projects include:

- the functionality required;
- the level of automation required;
- the types of systems employed.

Most primary pharmaceutical manufacturing processes can be classified as being either 'continuous' or 'batch' with a few, if any, being categorized as 'discrete' processes. This section focuses on the requirements of batch type operations.

The requirements of batch operations can generally be considered more onerous than those for other types of processing. Batch processing involves the sequential modification of process conditions through a predefined regime rather than maintenance of established 'steady-state' conditions.

Batch operations essentially consist of a series of phases that are executed sequentially. The execution of a phase is usually dependent on process conditions established in a preceding phase; therefore any fault that interrupts the execution of a phase may require the processing to be resumed from a point in the operation sequence other than that where it was suspended. The process automation system must be capable of executing sophisticated exception handling procedures. It may require the provision of facilities that enable the operator to intervene and manually adjust the point in the sequence at which processing is to resume.

System functionality

The functionality required of the system will principally depend on the processing objectives and the method of operation proposed. The plant equipment and its connectivity also affects the functionality; the following are some possibilities:

- single batch, single stream (one batch at any given time);
- multi-batch, single stream (more than one batch being processed at any given time);
- multi-batch, multi-stream, dedicated equipment trains;
- multi-batch, multi-stream, common equipment.

On plants where a variety of products are regularly manufactured, some form of automatic scheduling functionality may be desirable. When equipment is required to undergo Clean In Place (CIP) or Sterilize In Place (SIP) routines at regular intervals or at product changeover, the CIP/SIP operations may be considered as a 'product recipe' and scheduled accordingly.

The sophistication of the scheduling systems available vary from the basic, where queued operations (or batch recipes) are initiated when the necessary processing units become available (or predefined constraints are satisfied), to others which are capable of developing a production schedule from demand data transferred from ERP (Enterprise Resource Planning) or MRPII (Manufacturing Resource Planning) systems. The sophisticated systems are capable of queuing recipes, calculating the optimum batch sizes to complete a campaign, and making changes dynamically as 'demand' changes. (Some form of 'gateway' to control the transfer of data from ERP systems is recommended to prevent disruption of manufacturing operations by sudden changes in demand). Other factors that complicate scheduling include the following:

- resources that can be simultaneously allocated to more than one process (e.g., cooling fluid circuits, ring-main fed utilities);
- number of streams in the system;
- selection of the best resource to use when several (shared) non-identical units are available (requires knowledge of what will happen next);
- operations that are dependent on activities/equipment controlled by external systems (which may result in the duration of the operation being unquantifiable).

The recipe handling requirements of the process control system are affected by the type and configuration of the plant. The recipe system may also need to be able to cater for variations in the properties of raw materials, which may result in a requirement to modify the processing parameters. Any variation in the processing parameters/formulation, whether for a campaign of batches or for an individual lot, needs to be recorded and the appropriate mechanisms and facilities need to be provided to enable this.

As well as the quantity and complexity of the recipes that need to be executed, the number of recipes that can be simultaneously active in the system (on the plant) needs to be considered. In 'multi-batch' situations, the process control system needs to be able to report the impact of a malfunction or process deviation on other concurrent activities.

The exception is handling facilities that are critical to the successful operation of a batch plant. In the event of a deviation from the expected pattern of occurrences, the operator should be informed and appropriate action should be taken promptly. A minimum of three categories of operator message are recommended:

- critical alarms generated when there is risk to equipment or personnel;
- process alarms caused by deviations from the expected conditions;
- events which keep the operator aware of actions being performed.

In the case of critical and process alarms, the process control system will normally be expected to take action to put the plant in a safe condition automatically. Facilities are also needed to enable the system to restore the plant to its prior state as effectively as possible. A good understanding of both the process and the control system are required in order to develop the necessary procedures and phases.

The production data, exception reports and alarm information generated need to be associated with the appropriate batches and stored to satisfy operational as well as regulatory reporting requirements. As in the case of the process control software, the definition of the reports requires knowledge of operational procedures and company standards.

The recording and storage of data should be clearly differentiated from the reporting function. Justification should be provided for all data that is to be recorded because, while it is true that data not recorded is lost forever, recording excessive quantities of data can have severe drawbacks. Some systems enable data recording to be triggered by events; this enables data collection to be restricted to critical phases of an operation (such as during an exothermic reaction).

It is important that the recorded data is stored in a format that allows it to be manipulated in the manner required. While the control systems use a variety of data compression algorithms to facilitate the storage of large quantities of data, this can prevent data export and restrict the processing and manipulation of the information to the control system with the consequent limitations.

Interfaces and communication facilities with other systems also need to be evaluated when identifying the functionality required of the system and this is addressed in a later section.

Automation levels

In a processing environment automation should be aimed at removing the mundane and repetitive tasks from the operators, freeing them to add further value. The numerical processing capabilities of modern control systems enable advanced control strategies to be employed to improve efficiencies.

All areas of the plant will not require the same level of automation. There is also a trade-off between the manning level reductions available through automation and the flexibility available from lower levels of automation. In certain areas, such as raw material tank farms, a 'basic' level of automation can result in a far more effective system, while other areas benefit from all the sophistication available. In the main processing area, manual intervention may be restricted to critical operations where heuristic judgment is required or those aspects where the necessary facilities to allow automatic execution have not been provided.

As part of the development of the control philosophy, each area of the plant should be reviewed and the required automation level established. The basis of the justification for automation will vary and could include conditions within an operating environment such as physical aspects of the nature of the task to be undertaken, the need for an automated record of activities performed, etc.

5.4 Principles for layout of bulk production facilities

Many examples of unplanned developments can be seen on pharmaceutical sites throughout the world. Production facilities have grown in many cases in a totally uncontrolled manner with decisions made based on the priority of the moment with no regard for the future. This has happened due to lack of thought, concern for cost and lack of information on the company's future marketing plans. The result is a totally random 'hotch potch' of buildings leading to inefficient operation, potential hazards, questionable use of land, and expensive future development of the site.

Two types of development will now be considered. Green field development involves the use of land on which there has been no previous commercial developments. Plans for such sites will not generally be restricted by previous buildings and existing operations. Brown field development may, however, have some restrictions due to past or existing operations and freedom of design may be curtailed.

In both instances however, at some stage of design, it is necessary to review the impact of the new development on the future use of the site. All these principles equally apply to secondary production facilities.

5.4.1 General considerations

In the pharmaceutical industry, sites may be laid out for primary production, secondary production, research and development, warehousing and distribution or administration and head office activities. A single site could cover any number of these functions. There is considerable dialogue on the advantages and disadvantages of multiple use sites, which will not be discussed in this guide, except to point out that all the above activities do not necessarily sit well together. Here the guide is aimed at bulk drug primary production site layouts only.

5.4.2 Green field sites

Site location

It is assumed in this guide that the new site will consist of multiple production units; the first of which is to be built at the time of developing the site infrastructure, with others following on at some later time.

When selecting the site, due consideration will have been given to its geographical location with specific attention to road systems, communications, ports and airports, availability of skilled labour and adjacent developments. Any special environmental requirements and full information on the availability and capacity of public utilities will also have been investigated. Discussions with all appropriate planning and statutory bodies will have been carried out to determine if there are any requirements that would prevent the development of the optimum design for the site. It is also necessary to ensure that any adjacent developments in the planning stage are compatible with a bulk drug operation. For example, an open cast mining site adjacent to a plant manufacturing high cost pharmaceuticals would not be ideal.

It will be necessary to carry out full topographical and geotechnical surveys to determine the surface contours and the load bearing characteristics of the land. These surveys will provide information on underground obstructions, mine-workings and geological faults. Such information could influence the positioning of buildings or indicate the need to carry out specific rectification work. The land should also be checked for ground contamination. Information on the ambient climate of the site, including prevailing wind directions, is also necessary at this stage. The majority of the above data should be obtained prior to the purchase of the land. The above requirements are not exhaustive but do indicate typical actions which are required prior to finalizing on a particular site.

Conceptual design

The project may be divided into two parts. The first part covers site infrastructure, including:

- offices and administration buildings;
- operator and staff amenities;
- control and test laboratories (if not in the production plant);
- engineering workshops and stores;
- central warehousing;
- on-site utility generation;
- gate house and security fencing;
- utilities and services distribution;
- roads, road lighting and car parks;
- underground utilities;
- site grading and landscaping.

The second part will cover production facilities. This, as mentioned previously, may be the only production unit or may be the first of a number. In this guide it is assumed that the site is to be laid out to accommodate a phased development and the design must ensure that future construction will not cause interruptions to production. This second part typically will include:

- the main reactor and process facility;
- special hazard production units;
- environmentally controlled finishing units;
- bulk raw material tank farm and drum store;
- effluent treatment final conditioning unit;
- control room for the production processes;
- production offices;
- on plot generated services;
- switch rooms and transformers.

The split of the project into two parts can be advantageous commercially. The infrastructure is mainly civil and building engineering and the production unit is mainly process engineering. More suitable contracts can be negotiated if this difference is understood.

Based on these various elements, it would be normal to look at a number of possible layouts to finalize the overall concept before proceeding with detailed design.

Generic production plot layout

Before proceeding with the layout of the site, it is advantageous to give some consideration to possible plot layouts. It is anticipated that the production units, which will eventually be constructed on the site, will produce a number of products that may benefit from a custom design approach. If the plants are to be of a multi-product design then consideration should be given to the maximum numbers of reactors to be included in one plant.

Regardless of the style of production unit, the fully developed site is likely to have a number of production buildings each with associated control rooms, on-site utility generation, offices and tank farm etc.

Based on the first production unit to be developed, it is advantageous, before considering overall site layout, to develop an outline plot layout that can be the basis for all plants on the site. This does not mean that all plots will be identical but the main principles will have been identified at this early stage and will have some influence on the ongoing development of the site. Typically control room positioning, spacing of on-site tank farms, policy for facilities for hazardous operations, position of on-site switch room and electrical transformers should be identified.

Whilst the brief for the first production unit may be well defined, subsequent developments may be unknown at this stage. It is essential to recognize this and to incorporate flexibility into the eventual site layout and to identify which production plot parameters could possibly change. Site master plans should not be written in tablets of stone but should be reviewed with each new development. They should not, however, be changed by default.

Site layout – master plan – zoning

The term 'Site Master Plan' has been introduced in the previous paragraph. In green field development this is likely to start with an area of land that has no structures or building on it. It could be a cornfield, an area of heath land or a cleared and level site recovered from some defunct industry. There are likely to be several ways to lay it out and the first exercise is to decide on a concept. As stated before, there may only be information on the first production unit but the positioning of that unit will have a critical influence on the success of the site in the future. It is essential to look ahead and prepare a conceptual image of how this site could look when fully developed to allow a logical expansion of the site in future years.

The first consideration of the master plan is associated with zoning of the site — which areas will be allocated to offices, amenities, warehousing, utility generation, workshops and production plants. Zoning plans also contribute to solutions for the most efficient utilities distribution design and are the first stages of development of site logistics.

Master plan – landscaping

Having zoned the site, the overall site landscaping strategy can be developed. This will be very much dependent on company policy and any particular need to screen the plant. The outline site contours will have been decided and any necessary planting schemes can be worked out.

The master plan

Once the site has been zoned, a generic plot plan has been developed and outline landscaping has been decided, it is then appropriate to proceed with the overall master plan. The purpose of master planning is to look at how the site could be when it is fully developed and then only build the part that is required in the first instance — this ensures that what is actually built will fit into a logical site development. The master plan should be revisited at the time of each future project and modified if necessary to keep in line with changing requirements.

On-site roads

Discussions with the statutory authorities will have already identified the approved entrance and exit from the site, but the on-site road system should be developed based on the zoning plan. This must take into consideration gate house procedure, off-loading facilities, car parking, restricted access areas, emergency access, road vehicle access, forklift truck access and pedestrian circulation. The road system must also be capable of progressive development as the site expands without disruption to operations.

Car parking policy can often present major problems. By the very nature of the site operation, the site is likely to be away from built up areas and operator car parking space is therefore essential. The safest practice is to provide it outside the main operational site boundary, but this may not be a popular choice on large sites in geographically exposed locations. The main emphasis must however be to ensure that private vehicles cannot get within recognized safety distances of operating units. Road system designs must recognize this requirement.

Public utilities and site generated utilities

Public utilities are likely to include towns water, electricity, natural gas and sewage. Earlier discussions with the supply companies should have identified where, on the site boundary, these utilities will be available. It is now necessary to decide on the appropriate site interface. In most cases a control booth is constructed for piped utilities and a transformer house and switch room for electricity is constructed adjacent to the boundary.

On-site centrally generated utilities will normally include steam and compressed air. Refrigeration and recirculated cooling water is normally generated on each production plot.

Utilities, liquid raw material and interplant transfers can be distributed in several ways:

- **above ground:** this will normally involve a pipe bridge and is possibly the most convenient way of distribution in that it does not interfere with traffic and pedestrian circulation at ground level. An access platform should be fitted to the bridge for maintenance purposes;
- **below ground (in an open culvert):** the culvert walls may be inclined or vertical. This has the advantage of easy access for maintenance, but has to be bridged at each road crossing and is difficult to keep clean;
- **below ground (in closed trench):** this is not favoured for bulk drug sites because of possible hazards to operating staff and difficulty in maintenance;
- **surface run:** this method causes problems to traffic and operator circulation.

The design of the distribution system must allow for future expansion in both layout and capacity. The question of ring main capability, which may be required in the future if not initially, must be examined. The master plan must reserve space on the site for the extension of possible bridgework in the future. This design will require an estimate of peak and average usage of utilities when the site is fully expanded. This, together with forward assessment of future marketing forecasts, will allow an informed decision on the initial sizing of the distribution system.

Site offices, gate house, amenities, laboratories, warehouses

It is assumed that the site being discussed is for production only. Based on this, the general administration offices are likely to be small and can possibly be sited in the same building as catering and possibly laboratories, although this will depend on the nature of work being carried out in the laboratories. The building should be sited adjacent to the entry gate to the site, thus limiting the need for visitors and office staff to go through any operational areas. The catering facilities are likely to be used by day staff as shift staff associated with production operation are likely to have their own facilities within the control room building of the production unit. The office building will be positioned in an unclassified area of the site.

The procedures for receiving road transport arriving at and leaving the site will determine the layout of the gate house area and the final positioning of the gate house. Appropriate lay-bys for lorries and weighbridge facilities may need to be incorporated in the layout.

It will always be good practice to minimize vehicular access to the vicinity of the operating units. The site warehousing policy will influence this considerably. Each production plot can have its own warehouse for raw materials and finished goods. This would of course require road transport to have access to loading and unloading docks near to operating units. In addition the storage of high value, finished products adjacent to chemical reaction operations could give rise to a potential financial risk in the case of a hazardous incident occurring. It is not possible to generalize on recommendations for positioning warehouses but if possible the main warehouse should be positioned in the unclassified area of the site and the specific production units could have a small storage capacity for finished goods under test and possibly one or two day raw material storage. The production plant stores would be supplied by on-site forklift trucks.

Engineering workshops

Engineering workshops may be directly associated with each production unit or may be a site centralized facility — the size of the site will influence the choice. In medium to large sites it would be normal to have both a central workshop and satellite workshops on each of the production units. Certain engineering operations can only be carried out under flame permits or in workshops in unclassified zones.

The production unit

The discussion on generic production plot layouts identified a number of considerations for the individual production plot. The plot will generally house the buildings and facilities identified above, but there are no hard and fast rules and the requirements for specific products may differ greatly. For the purpose of this guide, it is assumed that automated batch reactor plant are being dealt with that carry out potentially hazardous processes. Processes that could result in explosions and/or use or produce highly active chemicals should be housed in a special hazard unit in an isolated area of the site.

106

The site layout

With due regard to the above considerations, it is possible to draw up a site master plan based on typical processing requirements and information from marketing and research and development departments. This can entail some guesswork but it gives more logic to the development and hopefully prevents, for example, the construction of the site boiler house on the area that might be required for a future production unit. The data for the plot layout for the first production unit should be available but maybe not those for future units. It is normal, however, for a company to be involved in specific types of chemistry and this may allow the concept of a typical plot layout to be developed, although the concept is unlikely to satisfy the detailed requirements of the next factory. The flexible parameters of this master plan are discussed in more detail in the next section.

The master plan suggests that on the area of land under consideration it is possible to construct up to, say, five separate production units of a size applicable to normal bulk drug facilities. Each unit would have the necessary on-plot facilities including a bulk liquid tank farm, the relevant on-plot utility generation, a control room and management offices. Depending on the design of the main reactor building there could be reactor capacity up to 96,000 litres using a variety of reactor sizes. The site infrastructure possibly includes central site generation of steam and compressed air and space has been identified for engineering workshops, special hazard operation and effluent treatment and conditioning. A number of these buildings may be developed in a phased manner as the site expands.

The plan gives a basis for future expansion and allows a logical development that is not too restrictive.

5.4.3 Brown field sites

There is a wide range of brown field projects — it could begin with a cleared area within an existing production site that can be fenced off from adjacent operational areas or an area of an existing building that has been cleared for a new production unit or it could even be the installation of additional equipment in an operating factory. They all have one thing in common — they will all be influenced by what is already there. The cleared plot will have to take into account the existing site infrastructure; the cleared building will have to take into account the potential limitations of the existing structure; the additional equipment project will have to recognize the existing utilities and the impact of ongoing production operations within the building. For the purposes of this guide the discussion will be limited to the cleared site.

It is likely that the brown field project will be equivalent to the production plot concept described in green field section. The site boundary will be equivalent to the green plot boundary and it should be anticipated that the necessary public utilities and centrally generated site utilities would be made available at the boundary. The project may or may not include the augmentation of these utilities. Considerations for the layout will include:

- process buildings;
- control rooms;
- on-site utility generation;
- tank farm and drum stores;
- switch rooms and transformers;
- warehouse;
- offices and operator amenities.

In most cases the approach will be similar to a green field production plot except for the impact of the surrounding existing site and the restrictions it might introduce, both to design and construction activities.

In some instances integration with the existing site road systems might require substantial modification to the existing system. In other examples, the new production facility may be required for operation under GMP standards when the rest of the site is manufacturing a non-pharmaceutical product.

The overall approach to the layout of brown field site should follow the same general principles as described for green field sites. The overall picture should be considered before settling on the layout for the specific plot in question.

5.4.4 Layout specifics for biotechnology facilities

Personnel and material flows have to be carefully designed to allow an orderly progress of product from fermentation through purification to finishing whilst minimizing the risk of cross-contamination. Other factors that are important to facility design include constructability, operability and maintainability. The latter covers accessibility to equipment for maintenance purposes especially in clean rooms; services access can be provided via the interstitial space above ceilings or via voids in the walls connecting onto corridors. All these factors should be optimized to maximize space utilization and minimize facility cost.

Due to the changing nature of the biotechnology field, it is important to incorporate features into the design to enable expansion, re-use of existing space and re-use of equipment. Some of the methods available include:

- mobile vessels that can be moved easily to provide flexibility;
- centralized buffer solution preparation areas;

- centralized cleaning areas for mobile vessels, etc.;
- centralized kill systems for liquid/solid wastes.

However, these methods would have to be reviewed carefully to obviate any possibility of cross-product contamination.

5.5 Good manufacturing practice for BPC

5.5.1 Regulatory framework

The manufacture of any pharmaceutical product is subject to regulations dependant on the country in which the product is sold. In the case of BPCs, the main regulatory body is the Food and Drug Agency in the US. They expect BPCs to be manufactured in accordance with the rules laid down in the Code of Federal Regulations title 21. Within the EU the manufacture of pharmaceutical material is regulated by EU Rules for Pharmaceutical Manufacture, Volume IV.

Current thinking from the FDA is that they expect manufacturers to 'control all manufacturing steps, and validate critical process steps'.

A critical step is not necessarily the last step in manufacture but may be one which:

- introduces an essential molecular structural element or results in a major chemical transformation;
- introduces significant impurities into the product;
- removes significant impurities from the product.

Further information on this topic can be found in Chapter 3.

5.5.2 Good manufacturing practice (GMP)

The manufacture of BPCs in accordance with GMP ensures that the product has a high degree of assurance of meeting its predetermined quality attributes. GMP for a BPC is concerned with the manufacturing process, the equipment and facility in which it is carried out.

GMP is all about protecting the product from anything that can cause harm to the patient. This covers the processing itself and the avoidance of any contamination.

Modern BPC manufacture is generally carried out in closed process equipment so the potential for contamination is greatly reduced. Special attention is paid to activities that involve exposure of the product or its raw materials or intermediate stages. This involves protection of the operator and the process when dispensing, reactor charging, sampling and product packing.

GMP is also concerned with cross-contamination from other sources and linked systems. Special attention is paid to hold up within process systems, cleanability and the use of Clean In Place techniques, interactions with shared systems such as nitrogen and vents.

GMP is involved with the operating method. Any instruments that record critical data have to be calibrated and validated to ensure the integrity of the data. The process must be well understood and capable of being controlled.

5.5.3 Validation

The validation for BPC follows the same concepts and requirements to those detailed in Chapter 4. The main difference for BPC production is the concept of a critical step, and the point at which validation and pharmaceutical quality assurance have to be applied.

Secondary pharmaceutical production

6

JIM STRACEY and RALPH TRACY

6.1 Products and processes

6.1.1 Introduction

The selection of manufacturing methods for pharmaceuticals is directly related to the means by which the active substance is brought into contact with the agent responsible for the illness.

The obvious administration route for the delivery of drug therapy has long been via the mouth, perhaps on the basis that the ailment under treatment was probably caused by the assimilation of some hostile agent via the same route! More localized treatments involving the application of agents to the skin, or the insertion of medicament-containing substances into the various body cavities was a logical development of oral entry.

These methods, with enhancements and improvements, remain with us today and are still the most widely used, but they have been joined by injectable and other transcutaneous routes, inhalations and transdermals. A brief description of each of these, together with their associated manufacturing procedures, is outlined in the following sections.

6.1.2 Pills

One of the earliest forms of oral-dose treatment took the form of manually-rolled gum-based pills. Thomas Beecham, one of the pioneers of pharmaceutical formulation, sold his original 'Pills' in a market at Wigan. These wonders were originally produced by mixing the gums with herbal extracts known to have pain-killing or laxative properties, and were sufficiently popular that the initial production methods needed to be updated and mechanized quite soon in the products' life history. Thus, the pill-rolling machine was produced, followed by the introduction of quality control in the form of a device which ensured that the individual pills were as perfectly spherical and of equal size as the rolling machine could produce — rejects being recycled for further processing.

6.1.3 Tablets

Although a successful formulation, the pill suffered from production output restrictions and was overtaken by the modern tablet — produced by mechanically compressing suitable mixtures of drug substance and excipients held in a cylindrical cavity, or die, by the action of piston-type tools.

During the early development of the tablet, it was quickly realized that in most cases the active drug substance did not lend itself to the formation of a reliable compacted entity merely by the application of pressure. The addition of binding agents was found to be necessary, together with other excipients offering enhanced powder flow, and the following characteristics of well-made tablets were soon established as important:

- the ability to withstand mechanical treatment (packaging, shipping, dispensing);
- freedom from defects;
- reasonable chemical and physical stability;
- the ability to release medicaments in a reproducible and predictable manner;
- the drug and excipients are compressible.

6.1.4 Granulation

The process of tablet making using modern machinery involves the blending of the drug substance with binders, fillers, colouring materials, lubricants etc., followed by a series of operations designed to increase the bulk density and uniformity of the mixture and prevent segregation of the drug. These operations are known as granulation, and are an important part of modern pharmaceutical product manufacture, notably for tablets but also for other products. The granulation process is a critical step in reliable drug manufacture, as it often involves the relative 'fixing' of several ingredients and must therefore be carefully designed and controlled. Regulatory pressures, demanding as they do a strict equivalence of product performance before and after development scale-up, ensure that during drug research and development the selection of granulation methods must be made carefully. This selection, including the choice of individual equipment types, can be difficult and costly to change, owing to the need for the validation of continued product performance.

The desired increase in bulk density and uniformity can be achieved by compression methods followed by milling, a process known as dry granulation. The techniques used for compression include 'slugging', a process not unlike tablet making, and roller compaction, which involves the feeding of material between a set of closely spaced steel rollers. The former produces tablet-like structures, which can then be reduced to granules by milling, whereas the latter gives rise to a flake-like compact that is first broken into smaller pieces and then reduced by milling. In either case, the forces and friction involved are such that a lubricating material (such as magnesium stearate) is necessary. To ensure good material flow, a material such as Cab-o-Sil (silicon dioxide) is often used.

Figure 6.1 shows a flow diagram for a dry granulation process.

The dry granulation process is not very easy to contain in terms of dust emission and available equipment suitable for pharmaceutical applications is

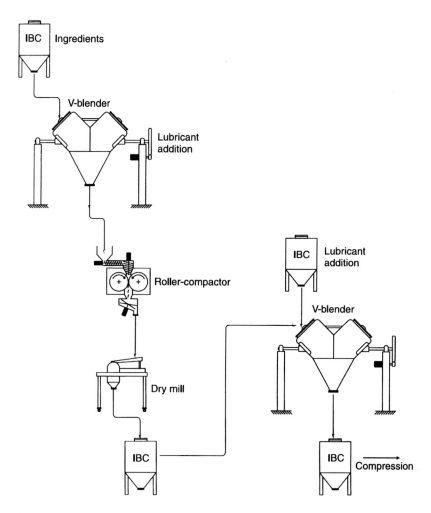

Figure 6.1 Typical dry granulation process

not common. This is mainly due to its greater use in heavy chemical, food and fertilizer manufacture. However, all formulation departments will attempt to formulate a dry process, as it is cheaper in capital equipment and a simpler process.

Therefore, the process most often used is wet granulation. This operation takes the blended materials, adds a suitable wetting agent, mixes the combined materials, passes the wet mass through a coarse screen, dries the resultant granules using a tray or fluid-bed dryer, and finally reduces the particle size of the dry material by passing it through a finer screen.

Figure 6.2 (see page 115) shows a typical flow diagram for a conventional wet granulation process.

The increasing potency of drug substances has encouraged manufacturers to seek granulation methods that are enclosed and free of dust emissions. Thus, a number of process equipment manufacturers have developed systems for enclosed processing which incorporate several of the granulation steps in a single unit.

The most common of these is the mixer-granulator, which combines the powder mixing, wetting, wet massing and cutting operations. These efficient machines can perform this set of processes within a matter of minutes, and discharge a wet granule which requires only drying, milling and final blending with lubricants to produce a tablet compression mix. In most cases, however, the discharged wet granule will be further reduced in size by passage through a coarse-screen sieve prior to drying, in order to improve drying rates and consistency.

The key to mixer-granulator operation is the combination of high-shear powder mixing with intense chopping of the wet granule.

Figure 6.3 (see page 116) illustrates a typical mixer-granulator.

The process steps employed in mixer-granulators are as follows:

- mixing of the dry ingredients with the main impeller and chopper rotating at high speed ($15 \, \mathrm{m \, s^{-1}}$ impeller tip speed and 4000 rpm chopper speed) for, typically, 3 minutes;
- addition of a liquid binder solution by pumping, spraying or pouring it onto the dry material with the impeller and chopper running at low speed ($5 \, \mathrm{m \, s^{-1}}$ and 1500 rpm) for around 2 minutes;
- wet massing with impeller and cutter running at high speed (2 minutes);
- discharge of the granulated material through a coarse sieve or directly to a dryer.

The step times indicated will vary according to the product involved, and are generally critical in relation to granule consistency.

There are a number of advantages that combined-processor granulators have over conventional methods, as follows:

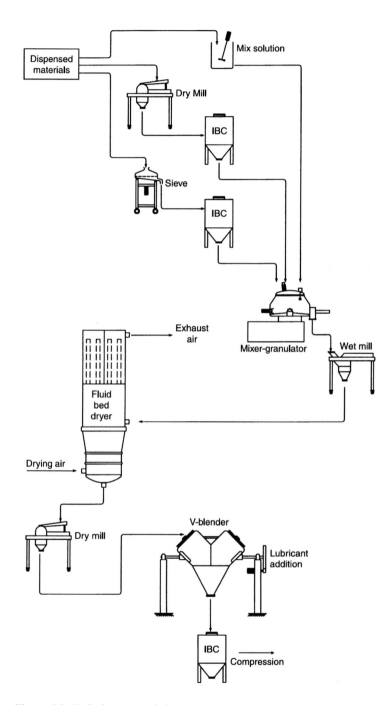

Figure 6.2 Typical wet granulation process

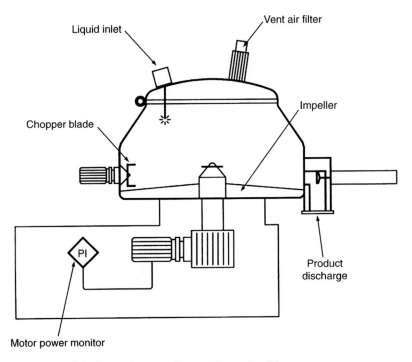

Figure 6.3 High shear mixer-granulator with opening lid

- the granulation steps are enclosed in a single unit that can integrate with subsequent-stage equipment, thus minimizing dust emissions;
- the process is rapid;
- binder liquid volumes can be reduced;
- granule characteristics can be adjusted easily by changing step times and binder addition rates;
- inter-batch cleaning can be performed easily, and can be achieved by use of automatic Clean In Place systems.

However, disadvantages do exist, mainly associated with the high speed and energy input provided by the agitators. This can give rise to mechanical breakdown of ingredient particles, over-wetting due to compaction producing over-sized granules, and chemical degradation of sensitive ingredients due to temperature rise.

Developments of the mixer-granulator include jacketed and heated or cooled mixing bowls, which avoid over-heating of the granules or assist in their drying, and the use of vacuum to reduce drying times and temperatures. These 'single-

116

pot' units aim to provide an efficient and contained operation covering as many granulation steps as possible in a single unit.

Single-pot mixer-granulators using vacuum and heated jackets, but employing slightly different configurations of impeller and chopper, include the Zanchetta Roto granulator/dryer, which uses a vertical-axis retractable chopper. This machine also operates slightly differently in that the bowl is pivoted so that the effective heat exchange surface can be maximized for reduced drying time. The planes of shear within the powder mass can also be altered at each stage of the process for optimum mixing and final size reduction.

The application of microwave energy for granule drying in-situ has been pioneered by Aeromatic-Fielder. The magnetron generators are situated on top of a mixer-granulator that operates under vacuum and are energized at the end of the wet massing/chopping cycle.

Figure 6.4 (see page 118) shows a flow diagram for a combined granulation process.

Spray granulation

A different and somewhat unusual granulation technique is the use of the spray dryer.

Spray granulation requires that all ingredients are soluble or dispersible in a common solvent and can be crystallized/combined from that solvent at a suitable temperature. The solution or suspension feed stream is passed through a nozzle inside the spray dryer chamber, where it immediately comes into contact with a co-current or counter-current gas stream at controlled temperature. The solvent evaporates rapidly and the resulting solids are separated from the air stream by cyclone separators and filters.

Spray granulation offers a number of advantages over mixer-granulation systems. The feed, being a homogeneous liquid, removes concerns over blending of liquid binders into dry solids. The resulting granules are homogeneous and, regardless of size, contain uniform proportions of the ingredients. Temperature control is also more consistent, thus eliminating problems of heat-degradation. Finally, the absence of mechanical moving parts generally improves cleanability and reduces contamination risks.

A recent example of this principle is the Spinning Disc Atomization system being developed in Switzerland by Prodima SA and EPFL. In this system a suspension of the product or a polymer melt is passed between rotating concentric conical discs and is released into the gas stream as fine uniform droplets, which dry or solidify to produce very spherical and similar granules.

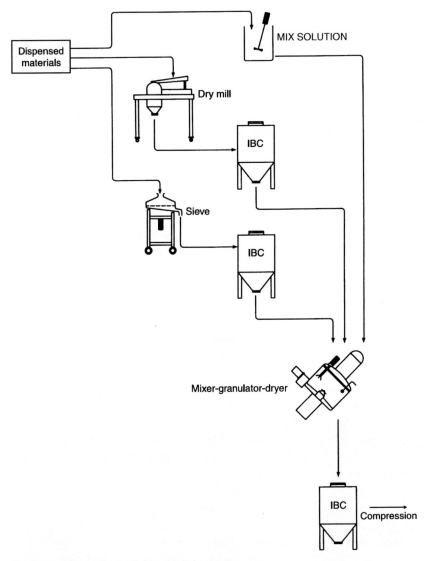

Figure 6.4 Typical combined granulation process

Fluid-bed granulation

A related process for achieving granulation by spray techniques utilizes the mixing action of a fluidized bed to mix powder ingredients in an otherwise conventional fluid-bed dryer. The mixture so created is then subjected to a sprayed-on binder solution, the evaporation of whose solvent produces an intimately-mixed granulate which is then dried by the fluidizing air stream.

Direct compression

Some drug substances have characteristics that allow them to be compressed without prior granulation, using a process known as 'direct compression'. This process avoids the cost and inconvenience of granulation, but often requires the use of special binding agents to avoid segregation during mass flow of the mix in the tablet compression process.

Figure 6.5 (see page 120) shows a typical flow diagram for direct compression.

6.1.5 Tablet compression

The basic principles of the tablet compression process have remained unchanged since their inception. The tablet press compresses the granular or powdered material in a die between two punches, each die/punch set being referred to as a station. Although many alternative methods have been tried, the principle of filling granules into a die and compressing them into a tablet between two punches is still the primary method of manufacture for all machines used in pharmaceutical manufacturing.

Developments utilizing a slightly different configuration of punch and die are under current examination in Japan and Italy. The primary incentive of these developments is to produce an arrangement which can reliably be cleaned-in-place, rather than relying on the time-consuming process of dismantling the machine to remove product-contact parts for cleaning with its attendant risks of operator exposure to active products.

Tablet machines can be divided into two distinct categories:

- those with a single set of tooling — single station or eccentric presses;
- those with several stations of tooling — multi-station or rotary presses.

Figure 6.6 (see page 121) illustrates the principles of tablet machine operation.

The former are used primarily in the small-scale product development role, while the latter, having higher outputs, are used in production operations. Additionally the rotary machines can be classified in several ways, but one of the most important is the type of tooling with which they are to be used.

There are basically two types of tooling — 'B' type which is suitable for tablets of up to 16 mm diameter or 18 mm length (for elliptical or similar shapes), and 'D' type which is suitable for tablets with a maximum diameter or

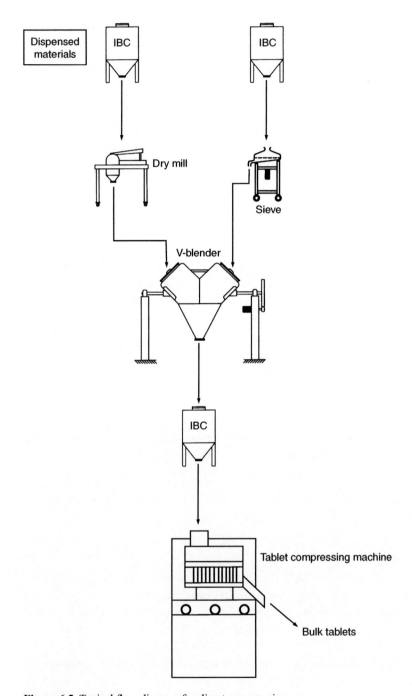

Figure 6.5 Typical flow diagram for direct compression

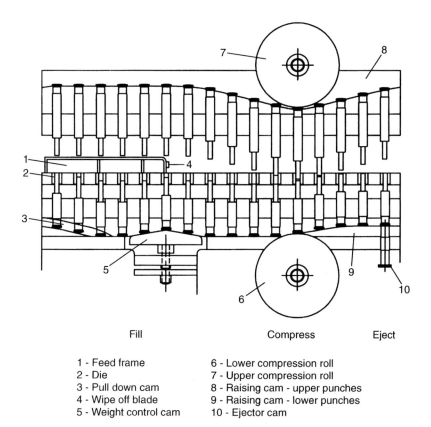

Fill   Compress   Eject

1 - Feed frame	6 - Lower compression roll
2 - Die	7 - Upper compression roll
3 - Pull down cam	8 - Raising cam - upper punches
4 - Wipe off blade	9 - Raising cam - lower punches
5 - Weight control cam	10 - Ejector cam

Figure 6.6 Rotary tablet compression machine operation

maximum length of 25.4 mm. The 'B' type punches can be used with two types of die; the small 'B' die is suitable for tablets up to 9 mm diameter or 11 mm maximum length, and the larger 'B' die is suitable for all tablet sizes up to the maximum for the 'B' punches. Machines can, therefore, be used with either 'B' or 'D' tooling, but not both.

Machines accepting 'B' type tooling are designed to exert a maximum compression force of 6.5 tonnes, and machines accepting 'D' type tooling 10 tonnes. Special machines are available which are designed for higher compression forces.

The maximum force that can be exerted on a particular size and shape of tablet is governed by the size of the punch tip or the maximum force for which the machine is designed — whichever is smaller.

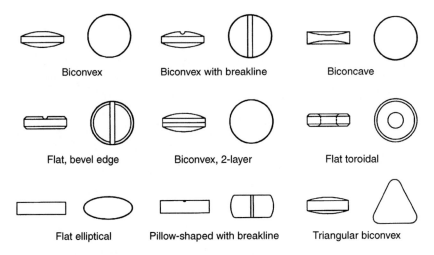

Figure 6.7 Some tablet shape possibilities

Tablets are now available in a range of diameters and thicknesses to suit the proportion, active dose and characteristics of the drug substance. Figure 6.7 shows some examples of tablet shape possibilities.

Formulation has enabled the production of tablets with special characteristics such as:

- effervescent;
- chewable;
- multi-layer;
- delayed or sustained release;
- bolii for veterinary use.

These examples indicate the extent to which development of the tablet has continued since its original introduction. Much effort was expended during the first half of the 20th century in establishing the best particle size of the active drug and the range and rheology of excipients needed to produce a reliable tablet with acceptable dispersion and absorption characteristics. However, the technology of tablet compression did not advance significantly during this period; reliable and robust machinery was produced and its performance and output were considered suitable for the demands of the time. Subsequently, improved excipient development by the pharmaceutical industry, based on enhanced glidants and micro-crystalline cellulose binding agents, and the introduction of reliable sensors coupled with electronic control systems have allowed compression technology to advance.

Whereas the manufacture of a single tablet is simply a matter for formulation development, the production of such products at machine speeds in excess of 300,000 tablets per hour raises additional challenges. The critical stage here is the delivery of the granulation into a die on a high-speed rotating disc accurately, so that tablets of minimum weight variation can be produced.

Very high-speed compression machines are now available with built-in tablet weight and thickness control and the ability to be self-monitoring from an output and quality standpoint. Hence, it has become possible for continuous, unmanned operation of the tabletting process to be carried out (the so-called 'lights out' working).

More recently, the greater impetus to improve has come from regulatory pressures, under which the need for uniformity, consistency and reliability has become paramount. The principles of current Good Manufacturing Practice (cGMP) and validation have greatly influenced the development of the tablet manufacturing process and the materials and methods used therein.

6.1.6 Coated tablets

Many tablet products contain active materials that require taste masking or a controlled release rate, and a variety of methods have been developed to achieve these objectives. A careful choice of excipients can mask the unpleasant taste of certain compounds, but a more reliable procedure is to coat the tablet with a barrier material. Such coating can be achieved by forming a compressed layer around the basic tablet, or core. There are compression machines that can accept a previously formed core and surround it with a layer of excipient material. An additional and similar use of compression can produce layered tablets.

The traditional method of taste masking is to apply a sugar coating to the core, and although this method has largely been superseded by film-coating techniques, it is still used. Originally the sugar coating was applied by pouring a sugar syrup, usually coloured, onto a bed of pre-varnished tablet cores rotating in a steel or copper pan into which warm air was blown. The skill required to achieve a successful application of the sugar coat was such that the true art of tablet making/coating resided in the hands of a small and respected elite. A key feature of the sugar coating process was that the tablet weight increased significantly with the sugar coating accounting for typically 60% of total tablet weight.

Subsequently this skill has largely been replaced by a more-automated system using mechanized spray/jets of sugar syrup applied in a pre-determined and controlled manner to a bed of tablets rotating in a perforated drum and warmed with pre-heated air.

A logical development of automated sugar coating was the introduction of non-sugar coating materials, based on plastic film-forming solutions/suspensions. This 'film coating' process has largely replaced the original sugar coating technique, although the method of application is basically similar. Advantages are the removal of food-type materials, a higher speed of through-put and a small increase in tablet size/weight, with consequent reductions in packaging cost.

Initially, most film-coating formulations included the use of flammable solvents for coating solution/suspension manufacture, and given the relative toxicity and safety risks associated with these materials it is not surprising that much effort has been expended in developing aqueous-based alternatives. The latter now make up the majority of film-coating formulations.

Figures 6.8 and 6.9 (see pages 124 and 125) are flow diagrams showing the stages of the film and sugar coating processes.

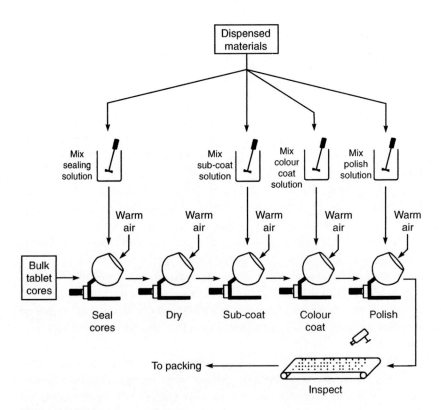

Figure 6.8 Tablet sugar coating

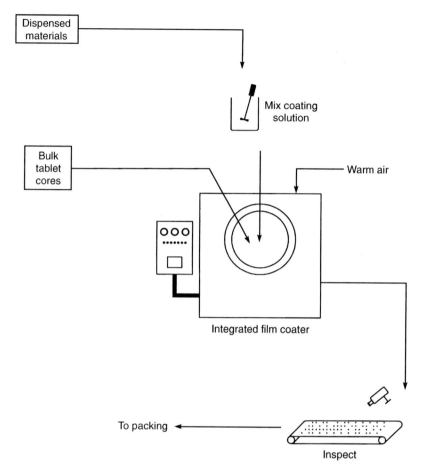

Figure 6.9 Tablet film coating

6.1.7 Capsules

The encapsulation process is an alternative to tablet compression, which also masks unpleasant-tasting actives. It can also have advantages where compression could result in a compacted tablet with unacceptably long or short dispersion time in the upper alimentary system. As with tablets, the gelatin barrier can be further coated with 'enteric' materials which ensure dissolution or dispersion only in that part of the system where optimum effect is produced.

Capsules are generally of two types, made with either hard or soft gelatin.

Hard gelatin capsules

Hard capsules are manufactured from bone gelatin and are produced as empty two-part shells supplied to the pharmaceutical manufacturer for filling. The capsules are produced in a number of standard sizes designated 5 through 000, with larger sizes available for veterinary applications.

Although originally filled by hand, and later by devices that allowed multiple cap/body separation, volumetric filling and reassembly, they are now filled on automatic machines. These separate the two parts, fill the body with powder, granules, pellets or semi-solids as required by the formulation to a controlled level, and reassemble the two parts prior to discharge. One disadvantage of the hard capsule is that a number of systems for dosage control have been developed by different filling machine manufacturers, so that (unlike tablets) the capsule has no standardized filling system.

The original hard capsule type, which was conceived as long ago as the 1840s, consisted of two plain-sided cylinders with hemispherical ends, one of larger diameter, so that one formed the body and the other the cap. Tolerances during manufacture (by dipping pins in molten gelatin) ensured that the cap/body clearance was minimized to prevent the possibility of powder leakage. Originally designed to deliver powder products, improvements in formulations and capsule tolerances have allowed the use of this dosage form for delivering oils and pastes.

Where fine powder escape or simple separation of the two parts proved problematic, these capsules were sealed by the application of a band of molten gelatin at the cap/body joint. This was achieved using conveyor-type machines, which provided space and time for the gelatin band to set, and provided an opportunity for visual inspection of the capsules.

The introduction in the late 1960s of the self-locking capsule, coupled with improved dimensional tolerances, largely removed the necessity for band sealing.

After the initial establishment of hard-shell capsules as a dosage form, machines were developed to increase the production rates of filled shells. One of the first types, developed by Colton and by Parke-Davis, consisted of a two-plate device that simply separated the two halves of the shells, filled the bodies volumetrically, and allowed recombination. One of the first commercially available machines to automate the process was developed by Höfliger and Karg of Germany, and filled at speeds of 150 capsules per minute. This machine used the differential diameter of the capsule cap and body to orientate and vacuum to separate the two parts, and an auger device to meter the product powders or granules and feed them into the capsule bodies. The caps and bodies were then re-combined prior to ejection.

Figure 6.10 (see page 127) illustrates a typical capsule filling process.

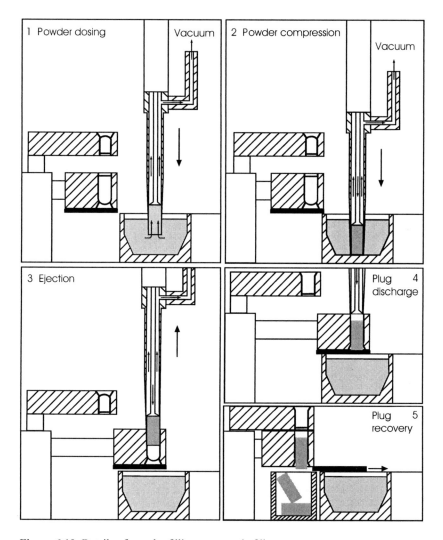

Figure 6.10 Details of powder filling on capsule filler

These techniques for capsule handling have basically been retained in later, higher-speed machines, but the dosing system has undergone a divergence in design. The original auger type filler is no longer used, mainly because it is not capable of high-speed operation without recourse to multiple stations, which would give rise to an unacceptably large machine.

The system developed in the 1960s by the Zanasi brothers in Italy, and still used today, employs a plug-forming method to produce the required dose.

A tube is plunged into a container of product having uniform depth, and the column of product so contained is compressed in-situ by the downward motion of a piston inside the tube. On withdrawal of the tube a cylindrical compact is retained within it, and this is then discharged into a capsule body by further downward motion of the piston. The dose weight and degree of compression (and subsequent dispersion) of the product is capable of adjustment by altering the depth of powder/granule in the product container and the extent of downward motion of the piston. One advantage of this so-called 'dosator' system is that the tube is quite small, so that a number of them can be arranged in a dosing module of modest dimensions to give increased output. Original machines worked with an intermittent motion, but later versions were designed to operate continuously by arranging the capsule feed/handling groups and the dosing units on separate rotating turrets, emulating to some extent the conventional tablet press.

To meet the challenge of the higher-speed dosator machines, Höfliger and Karg introduced their GKF range of machines, which utilizes the natural capacity of the capsule body for controlling product dosing. The capsule bodies, having been separated from their caps and fed vertically into cylindrical machined holes in a rotating disc, are moved so as to pass under a container of product powder/granule (not unlike the feed frame of a tablet compression machine), so that the product mix flows into the empty bodies. Before leaving the product container, the contents of the capsule bodies are subjected to compression by the insertion of pistons to a pre-determined and adjustable depth. After compression, the bodies are removed from the dosing zone by the rotation of the disc and reunited with their caps.

This system allowed for a significant speed increase compared with the auger type, but was disadvantaged in that the degree of dosage weight and compaction control was less than that allowed by the dosator system. A revised version was therefore introduced which included an intermediate dosing disc which allowed for the formation of a product 'plug', independently of the capsule body, which could then be transferred to the body after formation and compression. This development permitted the use of dosing discs of different thickness to control dose weight.

Again, the small dimensions of the Höfliger and Karg dosing arrangement made it possible to fill capsules at very high speeds of over 2500 filled capsules per minute.

Apart from size considerations, the key to high-speed capsule filling is powder flow, which in turn relies on consistent particle size and shape distribution. The bulk density of the filling material is of parallel concern, and must be uniform if reliable dosage weights are to be achieved. As with tablet compression, the conditions and processes employed for preparation of the filling mix have critical impact on performance. A typical capsule filling mix for a high-dose product may contain only the active drug and a lubricant (for example, many antibiotic products are formulated in this way), so the options for formulation adjustment are limited.

Products utilizing a lower active dose proportion may also contain a filler (such as lactose), flow-aid (for example, silicon dioxide) and surfactant (such as sodium lauryl sulphate) and may therefore have superior flow and output characteristics.

Soft gelatin capsules

Soft gelatin capsules, where the gelatin contains a plasticizer to maintain flexibility, were originally developed in France in the 1830s, and are generally used where the active product material is liquid or semi-solid, or where the most appropriate formulation is in this form. They were originally made in leather moulds, which provided an elongated shape and a drawn-out end which could be cut off to allow for the insertion of the product liquid, after which the end could be sealed with molten gelatin.

Although less popular than hard-shell capsules, their 'soft' counterparts satisfy a different set of product/market criteria, under which the total containment of the active principals is a key concern.

The manufacture of soft-gelatin capsule products is generally regarded as more specialized than that of other dosage forms and has been limited to a small number of producers. These companies have very much influenced the development of the technology employed in the production process.

R P Scherer developed the modern technology for automated soft-gelatin capsule production in the 1930s by designing the Rotary Die Process. The basic technique employed in soft-shell filling involves the melting of a gelatin/plasticizer mixture and the extrusion of this between the two halves of a mould formed by twin rotating cylinders, while the product liquid or solid is injected between the two half-shells thus produced. The continued rotation of the cylindrical moulds results in the closing and sealing of the resultant capsule and its subsequent ejection.

6.1.8 Pellets and other extrudates

A feature of capsules, which can have drug-release benefits, is that they can be filled with materials other than powder or granule mixtures. In addition to liquids and pastes, which are generally more suited to soft gelatin types, product in the form of large granules or pellets can be filled into hard-shell capsules.

Whereas 'large' granules can be prepared by the methods already described, pellets have their own production technology, based upon extrusion and spheronization. The spherical granules, or spheroids, have several advantages over conventional granules due to their uniform shape — they have superior flow properties, are more easily coated and have more predictable active drug release profiles. Dried spheroids may be coated and then filled into hard gelatin capsules to provide a sustained release dosage form capable of gradually releasing its active constituents into the gastro-intestinal tract over several hours.

The process of extrusion has been the subject of much scientific study in the polymer, catalyst and metal industries. It may best be described as the process of forcing a material from a large reservoir through a small hole, or 'die'.

Pharmaceutical extrusion usually involves forcing a wet powder mass (somewhat wetter than a conventional granulation mix) containing a high concentration of the drug substance together with a suitable binder and solvent, through cylindrical holes in a die plate or screen. Provided the wet mass is sufficiently plastic this produces cylindrical extrudates of uniform cross-section, not unlike short strands of spaghetti. These extrudates are loaded onto the 'spheronizer', a rotating scored plate at the base of a stationary smooth-walled drum. The plate initially breaks the strands into short rods, and then propels them outwards and upwards along the smooth wall of the drum until their own mass causes them to fall back towards the centre of the plate. Each individual granule thus describes a twisted coil pathway around the perimeter of the plate, giving the whole mass a doughnut-like shape. This movement of the granules over each other combines with the friction of the plate to form them into spheres.

A typical spheronizer arrangement is shown in Figure 6.11 (see page 131).

The basic core granules for the preparation of controlled release pellets for filling into capsules can be prepared by several methods, such as spray coating, pan/drum granulation, melt granulation, as well as spheronization. Core granules are then coated with a suitable polymer or wax to confer on them their controlled-release properties, either by spraying wax-fat solutions onto granules tumbling in pans or by spray coating them with polymers or waxes in a standard film coating machine.

The melt-granulation pelletization process is a fairly recent technique, based on high-shear mixer-granulator technology. In this process the core material (drug substance) is mixed with a suitable low-melting solid excipient (such as high molecular weight polyethylene glycol) in a high-shear mixer. The agitation is continued until the heat generated melts the excipient, which forms a wax-like coating around the core material. Under controlled conditions it is

130

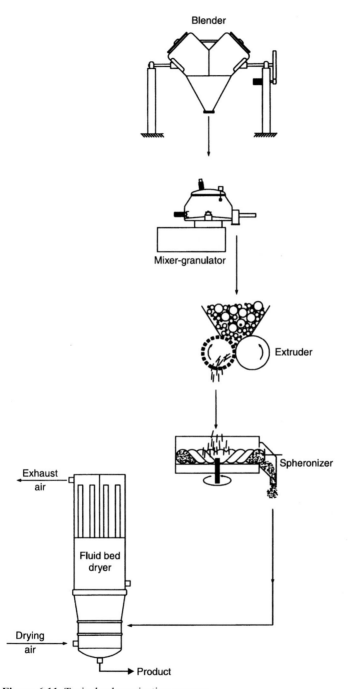

Figure 6.11 Typical spheronization process

thus possible to produce coated pellets of reasonably uniform size, which can exhibit dissolution or dispersion properties suited to the drug substance involved.

6.1.9 Syrups, elixirs and suspensions

These dosage forms are basically produced by the dissolution or suspension of a drug substance in a suitable solvent/carrier (usually purified water), together with appropriate sweeteners, flavours, colours and stabilizing agents.

The primary use of these products is in paediatric and geriatric treatment, where the patient may have difficulty in swallowing solid-dose medicines, although they are also valuable where the pre-dissolution or pre-suspension of the active drug can enhance therapeutic effect (for example, cough remedies).

The production of solutions is a relatively straightforward procedure, typically using purified water heated to a minimum temperature suitable for dissolution of the materials, with the addition of the active and excipients followed by a filtration to remove possible haze prior to filling.

The difficulties inherent in syrup manufacture are associated with product stability, for example dissolution and solubility, which may not be adequate at normal temperatures and taste masking, which is made more difficult when the drug is in solution.

Suspensions overcome some of these problems for suitable products, but other difficulties exist — notably maintaining the product in suspension. This latter challenge can only be met by the use of a high-shear dispersion system, or homogenizer, which utilizes wet-milling techniques to reduce particle size and enable reliable product suspension.

Elixirs are basically clear, flavoured solutions containing alcohols and intended for oral administration. Other ingredients may include glycerin, sorbitol, propylene glycol and preservatives. Quite high alcohol contents were common to ensure dissolution of certain drug substances, although products formulated in this way are becoming unusual.

The distinction between medicated elixirs and solutions is not altogether straightforward, the latter often containing alcohol (for example, up to 4% is present in some ephedrine-containing syrups).

6.1.10 Emulsions

An emulsion is a two-phase liquid system where one liquid exists in very small droplet form (the internal phase), suspended in another (the external phase); the two liquids being otherwise insoluble in one another. An emulsifying agent contained within the mixture acts on the surface active properties of the two liquids such that the emulsion remains stable for a sufficiently long period to serve its purpose. If necessary, the liquids may be heated in order to enhance the stable formation of the emulsion, by reducing its viscosity. The active pharmaceutical material may be a solid, which is added to the liquid/liquid system, or may be soluble in one of the components. The product is prepared by high-shear mixing to reduce droplet sizes, using submerged-head agitation devices which draw the mixture through a high-speed rotating impeller contained within a close-fitting housing, not unlike a centrifugal pump.

Most pharmaceutical or cosmetic emulsions contain water and oil as the two phases, and may be oil/water or water/oil, depending upon which is the internal and which is the external phase. It is possible for emulsions to 'invert'; a process in which is the internal and external phases change identity between the water and oil ingredients.

Although more usual in cosmetic topical formulations, pharmaceutical emulsions are prepared for topical, oral and parenteral use. Owing to their difficulty in preparation, pharmaceutical emulsions are used infrequently and only where they exhibit particularly useful characteristics such as drug solubility or specific absorption capability.

6.1.11 Creams, ointments and other semi-solids

Creams are basically similar to emulsions in that they are two-phase liquid systems; however, they exhibit greater physical stability at normal temperatures than emulsions and can thus be more useful for topical applications. The external phase is often water, while the internal phase is usually a high-viscosity oil or semi-solid oleic material.

Manufacturing involves the heating and stirring together of the two phases in the presence of emulsifying agents and other excipients (colour, stabilizers, perfume etc.) with the assistance of a high-shear mixing device (colloid mill, homogenizer or ultrasonic mixer). The operation is most often carried out at slightly elevated temperatures to enhance dispersion. If the active substance is a solid, it will normally be added to the stabilized mixture, followed by further agitation and homogenization.

Ointments are solutions of high melting point and lower melting point hydrocarbons, usually mineral oil and petroleum jelly. The active drug and other excipients are incorporated in much the same way as with creams with the semi-solid matrix being heated to assist dispersion of these additives.

An advantage of ointments over creams is that, when used as a base for sterile products such as ophthalmics, being solutions they can be sterilized by filtration after the addition of a soluble active or prior to the final addition of an insoluble sterile active ingredient. Cream bases would break down under microfiltration conditions.

133

Modern ointments based on polyethylene glycols (PEGs), which are available in a range of viscosities, have the advantages of typical ointments but are water miscible.

Pastes are similar to ointments except that they contain much higher insoluble solids content. They are prepared in a similar fashion, with the semi-solid base being added to the solids gradually with mixing until the required concentration is achieved and the dispersion is uniform. Pastes are used where a particularly high concentration of the medicinal compound is needed in contact with the patient's skin (such as for burns, prevention of sunburn or the treatment of nappy rash).

Gels are semisolid systems in which a liquid phase is held within a three-dimensional polymeric matrix consisting of natural or synthetic gums, with which a high degree of physical or chemical cross-linking has been introduced. Polymers used to prepare pharmaceutical gels include natural gums such as tragacanth, pectin, carrageen, agar and alginic acid and synthetic materials such as methylcellulose, hydroxyethylcellulose, carboxymethylcellulose and the carbopols (synthetic vinyl polymers with ionizable carboxyl groups).

6.1.12 Suppositories

The original suppositories were hand-formed pellets based upon white paraffin wax and containing active material and relevant excipients dissolved or dispersed in the melted matrix. Eventually the need for standardization resulted in the development of pre-formed moulds into which the cold product mass was forced by means of a piston and cylinder arrangement.

This slow process was later superseded for volume production by warming the mass to its melting point and pouring the liquefied material into split moulds, which were then solidified by cooling.

The early types were wrapped in greaseproof paper packaging and were successful except that any rise in ambient temperature would result in melting, with subsequent leakage and product spoilage; hence the introduction of plastic disposable mould materials which were closed with adhesive or heat-sealed cover strips. Initially the moulds were sold as pre-formed strips containing typically five moulds. Machinery was developed which filled these strips in rows, followed by cooling/solidification and the application of seal tapes.

These machines have relatively low output, but are suitable for the production rates often associated with this dosage form. Later form-fill-seal machines provide capacity for larger product sales, and involve the forming of moulds automatically on-line, followed by filling, cooling and heat sealing using a single packaging material. A feature of all fill-seal suppository machines is the need to allow for the shrinkage coincident with the cooling/solidification process. This requires that the filled moulds are cooled to allow solidification of the contents prior to sealing, and the machines are often quite long in size to accommodate the length of the cooling section.

6.1.13 Oral, nasal, aural drops and sprays

Oral medicines applied in drop form are usually neonatal versions of paediatric syrups and suspensions. They are filled into small bottles, often of a flexible plastic that allows the container to be squeezed so that the requisite number of drops of liquid can be exuded through the plastic dropper insert.

Nasal solutions are similar except that the formulation will usually be isotonic with nasal secretions to preserve normal ciliary action. The drugs used in such formulations include ephedrine, for reducing nasal congestion, antibiotics, antihistamines and drugs for the control of asthma.

Products formulated as aural drops, usually referred to as otic preparations, include analgesics, antibiotics and anti-inflammatory agents. They are usually based on glycerin and water, since glycerin allows the product to remain in the ear for long periods. In the anhydrous form, glycerin has the added benefit of reducing inflammation by removing water from adjacent tissue.

Sprays used orally or nasally, are similar in formulation to their equivalent drops, being simple solutions and suspensions traditionally applied to the mouth, throat or nose by bulb type spray devices. Modern formulations make use of plastic pump sprayers or simple flexible bottle/nozzle combinations to produce the required spray pattern.

6.1.14 Ophthalmic preparations

Two formulation types are generally used in ophthalmic treatment; ointments and liquid drops, which together provide for both water soluble and oil soluble active principals. They are produced in the same way as oral formulations in terms of the equipment and processes, although a higher level of cleanliness is required.

Products for the treatment of eye disorders have traditionally been manufactured under clean conditions, not least to avoid complications arising from the introduction of foreign particles to the eye (such as corneal ulcers or loss of eyesight). The need for medicines used topically on the eye surface to be aseptic was not originally thought necessary, owing to the fact that under normal conditions the eye's surface is in direct contact with the external environment, which contains many infective agents. Thus, like the alimentary system, the eye was thought able to cope with such challenges without additional protection. More recently however, it has become accepted that under many circumstances requiring medicinal treatment, the eye has an increased liability to infection by organisms such as *Staphylococci* or *Pseudomonas aeruginosa*, and should therefore not be exposed to any substance likely to give rise to such infection. It is now an internationally recognized pharmacopoeial requirement that ophthalmic preparations be prepared aseptically.

6.1.15 Injections

A potentially unwanted feature of orally dosed medicines is their introduction to the body's system via the route designed for digestion, a process more effective in decomposition of chemical entities than in their intact delivery to the remotest regions of human or animal physiology!

The mouth, throat, stomach and intestines contain a complex mixture of enzymes and acids, which will usually ensure that any orally-ingested medicine is, at the very least, altered before it can be absorbed into the bloodstream. It is the bloodstream that distributes the absorbed material and until the said material enters the bloodstream it is unable to create any effect beyond areas of immediate contact within the alimentary system.

Hence, if a medicinal substance has poor stability in acid solution or is easily broken down by digestive enzymes, it is of very little use in disease control as it will probably not reach those parts of the body's systems requiring treatment. A method of avoiding this effect and delivering the substance closer to the site of the illness or infection is via a transcutaneous injection. Although some drugs are unstable in body fluids including blood, the injectable route very much enhances the possibilities for overcoming instability problems.

The two most common forms of injection are intramuscular, where the substance is injected into tissue containing small blood vessels and therefore remains most effective local to the injection site; and intravenous, involving direct injection into a larger blood vessel, thus ensuring rapid transit around the body. A further procedure involves sub-cutaneous injection, used for the deposition of controlled-release formulations.

Whether for intramuscular or intravenous use, these products are liquids or suspensions, which are produced as a pre-sterilized material contained in ampoules or vials. The medicinal product may be based on aqueous or oil formulation depending on the relative solubility of the drug substance and/or the required release rate into the surrounding body tissue. Most injectable products are made as single-dose containers, although multi-dose systems are available for use in vaccination and in veterinary practice.

Additionally, drugs requiring sustained application via intravenous infusion over long periods are produced as large volume systems (typically 500 or 1000 ml).

Liquid products in solution can be filled under sterile conditions within suitable clean areas, the solution being itself sterilized by filtration using

0.2 micron porosity filters. However, the preferred manufacturing procedure is to ensure sterility by terminal sterilization of the filled ampoules or vials, by autoclaving or gamma irradiation. Only where such terminal sterilization techniques are likely to cause decomposition of the drug substance is it considered acceptable to rely only upon manufacture under sterile conditions to achieve the required standard. In such cases the extent of sampling for sterility testing of the final product will be increased.

Although sometimes desirable for the terminal sterilization of heat-sensitive suspensions, it should be noted that irradiation is not without problems. Apart from the obvious safety considerations, the effect of gamma radiation on the type of glass used for ampoule and vial manufacture is to cause brown discolouration, thus adversely affecting subsequent inspection operations. The generation of free radicals within product solutions is also a possibility, with consequent chemical deterioration.

Where the active drug is unstable in solution (such as for certain antibiotics) the product is filled into vials, under sterile environmental conditions, as a dry powder. Such materials are often very moisture-sensitive, and special arrangements need to be made to ensure a low-humidity environment in areas of product exposure. A key consideration here is that the products are themselves required to be sterile before the filling operation, which implies preliminary processing under sterile conditions.

The filling of powders into vials involves considerations not customary for liquid filling, such as the mechanism used for dosage weight control. Similar techniques to those used for capsule filling have been tried, but most suffer from excessive particulate contamination generation. Modern high-speed sterile powder filling machines utilize a vacuum/pressure technique which forms a temporary solid compact from the product powder prior to its ejection into the vial.

Although some powder products can be sterilized by gamma irradiation or heat sterilization, most cannot be treated this way. Methods adopted to manufacture bulk sterile products include spray drying, bulk freeze drying, and crystallization under sterile conditions.

An alternative technique for the manufacture of products exhibiting instability in solution is to prepare such solutions using non-sterile product material and sterilize them by filtration, fill them within a controlled time-span into vials in small batches, and freeze dry. This method ensures that a solution can be produced, sterile filtered and filled under aseptic conditions, then re-crystallized by sublimation within the vial.

Equipment for this process relies on the use of special vial seals or plugs which, when partially inserted into the vials, allow evaporation of the solvent during the drying phase. The drying is followed by the automatic full insertion of the plugs within the dryer chamber, under aseptic conditions. In this way the finished filled vials can be demonstrated to be equivalent to vials filled with liquid under aseptic conditions.

6.1.16 Sterilization techniques

Products intended for parenteral administration must not contain viable microbial organisms and their manufacture will inevitably involve one or more sterilization stages. Such stages may be used for the drug substance, the filling container or the finished product itself.

Even where materials are processed under conditions of strict asepsis, it is now required that the finished product should be subjected to a terminal sterilization process wherever possible.

A number of possible methods exist for the sterilization of products and materials, and the most appropriate method will be selected after careful consideration of the effects that the various alternative systems might have on those materials. Each method has particular benefits when applied to specific requirements.

The commonly used systems for sterilization include moist heat (autoclaving), dry-heat, chemical treatment, irradiation, high-intensity light and solution filtration. With the exception of the last one, all the methods rely on a combination of intensity and time to achieve the required reduction in microbial content.

Another factor to be considered is the possibility for pyrogens to be present in the sterilized material or component. Pyrogens are substances that cause a rise in the patient's body temperature following administration of the injectable pharmaceutical. They are in fact complex polysaccharides arising from the breakdown of bacterial cells, and are most likely to be present following moist heat sterilization or other lower-temperature sterilization techniques (such as irradiation).

Autoclaving

The most useful and longest-standing batch sterilization technique is autoclaving, which exposes the subject materials to saturated steam at a temperature/time combination appropriate to the stability of those materials.

Established effective sterilization conditions range from 30 minutes at 115°C, to 3 minutes at 134°C. Commercially available autoclaves are supplied with standard cycles that provide time/temperature combinations falling within this range. These standard cycles include specific time/temperature combinations and also the facility for cooling large-volume product solutions in containers at the end of the sterilization phase, by means of deionized or purified water sprays. The latter process includes the simultaneous application of cooling water and sterile compressed air to the autoclave chamber, in order to prevent high-pressure drops across the container walls and consequent breakages.

Provided that the steam in the autoclave is saturated and free from air, the different cycle temperatures may be attained by developing various specified pressures in the autoclave. It is preferable however to control the process by the temperature attained rather than by the pressure, as the presence of air in the autoclave results in a lower temperature than that expected under the correct conditions from the indicated pressure. In the case of porous materials, the air must be abstracted or displaced from the interstices in order to achieve sterilizing conditions, as the presence of residual pockets of air within the material may prevent contact between the steam and parts of the load.

The period of heating must be sufficiently long to ensure that the whole of the material is maintained at the selected temperature for the appropriate recommended holding time. The time taken for the material to attain the sterilizing temperature or to cool at the end of the holding time can vary considerably and depends on a number of factors, including the size of the container or object and the thickness of its walls, and the design, loading, and operation of the autoclave. It is necessary, therefore, that adequate tests are conducted to ensure that the procedure adopted is capable of sterilizing the material and that the material can withstand the treatment. Chemical indicators can be included in the autoclave load, which change colour after the specified temperature has been maintained for a given time. Reliance should not be placed, however, on chemical indicators except when they suggest failure to attain sterilizing conditions.

The process can be monitored by temperature-sensitive elements (thermocouples) at different positions within the load. Some indication that the heat treatment has been adequate can be gained by placing indicators at positions within the load where the required conditions are least likely to be attained (such as the chamber drain).

For the purposes of validating the sterilization conditions, the bactericidal efficiency of the process may be assessed by enclosing in different parts of the load small packets of material containing suitable heat-resistant spores, such as those of a suitable strain of *Bacillus stearothermophilus*. These are checked subsequently for the absence of viable test organisms.

It is common practice for autoclaves to be double-ended with access doors opening into a clean preparation area on the infeed side and an aseptic filling area on the outfeed, although single-door autoclaves are used in some applications.

Dry heat

Dry heat sterilization, often referred to as depyrogenation, uses high temperature conditions in the absence of moisture to destroy contaminating organisms
and eliminate pyrogenic material. It is particularly useful for sterilizing glass
containers (such as vials) or any other product-contacting material that will
tolerate the required temperature. Typical conditions for this process are 200°C
or more with a residence time at that temperature of 15 minutes, although
sterilization alone is achievable at lower temperature/time combinations. The
process can be operated on a batch basis using double-door machines (built into
barrier walls in a similar manner to autoclaves), which accept clean containers
on the non-sterile side and deliver them sterilized on the aseptic side.

Modern high-output filling lines use continuous tunnel-type sterilizers,
which include complex air-handling systems and deliver the cooled, sterilized
containers into the aseptic filling machine located within the aseptic area. The
validation of high-temperature sterilization techniques requires similar considerations to those applicable to autoclaving.

Heating with a bactericide

This process can be used for sterilizing aqueous solutions and suspensions of
medicaments that are unstable at the higher temperatures attained in the
autoclaving process.

In this process, a bactericide is included in the preparation at the recommended concentration and the solution or suspension, in the final sealed
container, is maintained at 98° to 100°C for 30 minutes to sterilize the product.

The bactericide chosen must not interfere with the therapeutic efficacy of the
medicament nor be the cause of any physical or chemical incompatibility in the
preparation.

Ambient chemical methods

Formaldehyde was once used extensively as a means of sterilizing spaces such
as aseptic production rooms and surgical operating theatres, but is now rarely
used owing to its high toxicity and relative corrosiveness. It is only an effective
sterilant in the presence of moisture; the process involves raising the ambient
room humidity by water spraying, followed by the sublimation on an electric
hot plate of paraformaldehyde pellets.

Peracetic acid has been used as an alternative to formaldehyde for the
sterilization of small spaces, such as filling machine enclosures, isolators,
together with their contents. Like formaldehyde, it is corrosive and toxic and,
therefore, is of limited application. It has been used in admixture with hydrogen

peroxide for the sterilization of isolators. Peracetic acid has the advantage that the sterilizing effect is (as with all chemical sterilants) dependent on concentration, which can be easily measured with suitable detection equipment.

Hydrogen peroxide has now largely supplanted peracetic acid for small-space sterilization, as this agent is far less likely to cause corrosion of equipment items. It is also used for sterilizing syringes, ampoules and other packaging materials.

Hydrogen peroxide is used at concentrations of 1000 ppm in air and is regarded as product-safe due to its decomposition products being water and oxygen. It has a melting point of 0°C, and its commonly used 30% aqueous solution has a boiling point of 106°C.

It is, however, toxic, having a time-weighted exposure limit of 1 ppm and an acute toxicity limit of 75 ppm. Another disadvantage has been the difficulty in monitoring accurately the concentration of hydrogen peroxide vapour under sterilization conditions, although in recent times suitable sensors have been developed. These sensors have relatively slow response times, making real-time analysis of hydrogen peroxide difficult, but it is now possible to reliably validate the sterilization process.

Various **alcohols** (ethanol, iso-propanol) can be used to decontaminate the surfaces of containers or equipment items, usually by swabbing. However, this activity cannot be relied upon to provide sterility in its own right and must be preceded by a validated sterilization process.

Ethylene oxide sterilization

Certain materials cannot be sterilized by dry heat or autoclaving for reasons of instability, but they may be sterilized by exposure to gaseous ethylene oxide. This process can be carried out at ambient temperatures and is less likely to damage heat-sensitive materials. It does, however, present difficulties in control of the process and in safety, and is currently only considered where it offers the only solution to a problematic sterilization requirement. It must be performed under the supervision of experienced personnel and there must be adequate facilities for bacteriological testing available. The most frequent use of the technique in the pharmaceutical area is for the sterilization of medical devices (such as plastic syringes).

Compared to other methods of sterilization, the bactericidal efficiency of ethylene oxide is low and consequently particular attention should be paid to keeping microbial contamination of subject materials to a minimum.

Ethylene oxide is a gas at room temperature and pressure. It is highly flammable (at levels as low as 3% in air) and can polymerize, under which conditions it forms explosive mixtures with air. This disadvantage can be overcome by using mixtures containing 10% of ethylene oxide in carbon dioxide or halogenated hydrocarbons, removing at least 95% of the air from the apparatus before admitting either ethylene oxide or a mixture of 90% ethylene oxide in carbon dioxide. It is also very toxic to humans (time-weighted average exposure limit 1 ppm) and has been demonstrated to be carcinogenic. For these reasons ethylene oxide sterilization is no longer frequently used as an industrial process.

There are two processes used for ethylene oxide sterilization, one at normal and the other at high pressure. The low-pressure process uses a 10% v/v concentration, a temperature of 20°C and a cycle time of around 16 hours. A suitable apparatus consists of a sterilizing chamber capable of withstanding the necessary changes of pressure, fitted with an efficient vacuum pump and with a control system to regulate the introduction of the gas mixture, maintain the desired gas pressure, adjust the humidity within the chamber to the desired level and, if required, a heating element with temperature controls.

The high-pressure process was developed to enhance output by reducing cycle times. It uses a more-substantial chamber design, suitable for the 10 barg operating pressure. The temperature is typically >50°C and the cycle time 3 hours.

As with any chemical sterilization process, the combination of time and sterilant concentration is the key factor. The sterilizing efficiency of the process depends upon:

- the partial pressure of ethylene oxide within the load;
- the temperature of the load;
- the state of hydration of the microorganisms on the surfaces to be sterilized;
- the time of exposure to the gas.

All these factors must be closely controlled for successful sterilization. The sensitivity of microorganisms to ethylene oxide is dependent on their state of hydration. Organisms that have been dried are not only resistant to the process but are also slow to rehydrate. Due to this, it is not sufficient to rely solely on humidification of the atmosphere within the chamber during the sterilizing cycle.

It has been found in practice that hydration and heating of the load can be more reliably achieved by conditioning it in a suitable atmosphere prior to commencing the sterilization.

Some materials absorb ethylene oxide and, because of its toxic nature, great care must be taken to remove all traces of it after the sterilization is finished; this is achieved by flushing the load with sterile air.

Irradiation

Sterilization may be effected by exposure to high-energy electrons from a particle accelerator or to gamma radiation from a source such as cobalt-60. These types of radiation in a dosage of 2.5 mega-rads have been shown to be satisfactory for sterilizing certain surgical materials and equipment, provided that precautions are taken to keep microbial contamination of the articles to a minimum. This method is not, however, widely regarded as a safe means of product sterilization, due to the possibility of chemical decomposition of many pharmacologically active substances.

This method can also be used for some materials that will not withstand the other sterilization methods. It has the advantage over other 'cold' methods of sterilization in that bacteriological testing is not an essential part of the routine control procedure, as the process may be accurately monitored by physical and chemical methods. It also allows the use of a wider range of packaging materials.

Control of the process depends upon exposure time and radiation level. It is important to ensure that all faces of the load are exposed to the required radiation dose.

Ultraviolet light

Ultraviolet light has long been known as a form of energy with bactericidal properties. It has particular uses in the maintenance of sterility in operating theatres and animal houses, and for the attenuation of microbial growth in water systems. Ultraviolet light exists over a broad wavelength spectrum (0.1 to 400 nm) with the bactericidal (UVC) component falling in the range 200 to 300 nm with a peak at 253.7 nm.

It is particularly useful for maintaining sterility in pre-sterilized materials and is used widely in isolator pass-through chambers to protect the internal environment of the isolator. It can also be used for continuous production sterilization of pre-sterilized components feeding into such isolators.

It can be used to sterilize clean materials in a continuous cycle provided that they are fully exposed to the radiation, but this is a relatively slow process requiring an exposure time of up to 60 seconds to achieve a 5-log reduction in viable organisms.

High-intensity pulsed light

A recently developed method of sterilization uses very short pulses of broad-spectrum white light to sterilize packaging, medical devices, pharmaceuticals, parenterals, water and air. It has been demonstrated that this process kills high levels of all micro-organisms. Each light flash lasts for a few hundred millionths of a second but is very intense, being around 20,000 times brighter than sunlight. The light is broad-spectrum, covering wavelengths from 200 to 1000 nm, with approximately 25% in the UV band. The latter component provides the sterilizing effect in short-duration high-power pulses, although the total energy required is quite low — an economic advantage.

High kill rates equivalent to 7–9 log reductions in spore counts have been demonstrated using a few pulses of light at an intensity of 4–6 joules cm^{-2}. Although the UV component provides the effectiveness of this method, it is considerably more rapid than conventional UV systems. Continuous in-line sterilization is, therefore, practical with this technology.

Pulsed light sterilization is applicable to situations and products where light can access all the important surfaces and also penetrate the volume. It will not penetrate opaque materials, but is efficiently transmitted through most plastics and may be used to sterilize many liquid products.

Filtration (liquids)

Liquids may be sterilized by passage through a bacteria-proof filter. This process has the advantage that the use of heat is avoided, but there is always a risk that there may be an undetected fault in the apparatus or technique used, and because of this each batch of liquid sterilized by filtration must be tested for sterility compliance.

Sterilizing filters can be made of cellulose derivatives or other suitable plastics, porous ceramics, or sintered glass. The maximum pore size consistent with effective filtration varies with the material of which the filter is made and ranges from about 2 μm for ceramic filters to about 0.2 μm for plastic membrane filters.

Particles to be removed in the sterilizing process range in size from 1 to 5 μm diameter, down to viruses of 0.01 μm. It appears at first sight that filters cannot remove particles smaller than the largest pore size of the filter. However, filtration occurs in a wide variety of mechanisms, including impaction, adsorption, adhesion and electrostatic effects, so that in practice particles much smaller than the interstitial channels may be effectively filtered out.

Filters for liquid sterilization have pore sizes of 0.2 μm, usually preceded by coarser pre-filters to remove larger particles. These filters are all fabricated as cartridges that are installed in leak-tight housings. For the filtration of liquids, hydrophilic forms of the filter material are used.

All standard filter types must comply with bacterial challenge tests performed by the manufacturer, which can be correlated with other integrity tests carried out routinely by the end-user.

144

Non-disposable filters must be tested periodically before use to ensure that their efficiency has not become impaired, using one or more of the following integrity test methods. Filters should be integrity tested after each sterilization and after each filtration. All integrity testing is performed on wetted filters. The tests depend on the principle that airflow through the wetted porous membrane is diffusive up to a certain pressure (the bubble point) and is a function of pore size and pressure. Above the bubble point, liquid is displaced from the membrane and bulk flow of gas occurs.

Bubble point test: In this test, air pressure upstream of a wetted filter is slowly increased. The pressure at which a stream of bubbles occurs downstream of the filter is the bubble point pressure. If a filter has a damaged membrane or an insecure housing seal, the test pressure will be below that specified by the manufacturer.

Forward flow test: A test pressure below the bubble point pressure is applied to a wetted filter. The diffusive airflow rate through the filter is measured. If it exceeds a specified value the filter is judged to be insecure.

Pressure hold test: A section of pipework upstream of the wetted filter is pressurized (below the bubble point). The rate of pressure decrease is measured. For a filter to be judged intact, this must occur below a specified rate.

Filtration is best carried out with the aid of positive pressure, as this reduces the possibility of airborne contamination of the sterile filtered solution through leaks in the system. If the filtration is likely to take a long time and the preparation is susceptible to oxidation, nitrogen or other inert gas under pressure should be used rather than compressed air.

Filtration (gases)

The uses of sterile air or inert gas in pharmaceutical sterile processing include the aseptic transfer of liquids using pressure, and blowing equipment dry after sterilization. In addition to these positive applications, air or gas also enters aseptic equipment during fluid transfers or cooling operations, and in all cases the air and gas must be completely free of micro-organisms. Air sterilization can be achieved by filtration with the required filter porosity being 0.2 μm as for liquids. Integrity testing also needs to be carried out in the same manner as with liquid filters.

6.1.17 Aerosols

The use of pressurized systems for the application of pharmaceuticals became common after World War II, when such methods were used for the topical administration of anti-infective agents, dermatological preparations and materials used for the treatment of burns. A logical development of spray technology, the aerosol relies on the propulsive power of a compressed or liquefied gas. The latter type have been of greater benefit, based on gases boiling at below room temperature (20°C) and at pressures ranging from zero to 120 psi above ambient.

Initial applications utilized flammable hydrocarbon gases, which were then largely replaced for pharmaceutical use by chlorofluorocarbons, notably for use in inhalation products. Recent developments have worked towards the replacement of the suspected ozone-depleting chlorofluorocarbons with hydrofluoroalkanes for environmental reasons.

A further method of avoiding the oral route for internal administration is to introduce the drug substance by inhalation.

Aerosol products for inhalation use first appeared in the mid-1950s and were used for treatment of respiratory tract disorders, based on the establishment of several key benefits:

- rapid delivery to the affected region;
- avoidance of degradation due to oral or injectable administration;
- reduced dosage levels;
- ease of adjustment to patient-specific dosage levels;
- avoidance of possible interactions with concurrently-administered oral or parenteral drugs;
- ease of patient self-administration.

The typical modern pharmaceutical aerosol consists of an aluminium container, a product (in powder, solution or suspension form), a propellant and a cap/seal incorporating a metering valve. The propellant provides pressurization of the container at normal temperatures, and expels the product when the valve is opened. The dose is controlled by the valve orifice configuration, which allows the release of a single shot of product liquid together with sufficient propellant gas to ensure production of an aerosol.

Continuous aerosol sprays for topical application use slightly different valve types that do not limit the dose size. Such products also sometimes utilize compressed gases to provide propulsion, including carbon dioxide, nitrogen and nitrous oxide.

The manufacture of pharmaceutical aerosols is complicated by the need to maintain a pressurized environment for the propellants during storage, mixing and filling. This includes the systems used for transporting the propellants from the storage location to the point of use, and is made more complex where flammable materials are involved.

The relatively complex nature of gaseous aerosol manufacture has led to the consideration of other methods for the delivery of drug substances by inhalation, including the creation of fine particles suspended in an air stream generated by the patient himself. Such powder inhalations utilize micronized powders delivered in unit-dose quantities, held in a device that simultaneously releases the fine material into air flowing through the device at the same time as that airflow is initialized by the user. By careful design using a multi-dose approach, a metered dose system providing relief of patient symptoms over a convenient time period is possible. Several such systems are currently available or under development.

6.1.18 Delayed and sustained release systems

The objective of any drug delivery system is to provide a specified quantity of the therapeutic agent to the appropriate location within the body, and to sustain the level of that agent so that a cure or symptom relief is achieved. In practice drugs are delivered in a broad-brush manner, which ensures arrival of sufficient drug to the body location needing it, but simultaneously provides the drug to parts not requiring treatment. This approach may ensure coverage but is somewhat wasteful and may engender unwanted reactions.

A targeted approach is therefore potentially valuable and there are a number of ways in which this can be achieved. The possible advantages of this approach are:

- improved patient compliance;
- reduced drug substance usage;
- reduced side effects;
- reduced drug accumulation;
- improved speed of treatment;
- improved bioavailability;
- specific delay effects possible;
- cost saving.

The objective stated above has two parts, namely the creation of a suitable drug level at the required site, and the maintenance of that ideal level for a period suited to the completion of treatment.

The first objective can be achieved by delayed release of the drug when taken orally, by localized application by injection, or by topical application local to the required site in the case of shallow-tissue disorders. Methods used for ensuring adequate levels of the therapeutic agent include sustained-release coatings for tablets and capsules, and formulations of injectable or topical drugs that allow controlled release of the active principal.

The combination of delayed and sustained release properties for orally dosed material can ensure, for example, that the drug is released, at a controlled rate, in the duodenum rather than the stomach. Such controlled-release is achieved with oral dosage by the formulation or coating of tablets and capsules so that the excipients (either internally or as part of the coating material) have a physical action on the drug dispersion or dissolution rate.

Injectable drugs in a suitable formulation can offer delayed or sustained release when delivered intramuscularly, as a 'depot'. Dissolving or dispersing the drug in a liquid medium that is not readily miscible with body fluids can reduce the rate of absorption. Oil solutions or suspensions are often employed for this effect, while aqueous suspensions can be used with insoluble drugs.

An alternative injectable route is the use of solid material injected subcutaneously, the 'depot' thus being formulated to ensure suitable release rates. The surgical implantation of drugs can be even more targeted, albeit at increased patient risk.

Topical drug application has a number of benefits, especially the opportunity to remove the material from the skin by washing, so reducing and ultimately stopping the rate of application. The absorption of drugs via the skin e.g. transdermal products, including intra-ocular routes involves the formulation of the actives in such a way that they can be released from the carrier material at the rates required. Such formulation can involve the use of microporous materials to which or within which the drug is applied or mixed, applied directly or attached to a substrate (such as adhesive plasters).

6.1.19 Microencapsulation

The process of microencapsulation involves the deposition of very thin coatings onto small solid particles or liquid droplets and differs from the technique of, for example, tablet coating in that the particles involved are much smaller — typically 1 to 2000 μm in diameter.

The benefits to pharmaceutical product development relate to the very small and controlled size of the particles involved. The technique alters the physical characteristics of the materials concerned to the extent that:

- liquid droplets can exhibit solid particle characteristics;
- surface properties are changed;
- colloidal properties are changed;
- pharmacological effects are enhanced or reduced by changing release patterns;
- the surrounding environment is separated from the active drug substance.

Although some similar effects can be achieved by alternative methods, the microcapsule can, due to its small size, be used in many product applications which would not otherwise be technically practical.

148

Methods available for manufacturing microcapsules include spray drying, pan coating and air suspension coating. The former is of particular value in the production of very small microcapsules (typically 1 to 100 μm in diameter), and has been used in protein-based product manufacture in which a protein solution is sprayed into a co-current air stream to form microcapsules. The co-drying of such materials with pharmaceutically-active substances is capable of producing particles of such substances coated with a protective or carrier layer.

6.1.20 Ingredient dispensing

All pharmaceutical manufacturing operations involve the use of one or more chemical materials in pre-defined quantities on a batch or campaign basis. Such materials are most often held in a storage location, in containers providing sufficient quantities of the material to enable the manufacture of more than one batch. These containers will be of such design as to afford the required level of protection of the material during the storage period and facilitate allocation to the dispensary.

The activity involved in the weighing of materials on a batch-by-batch basis is known as dispensing, and may be considered as the first step in the manufacturing process.

The sub-division of a bulk material into smaller batch lots inevitably involves the removal of that material from its original container. The environment in which this process is conducted must, therefore, be of a quality suitable for the intended use of the manufactured pharmaceutical product. For example, the dispensing of ingredients for the manufacture of oral-dose products will usually be conducted under class 100,000 conditions (to US Federal Standard 209e). The same operation for handling sterile ingredients for injectable products will usually be conducted under class 10 or 100 conditions, possibly using a glove-box.

Another key feature of dispensing is the need for assurance that the operation has been carried out correctly. This need will often be met by the checking of each weight by a second operator. With modern computer-controlled dispensing systems, the latter situation is most common, as the reliability of the dispensing process itself is such that only the potential for errors in transit to the production area need to be checked.

Containers

As indicated above, the 'input' container will be of such design as to protect the integrity of the material, and so too must the container used for transferring the dispensed ingredient to the manufacturing location. Where high-potency ingredients are involved, the latter must also ensure that subsequent handling can be performed without risk to operating personnel. Thus, a contained transfer system might be employed for this purpose (see Section 6.4 on page 176).

Incoming materials are likely to be contained in polyethylene-lined kegs (solids) or steel drums (liquids). These containers may hold as much as 200 kg of material and be transported on clean pallets. Space for the staging of such pallets adjacent to the dispensing zone is therefore required, together with handling devices suitable for positioning them conveniently for the removal of the required weights or volumes of ingredients.

Dispensed materials may be placed in similar containers to those used for incoming items. However, it is more usual for these aliquots to be transferred to manufacturing using dedicated sealable dispensed-material containers, often reserved for particular substances, and carrying provision for secure identification of the contents.

Weighing systems

As pharmaceutical ingredients are usually dispensed by weight (rather than volume), a suitable set of weighing scales is required. Scale sensitivity and accuracy usually diminish as capacity increases, so a two or three-scale arrangement is not uncommon. Thus, the active ingredients, which are likely to be of lower batch weights than the non-active or excipient materials, will usually be weighed-out on scales of higher accuracy. The three scales might, typically, have capacities of 1 kg, 10 kg and 100 kg respectively. The chosen scale capacities will depend on overall batch weights and on the weight of the active, or smallest, ingredient.

Electronic weighing scales are common in modern dispensaries, and these can be linked to computer-controlled dispensary management systems and to automatic identification and weight-label printers.

Operator protection and airflows

The protection of operating personnel from exposure to high-potency drug substances is as important during dispensing operations as it is in the subsequent processing. Hence, the arrangement of modern dispensing areas utilizes individual booths in which the ingredients for one product batch at a time are weighed and packaged. The operator must wear suitable protective clothing, which should include hair covering, long-sleeved gloves, dust mask, footwear and close-woven fabric overalls.

Modern pharmaceutical dispensing booths employ a ventilation scheme that seeks to separate the operator's breathing zone from the area in which product or excipient powders or liquids are exposed during dispensing. The basic principle relies on a downward sweeping of the ventilation air, from the ceiling above and behind the operator, to the lower edge of the booth wall facing them. Thus, any dust generated during scooping of materials into receiving containers is entrained in the air stream and kept away from the operator's head. A typical dust entrainment velocity is $0.45 \, \mathrm{m \, s^{-1}}$, and proprietary dispensing booths are designed to provide an operating zone in which the air stream moves at or above this velocity.

The air leaving the lower back wall of the booth may be filtered to remove entrained ingredient dust and recirculated, while supply air make-up and recirculated air will generally be filtered and conditioned to the environmental quality standard required by the product being dispensed, typically class 100,000 for oral-dose products. Figure 6.12 illustrates a typical airflow arrangement in a downflow dispensing booth.

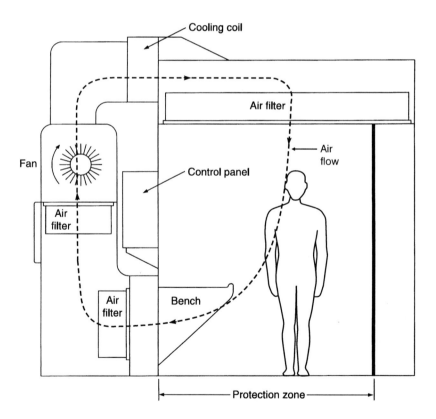

Figure 6.12 Sectional diagram of dispensing booth

151

Surplus materials

The disposal of surplus material remaining at the end of a product campaign in ingredient input containers generally poses a dilemma for dispensary managers. The options are to return the part-used container to the main raw material warehouse or to retain it as part of dispensing stock for later use. There is no universal 'best alternative', the decision being affected by such factors as the availability of space for storage within the dispensary area, the proximity of the main warehouse, the ownership of material stocks within the dispensary and warehouse, the sophistication of the materials management system, the level of security of the storage location etc. These all need to be considered when this issue is decided, but the overriding factor must be the security and integrity of the material itself.

Cross-contamination risks

In multi-product pharmaceutical manufacturing plants it is inevitable that the dispensary will be required to handle two or more products, probably at the same time. Thus, individual dispensing booths must operate in such a way as to ensure that there is no risk of materials from one product contaminating another. This is achieved by ventilation air pressure regimes that combine recirculatory air flow with slight positive pressure relative to adjacent access corridors and storage areas. By this means, dusts generated during the dispensing activity will be entrained and intercepted by the booth's extract filters, thus avoiding dispersion to the external environment. Meanwhile, any contaminant present in the adjacent spaces will be prevented from entering the booth by the positive pressurization.

Cleaning arrangements

One potential source of cross-contamination is the equipment and surfaces used during ingredient material handling. It is, therefore, important that all contaminated containers and utensils are removed from the dispensing booth for disposal or cleaning at the end of the operation, and that all working surfaces, including the fabric of the booth itself, are subjected to a validated cleaning procedure. Utensil and container washing is most effectively carried out in automatic washing machines, which should also incorporate a drying cycle. Open-sink washing of such items is unlikely to provide a validatable process, and should generally be avoided.

Operator clothing is a further source of contamination, and operators must change their outer garments when product changes are made, and in all cases should change their gloves between sequential batches.

Labelling

It is essential that all dispensed ingredients are reliably identified — including the batch number and name of the product batch that is to contain the ingredient, the item weight and material name. It may also include the identity of the dispensing operator and the time and date of dispensing. Although the manual generation of labels can be acceptable (assuming suitable checking systems exist and are in use), it is now considered worthwhile to arrange for these to be produced automatically by the dispensary management system. Thus, at the end of each weighing operation, the acceptance by the operator of the correctness of the weight and identity will initiate a bar-coded or alpha-numeric label being printed by a printer located adjacent to his workstation. Such labels, usually of the self-adhesive variety, will then be applied to the dispensed material's container.

Materials management systems

Modern dispensary management systems are computer-driven, with fully-validated batch recipe information held electronically. They are most often linked to business management systems such as MRP2, warehouse management systems, and intermediate specialist control suites which organize the flow of material throughout the production process and seek to prevent errors in material usage. The latter, which usually incorporates the dispensary management element, must comply with the principles of cGMP and must, therefore, be driven by fully-validated software — this makes such systems very specialized and potentially costly.

Materials management systems automatically update stock levels at each stage in the material pathway, including transfers of ownership between different departments (between warehousing and production, for example). They ensure that only approved material can be allocated for use, or indeed used, and that materials are consumed in accordance with normal stock rotation principles (such as first in, first out).

The specific role of the dispensary management system is to ensure that ingredients are weighed out in accordance with pre-programmed recipe information and in the correct sequence. Instructions to dispensing operators may be provided via a printed batch sheet or visually by VDU screen.

The systems often also include provision for printing of ingredient labels that provide identity, weight and batch code information, in either bar-code or alphanumeric form. Various add-on facilities may also be incorporated, such as programmed weigh-scale calibration routines, and authorized-operator identification.

153

6.2 Principles of layout and building design

6.2.1 Introduction

It has been said that the layout of a building can be designed in at least six different but equally acceptable ways. This may well be the case, although the degree of acceptability will vary depending upon the criteria applied by the accepting authority.

The criteria that give rise to the differences in pharmaceutical secondary production building layouts include, but are not limited to:

- safety/means of escape;
- complexity of the enclosed processes or activities;
- personnel level, type of occupancy and movement;
- ease of materials movement;
- specialized environmental classifications;
- type of partition construction;
- the structural design of the building.

6.2.2 Personnel safety

The primary safety consideration for all buildings is means of escape in the event of fire or other emergency. The issues are complex and covered by legislation and fire engineering principles, and will not be discussed here. However, the pharmaceutical engineer is well advised to take account of the basic considerations when planning the process-led layout of a building, and in doing so should seek the advice of a qualified architect at the earliest practicable point. Although failure to do so may not result in a potentially dangerous building, it will almost certainly involve time-consuming and costly reconsideration of the building layout during its architectural design phase.

Another important safety consideration relates to the product itself. In some cases the active materials involved in pharmaceutical manufacture are toxic in a high-exposure situation, and special precautions will then become necessary. These may involve modifications to the layout to accommodate specialist machines or environmental control equipment. The need for the use of flammable materials, although less common nowadays, may also arise and in such circumstances the design of the building may have to include the results of area zoning. This can be onerous, as construction materials may need different selections from those made elsewhere in the building, while the need for separate ventilation systems is also possible.

6.2.3 Process and activity complexity

Even simple pharmaceutical manufacturing and packaging processes must be carried out in areas with controlled environments. It is common practice to group final packaging operations, which usually involve the handling of products in a partially enclosed condition (such as filled and capped bottles, tablet blister packs) in a single room with limited spatial separation between linked groups of machinery, but with a common ventilation system. This is possible if the environment provides the required temperature (and sometimes humidity) to ensure product stability and that product cross-contamination is negated by the primary enclosure.

In the case of manufacturing operations, even where products require similar levels of product protection, separate environmental and spatial arrangements are usually necessary to prevent cross-contamination. It is, therefore, usual for manufacturing and primary packaging processes to be conducted in product-specific common environments and is essential for such processes where any degree of cross-contamination is hazardous to the product or patient to be separated physically, as a minimum.

It is possible for different products to share a common ventilation system, but only if that system allows for a single pass of the air supply, or if the recirculated air is passed through suitable filters. These filters must be of such porosity that it is possible to provide demonstrable evidence that any product dust passing through is of such low level as to ensure that products cannot become contaminated with one another at levels which pose measurable risk to patients.

Production processes involving specially clean conditions for product exposure (such as for parenteral, ophthalmic or inhalation products) add further complexity to the environmental and space planning activity. The transfer of materials between areas of differing cleanliness classification often involves a process such as sterilization by autoclaving or other means. Hence, the relative size, shape and position of the rooms on either side of the transfer process are important. In any event it is often considered necessary to separate such different areas by the insertion of air locks, in which decontamination of materials and equipment can be performed prior to transfer. This procedure will also be required during active product manufacture to prevent the possible spread of such material to adjacent areas.

In reality, the product mix in any production facility may be such that the above considerations demand dedicated spaces for different products. This demand inevitably impacts on the layout of the building, and it is for this reason that those responsible for facility space planning should understand the many and varying considerations.

6.2.4 Personnel occupancy level, type and movement

Although the use of machinery for manufacturing and packaging operations is widespread and increasing, pharmaceutical production requires the employment of people for the control of material selection, movement, processing and inspection, and it is unlikely that such employment will be eliminated in future.

A further consideration is that growing sexual equality coupled with a decreasing incidence of heavy lifting and movement means that both male and female production operators are equally likely to be employed on a process.

However, the numbers of personnel likely to be engaged on a given operation is relatively low by general industrial standards, so that this feature does not generally pose great difficulties in facility planning.

There is, nevertheless, a feature of pharmaceutical (and especially clean-room) operations that needs careful consideration. Just as air locks are a characteristic of material transfers, operator clothes-changing rooms are a common necessity in the protection of products from people. Clean areas generally need to avoid people-generated particles, while aseptic areas additionally need to be protected where possible from microbial contamination.

A careful selection of clean-room clothing, in terms of body coverage and particle-shedding ability, will significantly reduce both particle and microbial levels within the occupied clean room. Synthetic fibres generally shed lower particulate levels than cotton, and ceramic-coated synthetic materials are extensively used for clean-room clothing manufacture. However, fully covering synthetic-fibre clothing may enhance perspiration and thus microbial release, so high-specification clean areas should be operated at slightly lower temperatures to compensate.

The frequency of personnel movement within secondary production areas is not generally problematic, bearing in mind the relatively small numbers and the confined nature of the operations. However, the increased load on changing facilities at break times should not be overlooked, and neither should the ease of movement during any emergency condition. This is particularly important with clean areas, where many restrictions on movement occur such as the use of multiple doors, changing room step-over barriers etc., and where over-ambitious attempts to seal emergency doors to prevent dirt ingress have been known to result in blocked escape routes.

6.2.5 Materials movement

It is often the case that, along with personnel movement, material movement within pharmaceutical manufacturing facilities dominates the consideration of building layout. The separation of material and personnel pathways and the avoidance of cross-overs can consume a large amount of time during initial planning. While such considerations are important, the extent of such importance must be first established by the performance of a movement review, which in turn requires a full understanding of the operation of the facility and the type, size and frequency of movements. It will often be found that the problems envisaged are imaginary, and bearing in mind the low-volume nature of most pharmaceutical products this should not be surprising.

The pharmaceutical engineer must, therefore, play a key role in establishing the realities of building layout design and ensure that he/she provides advice to those with whom he/she is working.

Once the understanding of material movement is established, consideration can be given to key factors such as corridor width, door width and type, and the adjacency of related operational areas.

One key item sometimes overlooked in preliminary planning is equipment, both fixed and mobile. Care must be taken in layout design and equipment selection to ensure that larger equipment can be moved through the facility to its final operational position, and that routinely-mobile items have transport routes which have been planned with their movement in mind.

6.2.6 Specialized environments

Where products demand special environmental conditions, the building structure and layout should include separate spaces for their manufacture and/or storage.

In general, these special environments either have increased cleanliness, unusual temperature or humidity, or provide extra levels of separation from surrounding areas by virtue of high potency or other risks. For example, aseptic conditions are required for the manufacture of injectable forms, demanding higher standards of surface cleanability and ventilation air filtration efficiency.

These features must be used in conjunction with stricter operator clothing regimes and closely defined operating/handling procedures. Layout considerations must include provision for separation from lower-grade areas by means of air locks. Positive pressurization of the processing areas is of course necessary to prevent ingress of dirt and microorganisms.

An important feature of aseptic processing areas is the selection of structure and finishes. It is not uncommon in modern facilities to employ modular partitioning systems with close-tolerance self-finished panels. These have the advantage of providing crevice-free stable walls and ceilings which do not move or crack, even when the main building structure surrounding the area is liable to move due to thermal expansion/contraction. In conjunction with heavy-duty clean-area grade welded vinyl flooring systems, these modular clean rooms provide reliable and easily maintained surfaces ideal for aseptic operations.

They are, however, relatively expensive, and a lower-cost alternative is the use of steel-frame and plasterboard systems for walls and ceilings, coupled with vinyl flooring. This approach also provides a good-quality environment, but requires higher levels of maintenance attention due to joint-cracking potential and less-durable surface finishes.

Where products are especially temperature or moisture-sensitive, the rooms in which they are exposed to the operating environment need to be supplied with ventilation air which has been conditioned to the required levels. This requirement may not affect the layout of the area concerned, although air locks coupled with positive room pressurization may be included to ensure greater control of the special environment. However, it will demand changes to the air handling system, and this is typically achieved by localized heating, cooling or dehumidification of the supply air.

Care must be taken when humidity levels are unusually low (below 20% RH), as operating staff may suffer dehydrating effects such as sore throats and cracked lips, which may be avoided by reduced individual working periods in the areas concerned.

Where product materials of an active nature are exposed in-process, operating personnel may be protected by personal protective equipment, provided the exposure is of short duration (for example, during maintenance or product transfers). Alternatively, isolation/barrier methods should be employed to prevent such exposure. However, under either scenario it is possible that product dust may be emitted, and the rooms involved should be designed to take account of this possibility by the use of negative pressurization and the inclusion of air locks. It also requires consideration of room exhaust air filtration to protect the external environment, preferably sited at the room wall or ceiling interface. Such filtration systems should include a method by which the exhaust filters can be changed from within the room in a safe manner, personnel involved being protected by temporary personal protective equipment.

An additional desirable feature of active product processing areas is easy-clean surfaces for walls, floors and ceilings. This is essential to ensure containment.

6.2.7 Internal structure

Certain products and processes demand special consideration of construction and finishes. However, it is a general requirement for pharmaceutical production and storage areas that they should be easily maintainable in a clean condition, and walls, floors and ceilings, together with pipework, ductwork and electrical features should be designed with this in mind.

It is first necessary to consider the degree of product exposure at each stage of the storage, dispensing or production process, and to consider the risks to the product from such exposure. This analysis will provide a framework for the selection of surface finishes in each area. Thus, the movement and storage of materials that are always enclosed in sealed containers requires a very different selection of surfaces from that needed where sterile materials are filled into ampoules under aseptic conditions.

Provided that the need for cleanability in all areas of pharmaceutical manufacturing plants is ensured, a variety of surface finishes are available for selection. These range from painted blockwork walls, sealed concrete floors and insulated and plastic-faced liners to ceilings in warehouses, to fully sealed and crevice-free clean-room systems with coved interface joints in sterile areas.

In production areas it is generally wise to avoid the use of painted blockwork, in favour of a plastered and painted finish. It is also best to avoid suspended ceilings with lay-in tiles, as these do not provide effective barriers between processing areas and the technical/services areas above and may allow dirt ingress. In such areas it is also preferable to provide access to services distribution and plantroom areas, which does not involve direct penetration of the walls or ceilings of the operating spaces themselves.

6.2.8 Building structure

The choice of structural materials can affect the internal environmental conditions. An example of this is the effect of external environmental conditions on natural expansion/contraction of the building's structural material. Steel framed buildings will naturally provide greater potential for such movement than those fabricated from concrete or similar materials. In any case, these natural movements must be taken into account in the structural design of the building, and the presence of expansion joints in walls, floors and ceilings may be the consequence. Wherever possible, such joints should be avoided in manufacturing areas, except in the case of aseptic processing rooms where they *must* be avoided.

6.3 The operating environment

6.3.1 Introduction

As a consequence of the increasing regulatory pressures being exerted on the industry, the environment in which secondary production is undertaken has become progressively subject to greater inspection by the authorities. The 'environment' covers a number of issues, each of which is covered in the following sections:

- the avoidance of cross-contamination;
- product segregation;
- cleaning;
- environmental classification;
- ventilation systems;
- surface finishes;
- lighting selection.

The art of providing the correct operating environment lies in the selection of the systems that provide, *as a minimum*, no greater risk of contamination of the end product than has been accepted by the authorities during the drug approval process. This requires the engineer to select systems that meet this standard and are:

- economically justified;
- operable;
- maintainable (to the 'as-new' conditions);
- to cGMP standards.

6.3.2 Avoiding cross-contamination

All customers wish to receive exactly what they have ordered (or been prescribed, in the case of patients). Failure to do so can have unacceptable, even fatal, results. However, even if the product is correctly delivered, without proper controls it can be contaminated with another material. This potentially can have severe side effects, particularly if the patient suffers from a reaction to the contaminating material. Clearly, if the potential contaminant is another pharmaceutically active material or a viable organism, cross-contamination must be rigorously controlled.

The most likely sources of cross-contamination are:

- the operator;
- the previous batch of material;
- other materials in the working environment (such as paint, dust, micro-organisms, implements).

Sources of contamination can be identified and the level of risk determined for each product. However the industry has established a number of standard practices to reduce the contamination risks at all times. These 'standards' are commonly described as part of 'current' Good Manufacturing

Practice (cGMP) and are not always available in written form, although many guides have been published. The following paragraphs identify the main sources of cross-contamination in secondary pharmaceutical manufacture.

The operator brings to work several sources of contamination. External contaminants, such as soil, clothing fibres, etc. can be removed by the use of personal hygiene techniques on arrival at work and the wearing of non particle shedding clothing for production duties. Personal contaminants, such as dead skin scales and living organisms on the skin surface or in exhaled air, cannot be eliminated but risks from them should be reduced when these are known to be hazardous to products.

It is normal practice for all operators to change into clothing suitable for their duties on arrival at the manufacturing plant. Except in the lowest classes of operating environment, operators will change all their external clothing for 'coverall' man-made fibre working clothes, wear dedicated shoes and cover their hair and ears with a fine mesh hair cover. Those with beards/moustaches may be required to use a beard 'snood'.

As the quality of the environment increases, the standard of clothing and other protective coverings will increase. It is becoming standard practice, therefore, for manufacturing plants to have a series of change requirements to match the operating conditions. In the extreme — aseptic production — areas, operators have only their eyes exposed to the environment. Operators with infections are not permitted to work in these aseptic conditions, as the risk to the product is too high, even when protected by further containment methods.

Training is the principle method by which operators can learn to avoid the risk of creating cross-contamination. It is essential that they fully understand the need for absolute adherence to the Standard Operating Procedures (SOP), which have been developed to reduce the risk to the product in manufacture. Strict compliance with the clothing disciplines is required to avoid bringing contamination from one product to another on their clothes/skin. Learning to work at a pace that does not create excessive particulate disturbance requires skill and practice, particularly over exposed product.

It should never be forgotten that the human body loses particles of skin throughout the working day (see Table 6.1). These particles can become the chief source of contamination in a clean working environment.

The previous batch of material will always be a source of cross-contamination. Only when the previous batch is made from exactly the same components does this create no risk.

Segregation of products and the cleaning procedures required to avoid contamination are discussed in Section 6.3.3 (see page 163).

Table 6.1 Release of human skin flakes to the environment

Activity	Flakes released per minute
Sitting still	100,000
Moving limbs gently	500,000
Moving limbs actively	1,000,000
Standing up/sitting down	2,500,000
Walking/climbing stairs	10,000,000

Unless a process is undertaken in a totally contained manner, it can be assumed that the materials utilized in the manufacture of a product will be in the manufacturing environment. This is caused by many sources, but normally from particulate escapes, aerosols of liquids and from operators' clothing.

Methods of handling these materials can significantly reduce their discharge to the environment and the training of the operator is essential in the reduction of contamination risk from these sources.

Cleaning of the equipment and the surrounding areas can clearly reduce the level of contamination risk, but the need for excessive cleaning regimes should be avoided. Careful planning of production batches can reduce, or even eliminate, the need for cleaning between batches. Excipients (non-active ingredients) may be used in many formulations and, therefore, cleaning between batches of different products using the same active ingredient may be reduced in scope.

In summary, the risk of cross-contamination from a previous batch must be understood and reduced to an acceptable level.

Other materials can be present in the environment and not be caused directly by the operator or the previous batch of materials. A main source of such contaminants is the poor design of the premises in which the operations are undertaken. Information is given on surface finishes later in this chapter, but the particle shedding properties of all surfaces can be a source of contamination when the process materials are exposed to the environment.

Of more importance is the elimination of any surface on which contaminants can collect and later fall into the process. Flat surfaces should be replaced by sloping faces of easily cleaned materials; fixed equipment should be enclosed and ideally sealed to the ceiling; doors and windows do not require architraves; controls should be built into the walls or equipment; lights should be sealed to their surrounds; service outlets should be designed with minimum exposed surfaces.

Two further important sources of contamination should be considered in the design of all facilities with risks reduced to a minimum:

- the movement of air;
- the movement of process materials.

Most modern pharmaceutical premises are provided with air handling plants that supply a controlled volume of air to each process area. Correct specification and installation of the air system is essential to ensure an acceptable level of contamination of the air supply into a process area. Additionally, air movements between process areas can carry contaminating particles. This risk has to be considered for each process area and solutions found, usually by air locks, to prevent particulate movement between areas.

All process materials have to come from outside the process area at some stage. Liquids can be piped directly to a process without external contact, but dry materials have to be transported. If this transport involves movement between areas, the facility design and process operations have to assume that other spilt materials can contaminate the materials. Cleaning regimes on entry to a process area will need to be agreed at an early stage in the process design.

6.3.3 Product segregation

Product segregation is needed to avoid contamination by another product. This would ideally be by installing separate facilities for each product, but this can rarely be achieved due to prohibitive capital costs. The industry has, therefore, adopted a number of universally applied segregation techniques:

- Do not produce high-risk products in the same facilities as low-risk products. Antibiotics are always manufactured in facilities designed to produce only this type of product, as historically, patients have suffered reactions from cross-contamination of low-risk products by antibiotics. Hormonal products are normally manufactured in dedicated facilities for this potentially highly active material. Segregation allows specific cleaning and materials handling technology to be used in a dedicated manner as well as specific operator protection and training.
- Manufacture products requiring the same environmental standards in one area. Products at high risk of contamination, such as sterile products, require far higher quality environments and the cleaning regimes are more stringent. These areas should be kept to a minimum. Operators need special purpose clothing (to protect the product) and training to work in these areas.

- Dedicate an area to the production of one product at a time and ensure that the area and equipment are thoroughly cleaned before commencing the manufacture of a new product in the same area.
- Contain the production process, ideally within the manufacturing equipment. Where this is not possible, use airflow (laminar airflow or local extract) and enclosures to retain product spillage within the smallest possible area.
- Establish fully validated cleaning regimes for each product in each area/ equipment item. It is essential to know, and be able to demonstrate, that the production area and equipment is clean at the end of a production run. 'Clean', in this context, means that trace elements of the previous product left behind after the cleaning process are below acceptable limits.

Product segregation is therefore the practice of 'avoidance'. By avoiding the factors that cause cross-contamination between products, the risks are reduced to an acceptable level. For example:

- keep different products in separate locations;
- ensure that labelling clearly identifies the product and its components;
- never manufacture one product in the presence of another;
- prepare standard operating procedures that do not create a risk of cross product contamination;
- use clean equipment at the start of a new production run;
- identify the risks of cross product contamination (e.g. operator's clothing) and reduce these risks;
- train operators in the use of equipment and production processes;
- audit the production processes to ensure conformity.

6.3.4 Cleaning

Equipment

Emphasis has been placed on the need to avoid cross-contamination between products. The major source of such contamination, if not removed by cleaning, is the equipment in which the product is prepared, closely followed by sources outside the equipment.

It is not sufficient just to clean the equipment and assume that any risk of contamination has been removed. Every individual operator would use their own method of cleaning if they were not trained. Their individual methods will vary from time to time and there is no guarantee that any of the operators' methods will provide cleaning to the standards required to reduce the risk of contamination to an acceptable minimum.

It is critical to establish cleaning procedures that can be repeated consistently. Different procedures may have to be established for each product and all the cleaning procedures have to be validated for effectiveness.

Manual methods of cleaning cannot be guaranteed to be one hundred percent effective unless by 'overkill'. Mechanical means of cleaning, however, can be accurately reproduced on demand. For this reason, modern pharmaceutical plants are normally designed with 'in-built' Clean In Place (CIP) capability.

CIP technology, established in the brewing industry, is based on the combination of chemical/detergent action and mechanical action (from the effect of direct impact on, or flow of water over, surfaces). The sequence normally utilized consists of:

- initial hot or cold rinse to remove gross contamination;
- caustic detergent rinse to remove adhering materials;
- hot or cold water rinse;
- neutralizing acid rinse (if required);
- hot or cold water rinse;
- final water rinse of a quality equivalent to that used in the process.

Water quality is a critical factor in CIP systems and any possibility of contaminants being introduced by water from the cleaning process must be eliminated. For this reason, de-ionized water to USP23 or BP is normally used throughout the CIP sequence with a final rinse of Purified Water or Water for Injection quality if the process demands this standard of cleanliness.

CIP systems are normally controlled by automatic sequence rather than manual operation.

Large surfaces to be cleaned by CIP systems require the use of mechanical devices, such as spray heads, and an understanding of the 'shadow' effects created by internal fittings. Specialist companies supply both the equipment and 'know-how' for this technology.

Although cleaning by direct impact using spray heads can be designed into process equipment, the interconnecting pipework can only be cleaned by the flow of water and chemicals over the surfaces. Experience indicates that turbulent flow is required to provide maximum cleaning effect. This turbulent flow is normally created by flow rates at or above 1.5 ms^{-1} and the design of a CIP system should ensure that all process pipework is subject to this minimum flow rate.

The duration of flow of CIP fluids is determined by examination of the effects of the CIP process on the system. Access is, therefore, required to all cleaned surfaces during the validation of the cleaning process.

For this reason, most process pipework installations subject to a CIP system are designed to be taken apart on an agreed schedule to enable the cleaning procedures to be re-validated.

Materials of construction are frequently fabricated to a higher standard than is required by the process, to enable the cleaning procedure to be fully effective.

Contamination sources outside the equipment can be eliminated by the total containment of the process. For many reasons, the design of pharmaceutical processes cannot always permit this ideal arrangement and, in practice, many sources of contamination will exist that have to be controlled during production.

The following brief paragraphs aim to give an indication of some of the chief contamination sources that are created by normal operation of a process, and the techniques for avoiding these are outlined.

(a) Materials received into the facility from outside sources

These are expected to be contaminated by any material normally present during transport and materials handling operations. Normally all such materials are double wrapped (plastic linings inside outer containers) and are frequently over-wrapped by stretch film. Cleaning, other than gross contamination, will be left until the material is to be used.

(b) Sampling

This is undertaken of all incoming materials and requires the breach of the materials containment system. For this reason, sampling is undertaken within a sampling booth and the material containers will be cleaned externally before entry into the booth. The inner and outer containers will be resealed before return to storage.

(c) Storage and internal transport

These will not normally provide a severe risk of contamination, but all inner and outer containers must be kept sealed. Again, the outer containers will be cleaned before entry into the production area.

(d) Dispensing operations

This is naturally a dusty operation when dealing with dry materials. Contamination of other materials from this dust must be reduced to a minimum by cleaning the dispensary area. It should be noted, however, that cleaning between the dispensing of different materials for the same product is normally only on a limited housekeeping basis.

(e) Charging/discharging operations

Transfer into and from process equipment is normally dust free for operator safety reasons. Where, however, this operation is not dust free, the resultant dust spillage can be expected to contaminate all surfaces in the operating room as well as the operator. The operating area must be thoroughly cleaned on the completion of a production run, or at least once a week.

(f) The operator

The operator has freedom of choice in where to go and what to do. This freedom has to be strictly controlled, with high quality training provided and absolute discipline exercised to prevent the transfer of contaminating products between different process operations. Current practice indicates use of specific clothing for each production room and personal cleaning regimes on leaving the room. These cleaning regimes may be as limited as an external clothing change or as severe as air showers or water deluges, depending on the nature of the product and the company's policies.

(g) Processing equipment

This is normally selected to be non-particle shedding and, therefore, is not considered to provide a contamination risk. Care should be taken over new or maintained equipment that can be delivered with surface contamination invisible to the naked eye.

(h) Room fabric

This includes walls, floors, ceilings, doors, service entries, lights, etc. All have to be carefully chosen to avoid particle shedding characteristics and have easily cleaned surfaces. Ledges should be designed out of the room, but where unavoidable, should be sloped to prevent dust traps.

(i) Air handling systems

These bring a continuous source of replacement air to the operating environment. Care must be taken in the design of the air handling plant, equipment and, particularly, filters to prevent external contaminants being carried into the operations. The following sub-section provides information on the environments that have been found to be acceptable for pharmaceutical production.

6.3.5 Environmental classification

Pharmaceutical environments are classified by the number of particles of specific sizes contained in a measured volume of air, together with requirements for temperature and humidity. The information in this section is on European and United States requirements.

Table 6.2 United States Federal Standard 209D — air classifications

Class	Class limits in particles per cubic foot of size/ particle sizes shown (micrometers)				
	0.1 mm	0.2 mm	0.3 mm	0.5 mm	5.0 mm
1	35	7.5	3	1	NA
10	350	75	30	10	NA
100	NA	750	300	100	NA
1000	NA	NA	NA	1000	7
10,000	NA	NA	NA	10,000	70
100,000	NA	NA	NA	100,000	700

The most easily understood classification comes from US Federal Standard 209D (Table 6.2) and, although theoretically superseded, is still in extensive use. It is based on imperial measurements.

This Federal Standard has been updated to version 209E by conversion to SI units of measurement (see Table 6.3).

FS 209E permits the continuing use of 'English' terminology although SI units are preferred. Of particular importance in the Federal Standard is the need to specify and measure particle counts as either 'as-built' (no operators or equipment present), 'at rest' (equipment installed, but no operators present) or 'in operation' (equipment in use and operators present).

Table 6.3 United States Federal Standard 209E — air classifications

Class name		Class limits (volume units)									
SI	English	0.1 mm		0.2 mm		0.3 mm		0.5 mm		5.0 mm	
		(m^3)	(ft^3)	(m^3)	(ft^3)	(m^3)	(ft^3)	(m^3)	(ft^3)	(m^3)	(ft^3)
M1		350	9.91	75.7	2.14	30.9	0.875	10.0	0.283		
M1.5	1	1240	35.0	265	7.5	106	3.00	35.3	1.00		
M2		3500	99.1	757	21.4	309	8.75	100	2.83		
M2.5	10	12,400	350	2650	75	1060	30.0	353	10.0		
M3		35,000	991	7570	214	3090	87.5	1000	28.3		
M3.5	100			26,500	750	10,600	300	3530	100		
M4				75,700	2140	30,900	875	10,000	283		
M4.5	1000							35,300	1000	247	7.00
M5								100,000	2830	618	17.5
M5.5	10,000							353,000	10,000	2470	70.0
M6								1,000,000	28,300	6180	175
M6.5	100,000							3,530,000	100,000	24,700	700
M7								10,000,000	283,000	61,800	1750

In all cases, services must be functional.

There are a number of European Standards available based on national standards. The European Directives that created 'The Rules Governing Medicinal Products in the European Community' cover air classification systems for the manufacture of sterile products (see Table 6.4). These classifications are now considered as the established European standard and, for members of the EEC, are legal requirements.

In the 'Rules and Guidance for Pharmaceutical Manufacturers 1997' prepared by the MCA, a similar table is published for sterile production that gives further guidance between the 'at rest' and 'in operation' conditions (see Table 6.5).

For these airborne particulate classifications, the MCA also publish a table giving recommended limits for microbiological monitoring of clean areas 'in operation', (see Table 6.6, page 170).

Table 6.4 Air classification system for manufacture of sterile products

Grade	Max permitted number of particles per m^3 equal to or above		Max permitted number of viable micro-organisms per m^3
	0.5 mm	5.0 mm	
A Laminar air flow work station	3500	None	Less than 1
B	3500	None	5
C	350,000	2000	100
D	3,500,000	20,000	500

Extract from The Rules Governing Medicinal Products in The European Community. Note that class A refers to the air classification around the exposed product, whilst class B refers to the background environment.

Table 6.5 Airborne particulate classifications — MCA guidelines

Grade	Maximum permitted number of particles per m^3 equal to or above			
	At rest		In operation	
	0.5 mm	5.0 mm	0.5 mm	5.0 mm
A	3500	0	3500	0
B	3500	0	350,000	2000
C	350,000	2000	3,500,000	20,000
D	3,500,000	20,000	Not defined	Not defined

Table 6.6 Recommended limits for microbial contamination (average values) — MCA guidelines

Grade	Air sample cfu m^3	Settle plates (diam. 90 mm), cfu/4 hours	Contact plates (diam. 55 mm), cfu/plate	Glove print 5 fingers, cfu/glove
A	< 1	< 1	< 1	< 1
B	10	5	5	5
C	100	50	25	–
D	200	100	50	–

Note that all the above tables are published with comprehensive notes. It is important that these notes are fully understood before proceeding with the design of the environment.

The particulate classifications in use are normally referenced by either the FS 209D system (100, 10,000, etc.) or by the EEC rules (A, B, etc.). These two classifications correspond approximately and both are accepted by the regulatory authorities. In summary, Table 6.7 provides a brief check for the user.

It is recommended that the designer specify the particulate levels in the 'at rest' condition. In addition to the particulate levels, room operating conditions of temperature, humidity and pressure must be specified.

Humidity creates contamination risk to the product from condensation, absorption and human perspiration. It is, therefore, normal practice to maintain the operating conditions at 45% to 55% relative humidity.

Where a product is expected to absorb water from the environment, such as effervescent tablets, hard gelatine capsules, etc., the humidity has to be reduced. The humidity has to be controlled at a level that is acceptable

Table 6.7 Approximate equivalent international standards

MCA guidelines	FS 209D	FS 209E	ECC rules	Germany VDI 2083	UK BS 5295	ISO 14644 Part 1
1997	1988	1992	1992	1990	1989	Draft
	1	M1.5		1	C	3
	10	M2.5		2	D	4
A	100	M3.5	A	3	E or F	5
B	100	M3.5	B	3	E or F	5
	1000	M4.5		4	G or H	6
C	10,000	M5.5	C	5	J	7
D	100,000	M6.5	D	6	K	8

to the operator as well as avoiding risk to the product. In extreme cases, it will be necessary to provide the operator with a breathing air supply.

Temperature should normally be maintained at a level that permits the operator to work in comfort. The air supply temperature should allow for heat gains from all sources within the operating area. Many alternative methods of temperature control are available and the designer should seek expert advice. It is essential, however, to maintain the room temperature within the specified — and validated — limits over the full range of operational conditions.

Where production has to be undertaken at temperatures normally unacceptable to the operator, e.g., cold rooms, then protective clothing should be provided.

Pressure differentials are an essential part of the design of a clean room facility. To protect a product from contamination from outside sources, it is normal practice to pressurize the rooms in which the product is exposed to the environment. Where a sequence of operating rooms is installed, pressure 'cascades' are frequently used so that the most sensitive areas are at the highest pressure and the least sensitive at, or just above, atmospheric pressure. This situation is most frequently present in aseptic operations.

Where the product concerned is of high potency, negative pressure is used to contain the hazard to within the operating area. The risk from external contamination is usually reduced by surrounding the negative pressure room with other areas (e.g. changing rooms) at positive pressure.

The most commonly used pressure differential is 15 Pa.

6.3.6 Ventilation systems

Ventilation systems designed into any secondary pharmaceutical facility need to be carefully designed, installed, controlled and operated. The designer should consult with experts in this field to achieve the desired conditions within the process areas, but the following paragraphs give some general guidance.

The environmental standards specified within any operating area must be maintained to those standards at all times when process operations are active. At no time should the product, or the surfaces with which the product comes into contact, be exposed to environmental conditions that may cause unacceptable contamination. In practice, this means that ventilation systems will be fully operational for the majority of the time and only revert to night/weekend operation when all risks of contamination have been contained.

Assuming that the ventilation system has been correctly designed and installed, the system should not provide any significant source of contamination. This is achieved by both filtration of the air supply and monitoring and control of the pressure, temperature and humidity in each operating area.

Each area will have been commissioned against a specification that meets the environmental classification for the product being made and the area will be monitored on a regular basis for maintenance of this classification. Any deviation has to be reported and action taken. Significant deviation from acceptable limits will result in cessation of production.

To prevent this extreme situation, ventilation systems are normally designed to meet the following criteria:

Class A: Laminar airflow through terminal HEPA (High Efficiency Particulate Air) filters at a velocity of $0.45\,\mathrm{m\,s^{-1}} \pm 20\%$ at the working position (MCA guidance) with low-level extract. In all cases, operations at Class A should be contained within a purpose-designed workstation with no operator access other than gloved hands.

Class B: Downward airflow through terminal HEPA filters with low-level extract. The operator will be working and creating high particle counts in this area. Air volumes should be sized to ensure that particulate conditions for the 'at rest' state will be achieved in the unmanned state after a short 'clean-up' period of 15–20 minutes.

To ensure that the air movement is able to clean up the working area, current designs now utilize turbulent air movement delivered by purpose designed diffusers.

Class C: Airflow provided through (normally terminal) HEPA filters with air movement of sufficient volume to maintain the classification of the area. There is considerable debate on the use of low-level extract for Class C areas, but there is no specific requirement. Air volumes should be sized to ensure that particulate conditions for the 'at rest' state would be achieved in the unmanned state after a short 'clean-up' period of 15–20 minutes.

The higher cost of installing low-level extracts needs to be considered against the risks created by moving particles in the air stream over the entire working area when high-level extracts are used.

Class D: Airflow provided through filters (normally HEPA) with air movement of sufficient volume to maintain the classification of the area. High-level extract is the usual installation for this classification. Air volumes should be sized to ensure that particulate conditions for the 'at rest' state would be achieved in the unmanned state after a short 'clean-up' period of 15–20 minutes.

Where possible, air movement should be designed to flow downward over any exposed product to avoid particulate entrainment being carried over the product.

In areas where the majority of operations only require a minimal environmental classification, it is acceptable to provide higher local environmental conditions by use of air curtains. A good example of this method of protection can be found in many packing halls, where the general area will be to Class D, but local conditions around the product at the filling head will be to Class C.

HEPA filters are normally used to achieve the stated environmental classifications. Within Europe, the grades of HEPA filter are distinguished by the use of EU classifications, each of which has a known retention efficiency at 0.3 mm (see Table 6.8).

In the USA, HEPA filters are required to have efficiencies of 99.97% (EU12 and greater).

Not only is it essential that the filter specifications meet the requirements of the environment, but also that the installation does not compromise the filter integrity. This can be caused through damage to the filter medium, or through passage of unfiltered air between the medium and its frame, or between the frame and the air supply system. Assurance of the integrity of an installed filter system must be subject to an 'in-situ' integrity test.

6.3.7 Surface finishes

Throughout this section, emphasis has been placed on the avoidance of possible contamination of the product. Consideration has been given to sources of contamination from outside the operating environment but it is equally important to appreciate that the fabric of the area and the equipment in which the product is produced, can itself contaminate the product.

All materials of construction should be non-particle shedding. Traditional building materials must, therefore, be sealed by the application of a surface coating. Current practice is to use a two part epoxy coating (or equivalent) that provides both an abrasion resistant surface and a sufficient degree of elasticity to avoid minor wall movements opening up hair line cracks, thus permitting particulate escape.

The use of partition systems has become widespread and several alternative systems are available. These systems, although more expensive, eliminate the

Table 6.8 Classification of retention efficiencies of HEPA filters

Eurovent classification	Efficiency at 0.3 mm (%)
EU10	$\geq 95 - < 99.9$
EU11	$\geq 99.9 - < 99.97$
EU12	$\geq 99.97 - < 99.99$
EU13	$\geq 99.99 - < 99.999$
EU14	≥ 99.999

wet building trades and provide an acceptable pharmaceutical finish with no further surface treatment.

Joints in wall and ceiling construction are normally filled with a silicone sealant that permits some building movement without any crevices forming. For ease of cleaning, joints between walls and floors are always coved in any area in which product is exposed. Current practice is to cove at wall to wall joints in Class C areas and also wall to ceiling joints in Class A/B areas.

Floors present a more difficult choice, as they have to accept movement of heavy loads, building settlement and movement as well as possible damage from containers, etc. Currently, epoxy floor coatings up to 6 mm thick are proving successful, but their expense limits their use to the more severely loaded areas. Vinyl floor, wall and ceiling coverings are an acceptable solution — reserved for lightly loaded areas and are the material of choice in Class A/B areas for many manufacturers.

In selecting materials of construction for the building elements, thought must also be given to the damage that may be caused by the normal daily operations, such as trucks and pallets hitting walls. Where such damage would expose particulates, wall protection is usually provided.

The cleaning regimes in the production environment normally involve wetting the surfaces of the area. In the controlled environment, these conditions provide excellent sources for microbial growth and it is, therefore, important to ensure that surface finishes do not support microbial growth.

Process equipment comes into intimate contact with the product and, therefore, the materials of construction are of most significance. Non-corroding materials are essential, not only to prevent contamination of the product, but also to stop any damage to the surface finish of the equipment.

A poor surface finish harbours crevices that can support microbial growth and traces of previous products and cleaning agents. For this reason, emphasis is placed on the specification of surface finishes and the methods by which they are prepared.

The great majority of pharmaceutical process equipment is fabricated from 316 or 316L stainless steel because of the non-corrosive nature of the material for most products and the ease with which it can be given a high quality surface finish. The surface finishes are normally specified (as Ra — average roughness) in either micro inches or microns. The polishing medium grit size should not be used as an indication of the surface finish.

Individual producers of stainless steel equipment will use both mechanical and electro polishing methods. Electro polishing gives a higher quality look to the surface and provides a more rounded edge to the microscopic grooves in the polished steel. This more desirable finish is, however, more expensive than mechanical polishing.

The selection of the surface finish is determined by:

- existing standards within a facility;
- end user preference;
- the need for a reduction in crevice size to reduce microbial growth;
- cost.

Table 6.9 lists surface finishes specified for stainless steel equipment.

6.3.8 Lighting selection

Apart from the need to ensure a safe working environment, the regulatory authorities are interested in the lighting levels in a facility to ensure the manufacturing operations are undertaken without error.

Although many operations in the modern pharmaceutical production facility are now automatically controlled, the operator still needs to oversee these operations. Frequently his work requires him to read Standard Operating Procedures and the slightest risk of error caused by misreading the instructions, instrumentation and alarms is not acceptable.

Lighting selection must, therefore, ensure that the level of illumination is sufficient to read documentation, displays and instrumentation and that this does not cause operational difficulties from glare, reflection or too high an intensity.

Designers of pharmaceutical facilities are recommended to take expert advice in the illumination specifications to ensure that all working areas are well lit throughout.

Table 6.9 Polished finished on stainless steel sheet — Sillavan metal services

Description	BS1449 No	Approx. lm Ra value	Reflectivity %
Coarse grade 80 grit	3A	2.5	10
Coarse grade 180 grit	3B	1.0	10
Silk	3B	0.4	30
Supersilk	3B	0.35	30
Brush	3B	0.2	30
Bright buff	No 7	0.05	48/55
Bright polish	No 7	0.05	53/60
Mirror	No 7	0.05	58/63

6.4 Containment issues

6.4.1 Operator protection

Pharmaceutical manufacturing operations involve the handling of sophisticated chemical compounds, many of which can exhibit toxic effects on personnel handling them in concentrated quantities. Additionally, and often at the same time, pharmaceutical materials and products can suffer if exposed to the operating environment (for example, sterile products for injection).

Operator protection can be provided by means of personal equipment (gloves, overalls, masks), while the creation of suitable macro-environments can provide aseptic facilities for injectable manufacture. However, the validity of these methods is questionable, and the use of techniques which enclose the product materials in a smaller space and provide means of remote operator access have become commonplace. These techniques are known as isolator or containment technology. Although the application of these methods differs between operator and product protection requirements, there are similarities in the equipment involved.

6.4.2 Product protection

A second application of containment technology is its use for the protection of products from environmental contamination. This application applies particularly to the aseptic manufacture of injectable or infusion products, which has traditionally been performed in high-quality environments conforming to Class 100 or better (to US Federal Standard 209E). The accepted approach is for the equipment and operations involved to be sited in Class 100 clean rooms, with localized enhanced protection to Class 10 being provided by fixed or mobile air supply units. The latter are designed to provide airflow of minimum turbulence (effectively 'laminar' flow when the units are unoccupied) so as to minimize particulate pick-up by the air steam in areas where sterile product or product-contacting components are present.

This arrangement has been demonstrated over a period of twenty or more years to provide minimal validated risk of contamination, and this proven assurance has given rise to its use in the majority of modern pharmaceutical aseptic processing facilities.

However, two undesirable features remain:

- the construction and operation of facilities reaching Class 100 conditions is expensive;
- there remains the possibility of human operator contact with product materials, with consequent risk of contamination.

Hence, recent developments of isolator technology have concentrated on the use of such equipment to provide a reliable localized barrier between the product and the operator, with the isolator forming a separate sealed environment of Class 100 or better, within which aseptic manipulations can be performed, either by hand using glove ports or automatically.

Apart from the increased potential for reliable sterility, the use of isolators having a sealed high-grade internal environment has meant that the surrounding room space need not be to the same high standard. Current opinions differ on the desirable room environment quality, the regulatory view being based on Class 10,000, while some authorities among users and equipment manufacturers claim reliable validated operation at Class 100,000. Clearly, the capital and operating cost of such environments is lower than that of a Class 100 suite.

The isolator equipment commonly used for aseptic processing is sophisticated and by no means low cost, but it does allow lower cost surroundings while supplanting the need for localized laminar flow units and often filling machine guards.

It is possible to link several machines for washing, sterilizing, filling, capping and sealing of injectable product containers within a set of linked barrier isolators or use a form fill seal technique

6.5 Packaging operations

6.5.1 Introduction

The early days of pharmaceutical product packaging saw predominantly manual systems involving, for example, the hand counting of pills or tablets which were dispensed to the patient in a suitable container, often merely a paper bag!

As demand and availability increased, the risk of mistakes became greater due to the wider range of products available and the frequency of dispensing. The same factors applied to the production of medicines, where centralization of manufacture led to multiple pack despatches. Increasing standardization led to:

- automated counting;
- pre-printed standard labelling;
- specific tested containers;
- secure capping/sealing;
- pre-printed cartons.

Much of this paralleled the growth of other consumer products, but the special security and safety requirements of medicines have extended pack features, which now include:

- tamper-evident closures;
- child-resistant closures;
- special protection against hostile shipping environments;
- security coding systems.

The early manual assembly of packaged products has given way to progressively more-automated methods. Machines for counting unit dose products (such as tablets) and discharging the correct number into manually-presented containers soon gave way to in-line counting, filling, capping, labelling and cartoning units linked by conveyors. These transport systems had gateing, accumulation and flow control elements built-in. Thus, the modern packaging line incorporates sophisticated handling and sensing equipment designed to minimize human intervention and eliminate human error.

As seen later in this chapter, the structure of healthcare management arrangements is leading to increasingly sophisticated and patient-dedicated packaging, which curiously is taking developments full-circle and returning the objective back to the days of direct patient-specific dispensing.

6.5.2 Tablets and capsules

The packaging of solid unit-dose items is generally carried out in one of two ways. These utilize multiple-item containers (typically glass or plastic bottles) and blister packs.

Bottle packs

This packaging type utilizes containers with screw or press-on caps, containing either a single course of treatment, or larger types intended to be used for dispensing from, in order to produce such single courses.

Methods of tablet/capsule counting range from photo-electronic sensing types to pre-formed discs or slats having a fixed number of cavities.

All counting methods have potential inaccuracy due to the non-symmetrical shape of tablets and capsules and the possibility of broken tablets giving false counts. Individual tablets or capsules have low weight in comparison with the container, so that container weight variation can be greater than the weight of an individual item. Thus, post-filling check weighing methods cannot be relied upon to detect missing tablets/capsules in a container.

As a result, modern counting machines are equipped with missing-item detection systems, utilizing infrared sensing or matrix camera technology.

Containers may be of either glass or plastic, but are increasingly of the latter as plastic materials with improved moisture-resistance have been developed.

Capping systems have been designed which prevent non-evident pilferage or which are resistant to the attentions of young children. These benefits do, however, become disadvantageous when used for arthritic patients, who may have difficulty in opening the packs.

Bottle packs have other disadvantages, namely:

- they offer no record of the dose having been taken;
- multiple-product treatment regimes mean the patient coping with several different containers;
- frequent pack opening may lead to product spoilage and risk of spillage;
- paper labels may become soiled, with risk of lost product identity.

However, they have two significant advantages, being generally cheaper to produce and of smaller size than the equivalent blister pack.

Blister packs

These are produced by a form-fill-seal process using PVC or similar thermoplastic material in reel form as the blister material. For products having enhanced moisture sensitivity, plastics such as polyvinylidine chloride may be used. The blister cavities are formed from the thermoplastic film using heated die plates or drums, with plug or vacuum assistance.

Tempered aluminium lidding foil with laminated plastic or an adhesive coating allows the two parts of the pack to be heat-sealed together. An alternative to plastic films for blister pack formation is the use of cold-formed aluminium foil, which can offer improved product protection from moisture ingress.

Blister forming methods include the use of continuous-motion cylindrical formers with blister cavities machined into them, or flat-platen types which cycle in a manner which matches the horizontal speed of the blister web, giving higher potential outputs.

The sealing together of the filled blister and lidding foil is achieved by the concurrent flow of the two material streams followed by the application of heat and pressure using heated rollers or platens.

Modern machines can operate at speeds of typically 400 blisters per minute, giving an equivalent tablet/capsule output of 4000 per minute for a ten-item blister.

A critical factor influencing machine output is the mechanism used for feeding the tablets/capsules into the formed blister cavities. Similar methods for detecting missing items to those used for bottle packs are employed.

It is not uncommon for finished blister packs to contain more than one blister strip. This packaging method requires the blister form-fill-seal machine to incorporate a stacking/counting unit for the blisters, prior to carton insertion.

6.5.3 Liquids

Liquid pharmaceuticals are packaged using either bottles or sachets, the latter being used for unit-dose applications.

Bottles

Early production systems for bottle filling were based upon manual dispensing from a bulk supply using a measuring container. As precision-moulded bottles became available and demand rose, methods of filling to a fixed level were established. Initially manual in operation, this approach was followed by a semi-automatic method in which the bottle was presented to a machine, which created a partial vacuum inside the bottle thereby encouraging the flow of liquid from a bulk tank or hopper. The liquid level rose in the bottle until it reached the height of the vacuum nozzle, when flow ceased. This vacuum method was developed for beverage production and is still used in some small companies.

Manually presented level-fill systems led on to automated bottle movement and presentation, with consequent increases in output. Indeed, the basic technology is still used in high-speed beverage production.

However, the fill-to-level method suffers from the disadvantage that the filled volume varies according to the accuracy of bottle moulding, making it relatively unsuitable for pharmaceutical product use.

In consequence, modern pharmaceutical liquid packaging systems utilize volumetric measurement, either by means of adjustable-stroke piston pumps, or by positive-displacement rotary lobe-type pumps controlled by rotation sensors.

High-speed dosing machines utilize 'diving nozzle' systems in order to reduce air entrainment and foaming problems (see Section 6.5.5 on page 182).

Sachets

Sachet packaging is mostly used for powders, which are then reconstituted with water or another suitable diluent by the end-user. However, a small number of examples exist of liquid-filled sachets. The pack is an ideal single-dose provision system. Sachets are formed from laminated foils, usually including a plastic inner layer with aluminium foil centre laminate and an outer layer of paper that provides a printable surface.

The sachets are formed as three-side sealed units prior to filling, and the final top seal is then applied, together with a batch/expiry date code.

Sachet packaging is more common for non-pharmaceutical products, where outputs can be as high as 100 sachets per minute.

The assembly/collation and cartoning methods of sachets are basically similar to those for tablet blister packs.

6.5.4 Powders

The powder is not a common finished dosage form for pharmaceuticals, but it is frequently used for granule or powder formulation products that have low stability in solution (such as antibiotic syrups/suspensions for paediatric use).

Products manufactured are typically in bottle or sachet form, the latter used for single-dose applications.

Powder filling systems can be either volumetric or gravimetric. The former is most often typified by auger filling machines, in which a carefully designed screw rotates in a funnel-shaped hopper containing the product powder. As the auger rotates, the number of rotations determines the volume of powder delivered at the bottom outlet of the funnel and into the container. Rotation sensors are used to control this number so that the volume and hence weight dose is also controlled.

A second volumetric system is the 'cup' type, in which a two-part telescopic cylindrical chamber is opened to the powder in a hopper and thus filled. The volume of this chamber is adjustable by varying its height telescopically. By rotating the position of the chamber between the powder hopper and a discharge chute, a controlled volume/weight of powder is discharged via the chute into the bottle or sachet. Automation of bottle or sachet feed allows relatively high output to be achieved.

A key feature of all volumetric systems is the control of powder level in the hopper, as the height of product powder above the infeed to the dosage control system affects the bulk density of the powder and hence the weight dosed.

A weight-dosing system can also be used for bottle filling. This method involves the automatic pre-weighing of the empty bottle followed by approximate dosing of typically 95% of the required fill weight (using an auger or cup filler). The partially filled bottle is then re-weighed and the weight compared with that of the empty bottle so as to allow calculation of the required top-up weight. The bottle finally passes under a top-up filler which delivers a calculated final amount to achieve the target weight.

The advantage of this approach is that the overall dosage accuracy can be greater, due to the finer control capability of the lower weight second/top-up dose.

6.5.5 Creams and ointments

These products are mostly filled into collapsible tubes, but occasionally into jars. The latter are filled and packed in much the same way as liquids. These semi-solids are also applied to impregnated tulles, although they are generally for burns treatment, where aseptically-produced versions apply.

Tubes

Tubes used for pharmaceutical preparations are either of the fully collapsible aluminium or aluminium/plastic laminate type, or are non-collapsible plastic. They are filled with product from the seal end before closing — the aluminium types being closed after filling by flattening and folding, while the plastic types are sealed by heat/impulse methods.

Filling machines are usually of the rotary plate type, with empty tubes inserted into holders fixed into this plate from a magazine by means of an automatic system. On low-output machines, tube insertion may be performed by hand.

The product is filled from a hopper via piston type dosing pumps through nozzles and into the tubes. These nozzles are often arranged so that they 'dive' into the empty tube and are withdrawn as the product is filled, a technique used to minimize air entrainment. The bulk product hopper is often stirred and heated, typically using a hot water filled jacket, in order to enhance product flow and uniformity.

Empty tubes are usually pre-printed with product information. This print includes a registration mark which allows the filling machine to sense the orientation of the tube, and rotate it prior to sealing so that the product name or details are conveniently positioned for user-reading.

Modern machines can also be equipped with code scanners that check a pre-printed bar-code, comparing this code with microprocessor-held recipe information, and reject or produce an alarm on any false codes.

6.5.6 Sterile products

It can be assumed that products manufactured aseptically arrive at the packaging stage in sealed containers that assure the integrity of the product.

The exceptions to this are items manufactured using integrated form-fill-seal systems, and impregnated dressings, where specific handling arrangements apply.

182

Ampoules and vials

Although some unit sterile products (both liquid and powder) are filled into pre-printed ampoules or vials, it is not uncommon that these components are effectively unidentified prior to labelling. It is, therefore, essential that filling controls are such as to ensure that the containers are held in identifiable lots, and that these lots are labelled with minimum delay or handling. It is thus usual for ampoules and vials to be labelled immediately following aseptic filling or terminal sterilization. In the latter case, they will be held in sterilizer-compatible trays that are used as loading cassettes for the labelling machine.

Wherever possible, manufacturers will arrange for unlabelled injectable product containers to have a form of product-specific machine-readable code. In such cases, the first task of the labelling machine will be to read this code and compare it with recipe information held in its control system.

As with oral-dose products, modern labelling systems use self-adhesive pre-printed/coded labels in reel form. It is common for these labels to use a transparent substrate such as polyester film to facilitate product visual inspection after labelling.

Pre-printed code checking is also included in modern labelling machine technology, and is again linked to control-system recipe information.

Syringes

Similar procedures apply to syringe packaging as for ampoules and vials, but the inconvenient shape of pre-filled syringes means that specifically engineered handling systems are required.

Form-fill-seal

High-volume production of single-dose and large volume infusion solutions is frequently performed using integrated-system technology. This approach is based on the use of high-quality thermoplastic materials (such as polypropylene) in granule form being heat-moulded in an enclosed system within a controlled-environment machine enclosure, to produce sterile empty containers, which are immediately filled in-situ with the sterile-filtered product solution. The filled containers, which may be single or multiple-moulded units, are immediately heat-sealed prior to emerging from the controlled enclosure.

This form of production requires sophisticated and expensive machinery but has high throughput and the possibility of locating the forming-filling unit in an area of lesser environmental quality. It is also possible to emboss product and batch code information onto the containers at the point of manufacture, thus enhancing identification integrity.

Creams and ointments

Such products are often filled aseptically into collapsible tubes using techniques similar to those employed for non-sterile products. These procedures are most often used for the manufacture of ophthalmic ointments.

Another application for semi-solid products is in the preparation of impregnated dressings. Although it is not a common product type, it has particular importance in the manufacture of material for the treatment of severe skin conditions, including burns. The technology involves the dosing of the medicated product onto a suitable substrate (usually tulle) in reel form, in a continuous or semi-continuous automated process carried out under aseptic conditions. The impregnated tulle is then cut into unit-treatment sections, which are packed into sachets, using a form-seal process. The sachet-forming material would consist of paper/foil/plastic laminates in reel form, pre-sterilized by irradiation.

6.5.7 Container capping and sealing

Solid or liquid products packed in glass or plastic bottles, jars or tubs require some form of lid closure to protect the contents. A typical bottle closure would be a pre-moulded screw-on plastic cap with a composite paper wad to provide a seal.

Such caps were originally hand-applied and tightened, but this action gave rise to unreliable seals and leakage, so mechanized systems were developed which provided a constant application torque, although the bottles were still hand-presented. As outputs increased the arrangement was changed to one of automatic presentation, application and tightening.

A small number of incidents of product pilferage occurred, so the consequent requirement for tamper-evidence led to various attempts to provide a 'pilfer-proof' feature. One such, for jars, involved the application of a plastic/aluminium foil laminate, heat-sealed onto the jar by means of heat/impulse sealer (similar to the system used for instant coffee jars). This solution provided the added benefit of enhanced product protection from moisture ingress.

Alternative tamper evident methods included the use of roll-on aluminium type caps, where the bottle thread is followed by spinning rollers that form the cap thread. These have also been utilized without the tamper-evident feature.

Plastics have been used successfully for many years as a material for both container and cap manufacture. These include both screw and press-on flexible plastic caps, the latter also being employed for glass bottles. Such flexible materials have the added possibility of including a press-on tamper-evident cap, which combines adequate product protection with ease of application.

6.5.8 Container labelling and coding

Early labelling systems used vegetable or animal-derived semi-solid glues manually applied to paper labels, which were then applied to the container.

This approach had many failings, notably:

- there were no reliable checks on label identity or batch code;
- the position of the label on the container was not fixed;
- there was no automatic batch coding.

Later systems, still used in many non-pharmaceutical applications, retain the use of wet glues but employ machine-application. Early versions of such machines employed automatic batch code printing, although the resulting print quality was not good.

Most modern pharmaceutical labels are of the self-adhesive type, which allows cleaner operation and reliable appearance. Automatic machines usually include product bar-code scanning and automatic batch coding, with alarm systems for integrity failings.

A long-standing feature of pharmaceutical packaging has been the use of market-specific labelling. This requirement gives rise to a potentially wide range of label alternatives, with stock holding and cost consequences.

A modern system has been developed to overcome this problem, utilizing plain self-adhesive label stock onto which all product, batch and expiry details are automatically printed in multiple colours using microprocessor controls. Recipe information held by the microprocessor system is fully validated to ensure correct output.

6.5.9 Cartoning

The placement of filled containers of liquid or unit solid pharmaceutical products into cartons was initiated for a number of reasons, including the need to insert leaflets providing patient usage instructions and, in the case of liquid products, the addition of a standard dose-measuring spoon. Such placements were initially performed manually.

As demand and output increased, automatic machines were introduced. This automation created a number of challenges to consider, for example:

- the importance of detail design, accuracy of cutting and assembly of blank cartons to ensure efficient mechanical erection and closing;
- the importance of humidity control during carton storage due to the effect of moisture on carton board making it less pliable and increasing friction — very significant for higher speed machines;

- the engineering design of cartoning machinery to allow smooth and reliable high-speed operation.

Modern cartoners may be fitted with automatic leaflet insertion, using pre-printed plain sheets, folded prior to insertion, or reel-fed leaflet stock. They may also be fitted with automatic batch and date coding and code scanning to determine correctness of carton type and overprinted information.

Automatic and semi-automatic cartoners are generally of four altern-ative types, which are characterized by method of motion indexing (intermittent or continuous) and by direction of container insertion (horizontal or vertical). Intermittent motion vertical (IMV) machines are used frequently in pharmaceutical packaging, not least because they can be operated in a manner that permits manual insertion of bottles/leaflet spoons at one or more operator stations. For high throughput, however, continuous motion horizontal (CMH) machines are favoured.

6.5.10 Collation, over-sealing, case packing and palletizing

The automation of 'end of line' operations within the pharmaceutical industry is not a universal practice, although it is becoming more commonplace for higher-output packing lines. Owing to the fact that, at this late stage in the production cycle, the product is fully sealed, protected and identified, the equipment required for final packaging does not generally need to be specia-lized. It is, thus, acceptable for it to be of the same type and source as that used for consumer goods packaging.

Collation of filled cartons and over-sealing with cellulose or polymer film is common for many medium-selling products. On low-output packing lines the collation is performed by hand and the over-sealing is performed using a semi-automatic heat-sealing unit with manual operation.

For higher-speed lines, typically over 20 cartons per minute, the collation of cartons and feeding into a wrapper/sealer is often performed automatically.

Automatic case packing and palletizing is not universally used, due to the relatively low outputs typical of many pharmaceutical products. However, it is not unknown, and once again, consumer goods equipment is employed.

One advantage of automatic final packaging is that it facilitates the automatic application and checking of outer carton labels.

6.5.11 Inspection systems

Modern pharmaceutical packaging systems rely heavily on inspection systems to verify the correctness of critical product parameters, including:

- fill volumes or unit counts;

- absence of contamination;
- container seal integrity;
- container label identity;
- label position and orientation;
- carton identity;
- outer container label identity;
- batch number;
- manufacturing date;
- expiry date.

In common with other consumer product industries, the pharmaceutical industry originally relied on human visual inspection to detect contamination and pack faults. Examples included the use of visual checking for particulate contamination in ampoules and liquid vials, container, label, cartons identity checking, and the monitoring of fill levels.

These procedures were known to be of limited reliability due to operator fatigue and attention-span limitations, and also suffered from slow and variable output rates, especially if inspection speeds were operator-controlled.

Initial mechanized systems, in which the containers were automatically presented to the operator's line of sight in an ergonomically efficient manner, were introduced. These still relied on operator visual acuity and attention, with benefits to output and reliability, but these were not significantly faster than a competent human operator, and remained less than 100% reliable.

A considerable amount of survey work was carried out in the 1960s and 1970s, especially in connection with injectable product inspection, and the data generated was used to compare performance with mechanical methods.

Camera-based systems were introduced during the 1970s by a small number of European and Japanese companies, and these provided benefits in terms of improved output rates to match similarly improved filling machine performance. Detection rates were improved and became more consistent, but the machines were limited in capability to a set number of reject types, largely due to limitations in the camera technology. These rejects were based upon physically measurable parameters (including volumes, counts, contaminants).

The introduction of digital matrix camera technology during the 1980s gave rise to an expansion in automatic inspection capabilities. These microprocessor-driven systems can be programmed to recognize deviations from standard shapes, the presence of contaminants, and even the correctness of components codes and batch and expiry-date numbers.

As with many advances in production technology, the improvements in inspection systems have arisen from the quality and output-led demands of the pharmaceutical and other high-volume product industries. These challenges have been met by the machinery and equipment manufacturing industry and the reader is recommended to approach these manufacturers for information on the latest advances in this fast-moving area of technology.

6.6 Warehousing and materials handling

6.6.1 Introduction

The storage of materials for pharmaceutical manufacture and the products themselves utilizes systems and procedures much like those employed in any high-volume consumer products operation. However, there are some special considerations applicable to pharmaceuticals resulting from the critical need to ensure the integrity of raw materials and products, and these affect the selection of storage systems, materials management systems and material transportation arrangements. The ultimate choice of system available in each of these aspects will be influenced by many 'normal' considerations, but ultimate pharmaceutical product security and integrity are the overriding factors.

6.6.2 Conventional storage

The extent of raw materials and finished product holding typical of pharmaceutical industry operations is not normally considered large. Hence, automated high-capacity storage systems are not always required or cost-effective. In these situations, 'conventional' warehousing, consisting of racking systems having, typically, no more than five pallets in the vertical direction and aisle widths between rack faces of around 2.5 to 3 metres, are common, assuming standard 1.0×1.2 metre size pallets.

The advantage of this arrangement is that the racking can be fully free-standing with no top-end fixing, and regular ride-on counterbalance fork lift trucks (which can also be used in a variety of non-warehouse duties) are suitable for stacking and de-stacking movements.

Although such arrangements are relatively low-cost, they do have certain disadvantages, notably that the pallet density per unit floor area is low, so that the area utilization is poor where site space is limited. A further specific disadvantage for pharmaceutical warehousing is that, being basically flexible and operator-controlled, the extent of automatic cGMP compliance in relation to material segregation is effectively zero, and adherence to procedures becomes the only method of avoiding mistakes in the selection of materials for production.

A solution to these deficiencies is the employment of automated systems (see Section 6.6.4).

6.6.3 High bay options

Where material volumes are high, in terms of total inventory and frequency of movements, conventional warehousing is inefficient, both in storage density and in speed of pallet insertion and removal. Where site space is limited, the storage density is especially significant.

High bay warehouses, having vertical pallet stacks of between 5 and 20 units, provide a solution to high-density storage requirements. They typically have narrower trucking aisles and special trucks which cannot be utilized for non-warehouse duties. The trucks can be of two alternative types — operator-controlled ride-on, or automatic crane. The former has many similarities with conventional systems, whereas the latter has no direct operator involvement and is controlled by a computerized materials management system. Many permutations are possible, and the selection will depend on material selection frequency, total capacity, number of alternative materials, etc.

Computer-controlled systems have considerable benefit in pharmaceutical warehousing duties, as quality assurance is enhanced by the automated nature of material selection and location (see Section 6.7).

These high racking configurations usually require structural bracing at the top in order to provide stability. Indeed, it is not unusual for very high warehouses to utilize the racking system as part of the building structure, with exterior cladding and roofing supported off the rack framework.

6.6.4 Automated warehousing

Some of the major international pharmaceutical companies have invested in automated production systems, including warehousing. The latter, based on high bay arrangements, utilize materials management systems for the control of material movement and usage, interfaced to warehouse control systems that handle the insertion, removal and security of raw materials and finished product. Such warehouses are typically un-manned and employ stacker cranes.

As there is no physical operator involvement in materials selection, it is possible for automated warehouses to be employed for the storage, in a single warehouse, of raw materials and finished products having 'quarantine' as well as 'approved' status. The selection of materials is controlled by the materials management system, which carries material status information and transmits simple location-only instructions to the warehouse crane.

This type of warehouse and management system may integrate with automatic production systems, where material movement within the manufacturing area is also mechanized, and where the production materials are always enclosed within the processing equipment or transfer containers.

'Islands of automation' arrangements are ideally suited for single-product manufacturing facilities, but have also been employed for multiple generic product manufacture. Their most significant challenges relate to the specification of control systems and their validation, and to the design of mechanisms for enclosed material transfer.

6.7 Automated production systems

6.7.1 Introduction

Earlier sections of this chapter refer to the application of automatic manufacture systems.

The adoption of automation in pharmaceutical manufacturing is driven by the need to minimize costs, and the desire to avoid the effects of human error. As labour costs increase, the reduction of direct manpower requirements makes economic sense. At the same time, the cost of pharmaceutical machinery is escalating as a result of enhanced technical sophistication and cost inflation, so that increased daily running times are necessary to meet return on investment criteria.

Although automated materials handling has been and continues to be utilized in pharmaceutical manufacture and warehousing, its application has generally been restricted to operations which basically involve a single-product type (such as tablets), or those where high-potency product containment has led to the development of enclosed systems.

The additional costs of fully automatic, or 'lights out' operation, are largely related to the inclusion of microprocessor-based monitoring and control systems, the hardware costs of which are steadily reducing in real terms. Hence the cost/benefit relationship is moving in favour of the adoption of automation.

In addition to these manpower and capital cost savings, automation can bring other advantages, including:

- improved product consistency and quality;
- enhanced adherence to validated systems;
- reduced services usage per unit output.

190

6.7.2 Process automation

Automatic semi-continuous operation of individual process units where bulk material input and product output systems are possible (including tablet presses, capsule fillers, inspection units) is achieving greater acceptance. Such units utilize automated sampling for off-line QC analysis, as well as automated measurement and feedback control of fill/compression weight, hardness and thickness. Self-diagnosis of electronic systems coupled with automatic switching of backup systems can also be expected to become common in the medium term.

Other less continuous processes can more easily be automated (such as granulation, drying, blending), as the number and range of control parameters are limited. However, automation of the product transfer arrangements linking these individual steps is perceived to be more difficult to achieve due to the greater separation distances involved and the need for connection and disconnection.

This perception can be answered by amalgamating unit operations within single areas, having 'permanent' connections between process steps, and using validated Clean In Place systems for inter-batch decontamination. This approach allows complete sets of linked operations to be run as 'continuous' processes. Applications of this nature are common in certain other industries and technology transfer is clearly a major opportunity.

Additionally, where scale of operation and product mix permit, Automated Guided Vehicle (AGV) systems for IBC movement with automatic docking facilities can be utilized. This is particularly attractive where bin movements and docking operations can take place within technical (non-GMP) areas.

6.7.3 Packaging automation

There is considerably wider scope for automation in pharmaceutical packaging operations, where higher unit volumes and repetitive tasks traditionally require the employment of large labour forces. Cost reduction and quality improvements have been achieved throughout the industry over the past 40 years by the use of automated operations and higher-speed machinery. There remains considerable scope for further automation of these activities, but factors determined by market and regulatory pressures are of great current interest (i.e., the movement towards original-pack dispensing and patient-specific production).

The following section of this chapter describes a pioneering approach to meeting these challenges, and provides useful information on the engineering aspects associated with packaging automation. The authors are grateful to Richard Archer of The Automation Partnership for agreeing to the inclusion of this section.

6.8 Advanced packaging technologies

6.8.1 Introduction

Compared to most other manufacturing sectors, the pharmaceutical industry occupies a unique position where the direct manufacturing cost of many of its products is a small proportion of the end user price. The major costs in pharmaceutical companies are the indirect ones in R&D, marketing and distribution, not manufacturing. In simplistic terms, it could be said that pharmaceutical manufacturing costs were not really important. If this statement seems contentious (which it deliberately is), consider the impact on respective company profitability of halving the production cost of a car compared with that of a tablet and how such a proposition would be viewed. For a car company, manufacturing costs are of paramount importance in achieving competitiveness, with the whole product design and development process geared to manufacturability and provision of maximum product features and choice at minimum cost. For the pharmaceutical companies, the primary emphasis is on discovering and launching increasingly effective molecules and therapies. Provided there is a method of manufacture that can be well controlled and monitored, the actual direct production cost is comparatively unimportant.

This unique situation has changed, however, as pharmaceutical prices have come under greater scrutiny from governments, healthcare providers, insurance companies and the challenge of changes in the selling and distribution of prescription drugs. Both the direct and indirect costs of production and distribution are under pressure, while the market is demanding greater choice, improved service, faster response and lower prices.

In many respects, therefore, the pharmaceutical industry is now having to face the same issues of cost and flexibility that most manufacturing sectors had to address decades ago. The industry is, however, unfamiliar with the key principles of truly flexible manufacturing and much of the available processing equipment is unsuited to rapid changeover and responsiveness. Too few pharmaceutical companies today recognize that the ultimate objectives of advanced flexible manufacturing are reducing indirect costs and generating new business opportunities, not direct cost reduction.

Packaging of solid dosage products is indicative of these aspects. The current equipment is comparatively high speed and is geared to long, efficient production runs in one pack format. Increasing pack variants and inventory reduction pressures have led to smaller batch sizes, but this then results in lines where changeover time often exceeds running time.

This section describes how a radically different approach to tablet packaging has been developed which seeks to address these new market issues. The objective, as with modern car manufacturing, is to reduce the viable batch quantity to a single product unit.

6.8.2 Conventional pharmaceutical packaging and distribution

Conventional drug packaging lines are geared to large batch quantities of single products, which are subsequently distributed through a complex internal and external chain of warehouses and distributors. (It has been suggested that it typically takes six months from packaging for a prescription drug to be received by the patient). It could be said that the inflexibility of the conventional packaging process is the cause of the current multi-stage distribution route rather than a consequence of it. Remember that the end user ultimately purchases one pack at a time; in other words, large batch quantities are a consequence of the existing packaging/distribution process not a customer requirement.

Traditional bottle filling systems are mechanically tooled and controlled, using tablet specific slats or pocketed disks to provide a pre-determined fill quantity. Tablet inspection, if used, is usually provided by eye. Changeover can take up to a shift to achieve and is primarily a mechanical technician task. Market data suggests that purchases of this type of filling system are declining markedly and that electronically controlled vibratory fillers are now selling in increasing numbers. These newer technology fillers, while theoretically slower, have fewer, if any, tablet specific components, use electronic counting methods and incorporate some basic automatic inspection of tablet area. Product changeover can be achieved in perhaps 1–2 hours. The trend to these new types of machine indicates that the industry is beginning to recognize that equipment flexibility is more important than absolute speed.

Aside from filling, the other areas of inflexibility in packing is the production and control of printed material, most particularly labels. Off-line printing techniques are used and the resulting materials are handled and released using control methods not dissimilar to those needed for producing banknotes. Nevertheless, labelling errors still cause around 50% of product recalls, with significant costs both financially and to product/company image.

While many pharmaceutical companies recognize the limitations imposed by their packaging equipment, it has been an area of relatively slow technology change. There are two related causes for this. Much of the pharmaceutical packaging equipment is produced by companies who, with few exceptions, are small relative to their customers. Not unreasonably, the equipment companies do not have the financial or technical resources to undertake major new product development programmes involving radically different technology and tend to concentrate on enhancing their existing products. In contrast, the pharmaceutical companies have the size and financial resources to develop new equipment but traditionally have not sought to develop their own packaging equipment and have sat back awaiting new offerings, preferably from well known vendors. It is not difficult to see how these two effects can lead to technology stagnation.

A further restricting factor is the relationship in pharmaceutical companies between marketing and engineering. Again taking car manufacture as a comparison, the linkage between these two departments in pharmaceutical companies is relatively small. Marketing would not naturally look first to areas such as packaging engineering for significant new business opportunities. It is typical to find internal 'new production technology' groups with no formal marketing involvement or, indeed, 'new market development teams' with no engineering input. Innovative in-house process technology developments have, therefore, to be justified against relatively small efficiency gains in direct labour reduction and material usage, rather than the substantial returns associated with new business generation. The end result is that where internal process innovation is pursued, it is often under-funded, has a low commercial priority and lacks a clear business objective and focus.

6.8.3 What does the market want?

In the last ten years the distribution channels for pharmaceuticals in the United States have undergone some dramatic changes and continue to do so. Pressure from corporate health programmes, medical insurance providers and government to reduce healthcare costs has resulted in new purchasing and distribution routes emerging. A key example is the explosive growth of companies who manage the purchase of pharmaceuticals on behalf of health plan providers. These companies act on behalf of the healthcare provider and negotiate substantial volume discounts with the drug producers against a restricted list of recommended drugs. These companies handle patient prescriptions at centralized semi-automated facilities and the packaged drug is shipped direct by mail to the patient. The conventional manufacturer/wholesaler/pharmacy distribution route is completely bypassed. A substantial proportion of the US population now receives many of its prescription pharmaceuticals in this way. Other organizations, such as hospitals and nursing homes, are now pursuing similar methods to obtain price benefits through centralized pharmacies. Whilst these are primarily US phenomena today, it would be naïve to assume that similar developments will not appear in Europe in due course once the financial impact of these programmes become apparent to government-funded health services.

194

There are a number of other market-related issues, all of which mitigate against conventional drug packaging methods. These include:

- 'globalization' of production by companies such that a single site may now produce all country and pack variants of a drug, requiring multiple label/language formats in the same facility with frequent changeovers;
- the requirement of the large supermarket-based pharmacy chains to have product identification and expiry date incorporated in a label bar-code to allow automated stock control. Conventional label production methods do not handle this need easily. Many chains are seeking their own branding on the label in addition to, or instead of, the manufacturer's name;
- label and insert data change frequently in response to new drug indications and side effects. Obtaining pre-printed material can delay the launch of a new or revised product by several weeks;
- direct management of retail shelf space by the supplier.

In summary therefore the market is demanding:

- increasing pack complexity, variety and customization;
- order delivery in a day with no intermediate handling and inventory costs;
- frequent pack design changes;
- single pack unit batch quantities;
- lower end user pricing.

The implication is that a make to order strategy is needed rather than make to stock. It is apparent that better management of, or enhancements to, conventional drug packaging lines will not address these new market needs, and that radically different equipment will be needed whose technology origins may be from outside the pharmaceutical industry.

6.8.4 New technologies

Other industries had to address the responsiveness/flexibility issues many years ago in order to survive. These manufacturers have had to take the initiative in stimulating the development and implementation of new manufacturing process equipment. Many of the principles and technologies that have resulted from this are equally applicable to pharmaceutical packaging.

Technologies that are relevant to an advanced tablet packaging system include:

- **'robotic' equipment design:** Whilst not necessarily using anthropomorphic arms, the underlying technology of electronically controlled actuation can give rise to machines that can switch instantaneously, under computer control, between different tasks and make intelligent decisions at high speed. That these machines may be both slower and more expensive than their less flexible predecessors should be neither surprising nor a problem, when the bigger commercial issues described earlier are taken into account.

- **image processing:** Machine vision is increasingly used for identification and inspection functions. The exponential growth in cheap computing power means that complex inspection and counting functions can be implemented in practical systems.

- **product identification:** A wide range of identification methods is available which allows product to be located and tracked by remote methods. Radio Frequency (RF) tags are extensively used in car manufacture to locate and route cars and components through variable process paths. These feature a short-range (50 mm) radio receiver/transmitter, memory electronics (typically a few Kbyte), and a battery in a compact, low cost format. All relevant product/process option data can be written to these tags and a complete process history recorded. On completion of the process these tags are reset and returned to the process start. These 'active' tracking methods have benefits over passive techniques, such as bar-codes, because they eliminate much of the need for large centralized tracking computers.

- **real-time computer control:** The use of smart machines depends on direct high-speed computer control. Whilst computer control of chemical processes is well understood in pharmaceuticals, it is comparatively uncommon to find computers used in this way in secondary processes. In general, computers are used only for scheduling, supervisory machine control and paperwork generation. The uncertainty of computer validation only leads to further caution over using direct computer control.

- **on-line printing:** Printing technology has been revolutionized in the last decade as sophisticated, low cost, high quality equipment has appeared, mostly for the office market. It is perhaps ironic that a packaging manager probably has more sophisticated computer power and printing technology on the department secretary's desk than on the packaging lines. Developments in ink jet, laser and thermal printing allow single, unique, high quality images to be produced rapidly and on demand. Technology developments for other industries will soon allow near photographic image quality to be achieved at line speed. Real time generation of unique single labels is already a practical proposition in both monochrome and colour.

Much of the necessary technology for an advanced, high flexibility, tablet packaging line already existed. The challenge was to select and configure it in an appropriate way.

6.8.5 Postscript technology

In 1991 The Automation Partnership ('TAP') began collaboration on a number of developments of novel manufacturing processes with Merck and Co. TAP offered a skill set in robotics, machine vision and computer control, while Merck recognized the need to take the initiative in developing radically different, advanced secondary process technologies. A number of these projects were aimed at line changeover time reduction, particularly the areas of tablet filling and on-line label printing. These early projects resulted in prototype production equipment which demonstrated that much higher levels of flexibility could be achieved for small batch, single product packaging, under GMP. These were still aimed at make for stock production.

These separate developments led subsequently to a concept, which became known as 'Postscript', for customer-specific packaging of tablets. With this, a customer order, down to a single bottle of tablets (such as a prescription), could be received electronically, counted, inspected, packed, uniquely labelled and despatched within a few minutes. Ideally, there would be little direct manual involvement in the process and a very high degree of integrity would be guaranteed by the system design. In principle, the line concept could receive, pack and directly despatch small end user orders within a day, eliminating all or most of the conventional distribution chain, large intermediate product inventories and the need for complex scheduling/forecasting systems. In other words, it would be closely aligned to the new market needs discussed earlier.

Not surprisingly, the concept was received with a mixture of technical concerns and business interest. It was decided that Merck would jointly develop, construct and demonstrate a near full-scale pilot line which would include all the essential novel elements and allow the feasibility and practicality of the new process to be assessed. The key functions and technology are described below; however, the concept's modularity allows a range of alternative configuration and capacities to be created for other specific needs.

6.8.6 Pilot plant configuration and equipment

The pilot plant line uses a U-shaped configuration with a conventional process flow involving empty bottles entering at the line start then progressing through filling, capping, labelling, collation and packing into shippers at the end.

For the purpose of demonstration, the pilot line was configured to receive small (hypothetical) electronic orders from customers, such as individual retail pharmacies, for a combination of differing product types. In this first case, up to four different tablet or capsule types were packed on the line simultaneously (although by adding a further four-channel filler modules this could be easily expanded to sixteen products or beyond). Orders comprised typically 20 bottles for a single customer with unique labels on each bottle showing the product identification, manufacturer, tablet count and the retailer's address. The bottles were packed into an order shipper at the line end, together with a dispatch label and order manifest. The system was, however, equally capable of packing a single patient prescription.

The key elements of the line were as follows:

(a) 'Puck'

The line had about two hundred identical 'pucks', which were used to carry individual bottles through the system. The base of the puck contained a proprietary RF tag, which allowed all relevant details of the order to be carried through the process with the bottle. The fingers on the upper part of the puck located the bottle while still allowing it to rotate for labelling. Specific finger designs allowed differing bottle sizes to be processed.

(b) Puck Handling Station ('PHS')

Four PHS's were used on the line to provide tracking and routing. The data on the puck could be erased, written or read at the PHS and the puck plus bottle could then be sent in alternative directions or rejected if faulty.

(c) Flexible filler

The filler was a novel patented design that used a vibratory feed, conveyor belt, imaging system and diverter to feed, inspect, count and divert tablets to the bottle. The filler consisted of four separate identical channel modules, each of which processed one single tablet type. Each channel could process between 500 and 1000 tablets per minute (dependent on tablet/ capsule size) and every tablet was automatically inspected for size, shape and colour. Damaged or rogue tablets were automatically diverted out of the stream and eliminated from the count. Tablet count was verified by two independent systems and any count discrepancies resulted in bottle rejection. The tablet count in a given channel could be varied for each successive bottle.

(d) Labeller

The labelling station used a conventional labelling machine but with a customized high-speed thermal printer. A specific label was printed on blank feedstock, in response to the bottle's puck data, and then applied. The label could also be verified by on-line print quality and character verification systems. The label incorporated a unique bar-coded serial number, giving each bottle a unique identity.

(e) Collation system

The order collator used multiple tracks and gates to assemble complete order sets. The puck determined the order routing. On completion of the order, the set was released and the bottles transferred from the pucks to a tote and then to the shipper carton.

(f) Control system

The system used multiple networked PC's to provide machine control, system monitoring and order tracking. System set up and running was through a touch screen. The system software was developed and tested under a structured environment suitable for validation.

(g) Ancillary equipment

The line used a conventional capper, and standard equipment, such as cotton and desiccant inserters, could be easily added as additional stations. The pucks were transferred on normal slat conveyors. The neck of the bottle, irrespective of its size, was always in the same position relative to the puck base. An overhead conveyor returned the empty pucks back to the line start. The pucks were reloaded with empty bottles using conventional unscramble/centrifugal feeder mechanisms.

6.8.7 Packing flow

The process flow is as follows:

- pucks are loaded with empty bottles fed from bulk and then queued on the conveyor;
- the first PHS erases all previous data on the puck and verifies a bottle is present;
- the filler receives a common train of empty bottles/pucks which feed the four channels as required;
- the filler receives data on the next bottle's fill requirement from the controller and then inspects and counts the correct number of tablets into that bottle. The puck receives all the data specific to the bottle while filling is in progress. Any errors in filling (such as a count error) give rise to an error flag in the puck data;
- the second PHS verifies the data on the puck and rejects any misfilled bottles. If appropriate, routing to alternate parallel cappers could occur at this point (e.g., choice of regular or tamper proof formats);
- the capper applies the cap;
- the third PHS verifies cap placement and reads the relevant data from the puck for label printing;
- the on-line printer produces a correct sequential stream of labels, which are then applied by the labeller;

- the final PHS reads the unique bar-coded bottle serial number on the label and correlates this with the serial number held on the puck. This ensures that the label is always correctly assigned to the right bottle;
- the collator uses the puck data to assemble completed orders. Note that several orders are processed in parallel — consecutive bottles on the line do not necessarily belong to the same order;
- successful completion of an order is reported back to the line controller. Parallel new orders are continually being initiated automatically.

6.8.8 System features

Particular features are:

- each filling channel operates asynchronously, i.e. the tablet fill speed and bottle rate through each channel will be different and may be zero at times depending on the content of individual orders;
- depending on tablet count per bottle, the throughput limit for each channel is determined by either the 500–1000 tablet/minute rate or the 20 bottle/minute rate. For example, typical limits for a four channel filler module would be 80 bottles/minute at 30 tablets/bottle or 40 bottles/minutes at 100 tablets/bottle;
- capsules and tablets can be packed simultaneously using identical channel equipment;
- the channels are physically isolated from each other and contained, with vacuum extraction to reduce dust generation and prevent cross-contamination;
- the product contact parts in a channel can be replaced within about ten minutes without the use of tools. There are no tablet-specific parts;
- the system can 'learn' the size/shape/colour profile of a new tablet design in about two minutes;
- labels can be designed off-line using standard software and then electronically downloaded into the system;
- on completing a run, the line automatically empties itself of orders;
- the system generates a separate computer batch record for every bottle processed, giving unparalleled traceability;
- an order can be filled, labelled, packed and ready for despatch within five minutes of receipt.

Overall the pilot system has demonstrated all the specified functions and performance, and has shown that the concept is valid and achievable. It has been subjected to an extensive validation programme.

6.8.9 Future developments

TAP is exploiting the technology more widely and is currently evaluating various applications in pharmaceutical packaging and distribution that might use a rapid pack to order approach. These include:

- direct supply to retailers;
- mail order pharmacy;
- clinical trial packing;
- product repackaging;
- hospital supplies.

Each of these would use the same core technology but in different line configurations. TAP is also exploring the opportunities for a similar concept for blister pack products for the European market. On-line, on demand, printing of blister foil has already been demonstrated at a prototype level by TAP and similar systems are becoming available from other suppliers.

6.8.10 Conclusions

The Postscript system has demonstrated that the concept of automatically packing a batch quantity is both feasible and reliable for solid dosage forms in bottles. Changing to a true make to order strategy from make to stock methods is, therefore, becoming a viable proposition. Whilst the system has unique elements, many of the principles and technologies have been successfully transferred from related applications in other industries. Perhaps the most fundamental conclusion, however, is that pharmaceutical product packaging can change from what some perceive today as a non-value adding process, to being an important strategic manufacturing technique that generates significant new business opportunities.

Safety, health and environment (SHE)

7

JOHN GILLETT

7.1 Introduction

This chapter briefly explains how risks to safety, health and environment (SHE) are managed in the pharmaceutical industry and how effective process design can eliminate or control them. The principles and practice of 'Inherent SHE', systems thinking, risk assessment, and compliance with legislation, are explained for the benefit of process designers and pharmaceutical engineers. Since this topic is too large to cover in a single chapter (see Figure 7.1), a useful bibliography is provided at the end for further reading. Specific pharmaceutical industry hazards that can be controlled by suitable process design are also reviewed.

Effective process design is an essential requirement for controlling risks to safety, health and environment (SHE) in pharmaceutical production facilities. Process design that results in robust, inherently safe, healthy and environmentally friendly processes, simplifies the management of SHE through the complete life-cycle of a pharmaceutical facility.

Fortunately, the considerable process design knowledge about SHE gained in the petrochemical, fine chemical, nuclear and other industries can be adapted and applied effectively in the pharmaceutical industry. Although, the pharmaceutical industry was slow to apply this knowledge initially, it has since expanded its use from primary to secondary production and other areas.

7.2 SHE management

The over-riding impact on SHE management over the last decades has come from societal pressure and legislation. Several major industrial accidents generated public concern and led to stricter legislation. Single-issue pressure groups raised public awareness, particularly concerning the protection of the environment, which led again to stricter legislation. As a result, the emergent requirement of recent SHE legislation worldwide is for auditable risk management based on effective risk assessment.

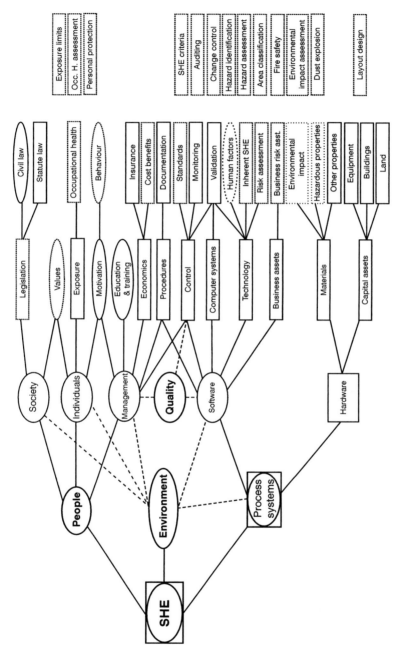

Figure 7.1 The safety, health and environment domain map

7.2.1 Integrated SHE management

Most pharmaceutical businesses adopt an integrated approach to managing SHE. In the past, safety, occupational health and environmental protection were usually managed as separate functions. The recognition that SHE was a line management responsibility that must be driven from the top to be effective converted the roles of SHE professionals from policemen to facilitators and enabled more effective use of SHE technical resources. It is well recognized that effective SHE management significantly reduces risks to product security and business as well as enhancing quality assurance.

As explained previously, SHE management has been driven by societal pressure and legislation to manage and assess risks effectively. However, the sheer urgency of business survival requires effective risk management — accidents cost money. Successful businesses give SHE management high priority from economic necessity. High quality and effective SHE management are also seen to go hand in hand. In successful enterprises, SHE is managed from the top to the bottom of the business organization with accountabilities and responsibilities clearly stated.

An effective SHE management system that is used in many successful businesses is shown diagrammatically in Figure 7.2.

The SHE management system described in Figure 7.2 consists of a cycle of activities with feedback to ensure continuous improvement of SHE

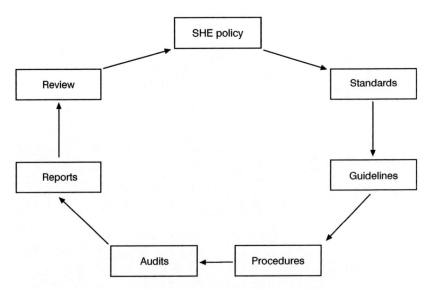

Figure 7.2 The safety, health and environment management cycle

performance. The cycle starts with a clearly stated SHE policy for all staff. This policy, together with more detailed SHE performance standards, is mandatory for all business areas. It is important to note that international business SHE standards must be written so that they can be applied to different cultures and legislative systems. The quality feedback loop is closed by compliance reports and SHE monitoring that provides the substance for a board level annual review of the SHE management system and performance achieved. In the example of Figure 7.2, the standards will define acceptable risk criteria and procedures for performing risk assessment in an effective and auditable manner.

This SHE management cycle is well suited to the pharmaceutical industry where similar quality assurance systems are well known and accepted. Most pharmaceutical businesses already have similar SHE management systems to that described. It is important that these systems include suitable hazard identification and risk assessment procedures and criteria so that SHE management is performed effectively.

7.2.2 Safety culture

Since the Industrial Revolution, attitudes to safety have changed considerably for the better. At the outset, injury and loss of human life were largely ignored in the drive for profit. However, several philanthropic industrialists and individual campaigners eventually persuaded the government of the day to pass legislation that required employers to provide reasonably safe working conditions for their employees and to record and report accidents.

The gradual improvement in industrial accident rates that followed was in four stages (see Figure 7.3a, page 206). The first stage was driven by legislation. During this stage, when there were numerous accidents, it was relatively easy to make simple improvements in procedures and protection to comply with the law. The second stage reduction in accident rates was driven by loss prevention and was largely due to improvements in process design and equipment based on quantitative risk assessment. The third stage was driven by effective SHE management and by recognizing the importance of human factors. During this stage, several major accidents due to poor management occurred and legislation became stricter. Some pharmaceutical businesses may still be at this stage of safety management, but others have already identified a fourth stage of improvement. The fourth stage improvement depends on the behaviour of the people in the business organization and a potent 'Safety Culture'. This is a topic that is outside the normal province of process designers, but must be borne in mind during risk assessments involving human factors.

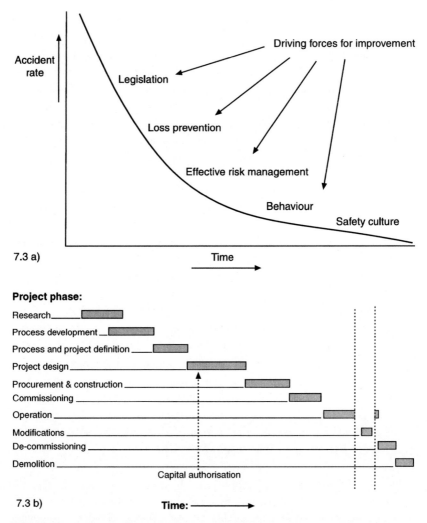

Figure 7.3 a) An accident rate reduction model. b) Life-cycle of a typical pharmaceutical product

7.2.3 Change control

Change is a natural phenomenon that occurs everywhere and is unavoidable. Change can be initiated deliberately to gain improvements or can occur unexpectedly. Whenever there is a change in a system, risks will be increased if there is no method of change control. Changes must, therefore, be controlled to eliminate or minimize risks.

206

There are two basic types of change. The most obvious type is change to hardware. Less obvious is the software change. Hardware or engineering changes are usually controlled on the basis of cost, although it is important to recognize that some inexpensive changes can, nevertheless, be very hazardous. Software changes are usually very easy to make and are often the most hazardous. (Software in this context includes not only computer software, but also procedural, organizational and people). It is extremely important that any system for managing change can identify whether risks are acceptable, regardless of the type or cost of the change.

7.2.4 Performance management
'You cannot manage what you cannot measure' is a well-known adage. Unfortunately, SHE performance is rather difficult to measure, particularly when it has been improved significantly. After the Industrial Revolution, the number of fatalities provided an easily recognizable and practical safety performance measure. As safety improved and fatalities became more rare, there were not enough to be able to determine trends easily, so major injuries were included to increase the event frequency. Eventually, as there were further safety improvements, minor accidents were included. The pharmaceutical industry has a good safety record, and even minor injuries are becoming too infrequent to be a reliable measurement of management control. Many organizations now record 'Near Miss' events as a more responsive performance measure. The measurement of SHE inputs such as training, auditing, documentation and human behaviour, are also used to provide more responsive and precise measures of performance.

7.3 Systems approach to SHE
'Systems thinking' is an extremely valuable tool in the pharmaceutical industry. This is because the industry involves a complex interplay between different people, organizations, cultures, processes, equipment, and materials. It is, thus, essential to consider the whole picture to take effective decisions. 'Systems thinking' must be at the heart of process design and management to control both SHE and business risks. The lateral thinking needed to obtain 'Inherent SHE' (discussed in Section 7.4) often stems from 'Systems thinking'.

7.3.1 Basic principles
'Systems thinking' or 'Holistic thinking' has been used widely by many disciplines to provide new and improved understanding of complex problems.

There are many definitions of the word 'System'. In the context of this book, a system is 'a whole' or 'a combination of many parts that work together towards a common goal'. The parts may be tangible or intangible, objective or abstract. Systems can be explained as a hierarchy. Every system exists inside a higher system called its environment. A system can also be divided into subsystems that can be similarly divided into sub-sub-systems. For example, an international pharmaceutical business will operate in many countries, and include research, development, commercial and manufacturing organizations. Each organization will have people, processes and equipment at different locations. At any one location there will be processes that contain equipment items. An equipment item will be made of several parts and each part will be made of several elements. 'Systems thinking' involves the whole system from the top of the business down to the last bolt connecting one of the equipment parts into the whole. Determining the correct balance between the depth of detail and the case of understanding a system is very important in process design and risk assessment.

7.3.2 System definition

It is not always possible to define a system with sufficient clarity to resolve a particular problem. This is usually because there is insufficient knowledge about the system elements or their interactions, or because the system is too complex to understand in its entirety. Systems that involve human activities are particularly difficult to model. Nevertheless, system models, even imprecise ones, can be constructed to improve understanding of the problem and, thus, guide improvements.

In general, the better the system definition, the easier it is to identify problems within the system. When systems definition is poor, problem solving depends on the investigative methods used to probe the system and a balance must always be struck between the effort spent on systems definition and that spent on system investigation. For example, hazard identification techniques need to be more powerful or time-consuming when studying ill-defined systems. This aspect of systems thinking is very important when performing risk assessments, as will be explained later.

7.3.3 Life-cycle considerations

Pharmaceutical manufacturing systems exist in time as well as in a complex and international environment. It is, thus, very important to consider the changes that could occur to such systems over their normal life-cycle. This is particularly true when performing risk assessments. A snap-shot in time may not identify hazards that could occur later.

A typical pharmaceutical manufacturing project life-cycle will last for several years and consist of at least ten distinct stages (see Figure 7.3(b) on page 206). The research stage precedes the development stage to determine the product and processes. A series of commercial and therapeutic assessments of the project feasibility leads to the process design stage. Engineering procurement and construction stages follow this, and then the commissioning and validation stages are completed prior to beneficial production. The life-cycle continues for several years, usually involving many modifications and system changes until the product or process becomes obsolete. The facility may then be decommissioned, and finally demolished. Each of these stages present different hazards that must be assessed at the project outset.

7.3.4 Business and commercial considerations

In the past, SHE was usually maintained as a separate function in many organizations. The realization that SHE had a significant impact on business performance arose from holistic approaches to business management. Insurance systems, quality systems and manufacturing systems interact with SHE in a complex manner and systems models have been used to indicate the SHE contribution. Such studies have resulted in considerable cross-fertilization of ideas and practices. Risk assessment is a particular activity that has been transformed from a basic engineering tool into a powerful business decision-making tool.

7.4 Inherent SHE

In practice, 'Inherent SHE' is the elimination of hazards by suitable process design so that processes are, by their very nature, safe, healthy, environmentally friendly, unaffected by change and stable. The more a process is 'Inherently safe', the less protective measures are needed, and the final result is then usually less expensive.

7.4.1 Basic principle

The basic principle of 'Inherent SHE' is to avoid hazards by suitable process design. Although the principle is simple it is, nevertheless, often overlooked, or used too late to implement. To apply the principle, it is essential to have sufficient time and flexibility to derive and assess the potential solutions that 'Inherent SHE' can suggest. This means that 'Inherent SHE' thinking must be started early in the project life-cycle. It is best employed during the research and development stages when fundamental opportunities for change are possible.

However, 'Inherent SHE' thinking needs to be continued throughout the project life-cycle, particularly when changes are being evaluated.

An ability to think holistically and laterally is very important when seeking an inherently safe solution to a problem. Several useful guide-words for 'Inherent SHE' are given in Table 7.1.

7.4.2 Inherent SHE examples in the pharmaceutical industry

'Inherent SHE' has been used effectively in the pharmaceutical industry both in primary and secondary production. Inventories have always been much smaller than those in the heavy chemical industry due to the relatively high activity and low volume of the compounds used. Cleanliness and aseptic or sterile operations have also driven pharmaceutical engineers to reduce capital and operating costs using 'Inherent SHE' principles.

In primary production, many of the crude production processes use hazardous chemicals. The production of hazardous chemicals such as phosgene in-situ is one example of inventory reduction. Other examples include the use of direct steam injection, direct nitrogen injection, 10 bar g milling, microwave

Table 7.1 'Inherent SHE' guidewords

Guideword	Principles	What to consider
ELIMINATE	Avoid using hazardous processes or materials	Process chemistry, heat transfer fluids, refrigerants, processing aids, location
SUBSTITUTE	Use less hazardous materials or processes	Process chemistry, processing aids, location
INTENSIFY	Reduce inventory, intensify or combine processes	Other unit operations or equipment, continuous rather than batch, faster reactions, hazard density
ATTENUATE	Dilute, reduce, simplify	Keep it simple. Moderate the operating conditions. Consider process dynamics: • high inertia hazards develop slowly • low inertia deviations can be connected quickly
SEPARATE	Separate chemicals from people and the environment	Containment. Layout. Drains. Services. Remote control robotics

drying, solutions rather than isolation as dusty powder, and spray drying to obtain free-flowing particles.

In secondary production, film coating was originally performed using flammable or environmentally unacceptable solvents. To overcome the problems that such solvents caused, aqueous coating processes were developed. To reduce operator exposure, multi-stage granulation processes to make fine active drugs free flowing for tabletting have been simplified, integrated, replaced by fluid-bed granulation, spray granulation, and occasionally by direct compression.

7.4.3 Inherent quality and product security

In the pharmaceutical industry, the principle of 'Inherent SHE' can also be applied to quality assurance and product security. This is particularly applicable to purification, formulation and packaging processes, discussed in the previous chapters. The aim is for robust processes that can be easily validated. All the guidewords described previously can be applied to achieve 'Inherent Quality'.

7.5 Risk assessment

The understanding of the word 'risk' varies considerably throughout society and has caused many communication problems. To avoid this problem, this chapter will use the Engineering Council (BS 4778) definition of risk as follows:

'RISK is the combination of the probability, or frequency of occurrence of a defined hazard and the magnitude of the consequences of the occurrence. It is, therefore, a measure of the likelihood of a specific undesired event and its unwanted consequences.'

Risk assessment is an essential activity in pharmaceutical process design and management. The risk assessment of therapeutic versus toxic effects of pharmaceuticals, research and development activities, clinical trials and business risks is not discussed here, although the same principles and methods can be applied.

Risk assessment is performed at several stages in the life-cycle and is exemplified by the 'six-stage hazard study' methodology that has been adapted and used in various different forms in the chemical and pharmaceutical industry (see Figure 7.4 on page 212).

The six-stage hazard study consists of Hazard Study 1 (HS1) to get the facts and define the system, Hazard Study 2 (HS2) to identify significant

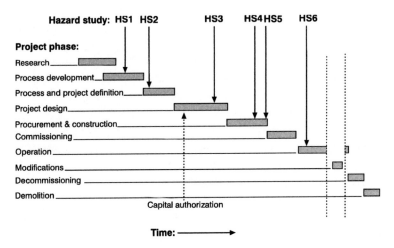

Figure 7.4 The six-stage hazard study methodology for a typical pharmaceutical product

hazards, Hazard Study 3 (HS3) to perform a hazard and operability study of the final design, Hazard Study 4 (HS4) and Hazard Study 5 (HS5) to check that the hazards identified have been controlled to acceptable standards, and Hazard Study 6 (HS6) to review the project and lessons learned. HS2 may be performed by several methods, including Preliminary Hazard Analysis (PHA). HS3 may also be performed in several ways, the most well known and powerful being Hazard and Operability Study (HAZOP) described later in Section 7.5.3.

7.5.1 Risk assessment principles and process

Risk assessment has been a human activity since men first walked on earth. People frequently perform risk assessment intuitively in their daily lives without realizing it. However, to present a logical and consistent approach to risk assessment, it is convenient to describe the risk assessment process as a series of separate activities. The risk assessment process is described in Figure 7.5 on page 213. The first activity is to perceive and define the system to be assessed. The second activity is to study the system to identify the hazards that it may contain. Each hazard identified is then studied further to estimate the consequences and likelihood of its occurrence. The combination of consequences and likelihood is then compared with a risk criterion to decide whether the risk is tolerable or not. These activities are described in more detail in the following sections.

212

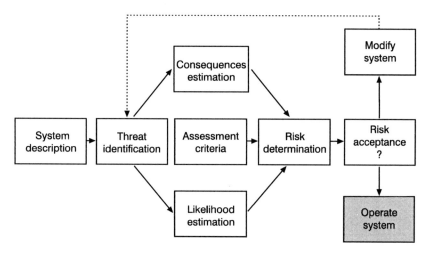

Figure 7.5 The risk assessment process

7.5.2 System definition

The first step in risk assessment is to define the system where the hazards exist. This step is crucial to the effectiveness of hazard identification. As explained previously, hazard identification in an ill-defined system will require more effort than in a well-defined system. It is, thus, important to try to model the system being assessed with as much detail and accuracy as possible.

In pharmaceutical manufacturing systems, it is important to define the software as well as the hardware. The software includes all the human systems, process and maintenance organization, controls, procedures, information, computer software and all the intangibles involved in manufacturing. The hardware consists of the tangible items involved in manufacturing such as the process materials, equipment, buildings, services and products.

It is advisable to start risk assessment by listing all the materials in the system to be studied. The materials' hazardous properties are then assessed, including their potentially hazardous interactions with each other. It is important to assess all the materials, including those that are used for services, cleaning, maintenance and activities supporting manufacture.

Having assessed the hazardous properties of the materials in the system, it is then possible to assess the manufacturing activities and production processes. Process flowsheets, piping and instrument drawings, engineering line drawings, activity diagrams, pictures, batch sheets, standard operating procedures and computer logic diagrams are typical pharmaceutical industry process system models that are used. The most powerful system models, however, often reside in the minds of the people who work within the system, so the selection of the risk assessment team is important.

7.5.3 Hazard identification

Effective hazard identification is best done by a carefully selected team of people and depends on two key factors — the accuracy of system definition and the method used to seek the hazards in the system. As explained previously, the better the system definition the easier it will be to identify the hazards within. A balance of effort must be struck between systems definition and hazard seeking. Hazards in a system that is defined completely and accurately in all its real or potential states may be obvious to the trained observer, but unfortunately this eventuality is rare. Since system definition in sufficient detail may not be possible, it is then essential to use hazard identification methods of increasing power, to generate deviations and ideas from the available system model and identify the hazards.

There are many hazard identification methods available to suit all types of system and system definition. In the pharmaceutical industry, the most used hazard identification methods are check-lists, 'What If?', Preliminary Hazard Assessment (PHA) and Hazard and Operability Study (HAZOP). These are briefly described in the following paragraphs.

Checklists

Checklists require little explanation as they are widely used as reminders in daily life for shopping, travel and household chores. The problem is that if an item is not listed, it will not be thought about! Checklists should be constructed and tested by the people with the most experience and knowledge of the systems that they are to cover. Regular revision of checklists is essential to maintain their effectiveness, although this often leads to the lists becoming longer and longer. Checklists are most powerful when used creatively to stimulate the imagination and raise questions. A slavish, mechanical application of ticks to a long checklist will rarely produce very effective hazard identification but can be combined with 'What if?' to overcome this problem.

Checklists are often used to identify hazards in plant modifications, proprietary equipment or laboratory activities.

'What if?'

'What if?' is a hazard identification method that uses the knowledge and experience of people familiar with the system to ask searching questions about its design and functions. Effective 'What if?' requires an experienced leader, since it is a brainstorming method and, therefore, not tightly structured.

When dealing with a large system, 'What if?' is best tackled by subdividing the system beforehand into specific subsystems. The study team performs a step-by-step examination of the best available system model from input to final output. Team members are encouraged to raise potential problems and concerns as they think of them. For each step, a scribe lists problems and concerns on a flip chart or notepad. These are then grouped into specific issues. Each issue is then considered by asking questions that begin with the words 'What if?' For example, 'What if the wrong material is added?' 'What if the next step is omitted?' and 'What if it gets too hot?'

The questions and answers are recorded and then sorted into specific areas for further study. 'What if?' is usually run in short sessions of about an hour per subsystem with a team of two or three people. Although the results of 'What if?' can be severely limited by insufficient team knowledge and experience, this method and its many variations have been used with apparent success for many years. There are now several computer software packages commercially available for assisting and recording 'What if?' studies.

'What if?' is often used at the research and development or feasibility study stages of the product life-cycle. It is also used for identifying hazards in plant modifications, proprietary equipment and laboratory or pilot plant activities.

Preliminary hazard assessment

Preliminary hazard assessment (PHA) was specifically developed to identify significant hazards during process development and feasibility studies. PHA is a variation of the checklist method that is enhanced by the creativity and judgment of a team of experts along the lines of a 'What if?' A list of specific subsystems is examined against a list of specific hazards to identify likely causes, consequences and preventive measures. Each hazard or hazardous situation identified is ranked in order of criticality to allocate priority for safety improvements. PHA is not a very searching hazard identification method, but is very useful for obtaining a structured overview of the hazards before resorting to more sophisticated and time-consuming methods later. PHA is a 'top down' method as it usually identifies the top events, such as loss of containment, which can then be investigated further down the chain of events until the prime causes are identified. It is a useful precursor to HAZOP.

Hazard and operability study of continuous processes (HAZOP)

HAZOP is one of the most powerful hazard identification methods available and has been well described in the literature. The imagination of a selected team is used to perturb a model of the system being studied by using a methodical process to identify potential accidents. The system is studied one element at a time, and is a 'top down' method. The design intention of each element is defined and then questioned using 'guide words' to produce deviations from the intention. The causes, consequences, and safeguards for each deviation are then discussed and recorded. Any hazards that require further action or information are listed for follow-up later.

HAZOP was originally developed for large-scale continuous petrochemical processes, but has been adapted and applied successfully to pharmaceutical batch processes. HAZOP of batch systems can be very time-consuming and requires an experienced hazard study leader to be completed effectively. The procedure for HAZOP of a continuous process is well described and many people have been trained in its use. Since the procedure for continuous systems is simpler than that for batch systems, it is described first (see Figure 7.6):

- study the system model and sub-divide it into its key elements (Nodes). If a Piping and Instrument Drawing (P&ID) is used as the model, look at the arrangement of the lines and decide how to divide the drawing into study areas;
- identify each element to be studied (Node) with a reference number. If a P&ID is used, number all the junctions that define the elements (Nodes) to be studied;
- select an element (Node) for study;
- state the design intention of the element (Node). This is an important step in the method and must be done carefully and precisely. The design intention

Obtain a Piping and Instrument drawing (P&ID) of the system

1. Study the system P&ID and subdivide it into **nodes** (discrete parts)
2. Identify each node with a **reference number**
3. Select a **node** for study
4. State the **design intention** of the node
5. Select a **parameter** in the design intention for study
6. Apply the first **guideword** to the parameter
7. **Identify** all **deviations** that could occur with causes, consequences and controls
8. **Record** all deviations that require **corrective action**
9. Allocate **responsibility** for completing the corrective actions
10. Apply the next guideword. Repeat 7–9 until all guidewords have been applied
11. Select the next parameter
12. Repeat steps 6–11 until all relevant parameters have been studied
13. **Mark** the node on the system P&ID to show it has been studied
14. Select the next node and repeat steps 4–13
15. Continue this process until all of the system has been studied

Figure 7.6 HAZOP of a continuous process

defines the processes or activities involved in the element and the boundary for examination. The intention will include details of the process parameters that can be changed in the element. Typical parameters stated in the intention are flow, temperature, pressure, level and time;

- select a parameter for study;
- apply the guidewords to the intention relating to the parameter selected and identify any deviations from the intent. The guidewords are listed with brief examples of typical deviations in Table 7.2;
- for each deviation identified, study the causes, the effects and the safeguards provided;
- decide whether the deviation requires a design change or corrective action;
- record the decision and allocate the action to a team member for completion by an agreed review date.

When using a computerized recording package, all the deviations are recorded and it is also possible to risk rank each deviation. This is useful for subsequent auditing of the study and for generating a project risk profile. When the study is recorded manually, it has been common practice to record only the actioned deviations, but this makes auditing difficult. It is recommended that all deviations studied be noted with suitable comments to explain actions taken or reasons for acceptance. A typical HAZOP Proforma for recording the study is shown in Figure 7.7 on page 218.

- once all the guidewords have been applied to the parameter selected, select the next parameter;
- repeat steps 6 to 10 for the second parameter;
- repeat steps 5 to 11 until all the parameters have been studied for the selected system element. Mark the element (Node) studied on the model (or drawing) with a crayon or highlighter to indicate that it has been studied;

Table 7.2 Hazard and operability study guidewords

Guideword	Example of a typical deviation
NO (NOT or NONE)	No flow in pipe. No reactant in vessel
MORE OF	Higher temperature. Higher level
LESS OF	Lower velocity. Lower bulk density
MORE THAN (or AS WELL AS)	Two phase flow. Contamination
LESS THAN (or PART OF)	Reduced concentration. Missing component
REVERSE (the complete opposite of the intent)	Valve closes instead of opening. Heat rather than cool
OTHER THAN (a different intent)	Non-routine operations maintenance, cleaning, sampling
SOONER/LATER THAN	More/less time. Operation out of sequence

Hazard Study 3: Report Form	Project:			Session:	Drawings:				Sheet......of......
HSL:............ Team:...........									Date
Node:	Parameter:	Intention:							
Guideword	Deviation	Causes of Deviation	Consequences	Safeguards	Actions to be taken	Ref. No.	By	Remarks	Date Completed

Figure 7.7 Hazard and operability study report form

- select the next element (Node) for study and repeat steps 4 to 12;
- continue this process until all the system elements (Nodes) have been studied;
- record all actions and file all associated documents in the project SHE dossier;
- the Hazard Study Leader (HSL) then reviews the study overall to prioritize the hazards identified. Depending on this overview, the HSL may then perform further studies such as a CHAZOP of the computer systems, or a Failure Modes and Effects Analysis (FMEA) of critical items;
- the project manager plans HAZOP action review meetings to ensure that the actions are implemented satisfactorily. The HSL appends remarks to the HAZOP report to check whether further hazard study of the changes made is required at these reviews.

HAZOP procedure for batch processes

Batch processes are more difficult to define and study than continuous processes because they are time-dependent, flexible, subject to changes of product and process and frequently involve multiple-use equipment. A batch process element can exist in any one of several different states depending on the batch process sequence. At a given time, a batch process element is either active or inactive. An active or inactive batch process element can also exist in several different conditions. An active element can be waiting for a previous batch step to complete, or for a subsequent step to be prepared. Active elements are also subject to sampling, inspection, batch changeover and other activities that are

218

governed by external factors. An inactive element may be undergoing cleaning, maintenance, product changeover or merely waiting for the next planned production campaign.

Another factor that complicates batch processes is human intervention. Most batch processes have stages that are controlled manually. Human reliability assessment of key operations may sometimes be essential to maintain quality and production efficiency. The use of computer control may alleviate some of the human reliability problems, but then generates additional complexity of a different nature. A hazard study of batch process computer systems will be required as an additional exercise.

The hazard study of batch processes is very demanding. The hazard study team needs to work very intensely and creatively to link all the diverse elements of the batch system together without missing interactions or deviations. It is always very difficult at the end of a hazard study to be absolutely sure that all the hazards in a batch process have been identified.

Effective HAZOP of a batch process depends on the HSL and the study team. HSLs experienced in the hazard study of batch processes all adopt similar approaches to the HAZOP methodology, but each will have different ways of running a particular study. There is no right or wrong way of doing HAZOP on a batch process. The method used must be tailored to suit the study. The following approach may be helpful:

- The team members discuss the batch system in general terms to get an overview. They use the available documents and drawings to get a clear understanding of the key problem areas and to agree on the level of detail required for the study.
- The team identify the main sub-systems in order to plan the study. A maximum of six or seven is a practical guide. These can then be sub-divided to provide the full detail when each is studied individually. There may be some duplication and overlaps, but this should not be a cause for concern. It is useful to identify a single key element to anchor the attention of the hazard study team. For example this might be a reactor with several sub-systems such as a heating/cooling system, a charging system, a services supply system, an effluent system, and so on.
- The team then construct an activity diagram for the batch process. This step ensures that the team understand all the batch process sequences and activities. Alternatively the team may decide to use the operating instructions for the same purpose.

- At this point in the study the HSL has to decide on the level of detail. The level of detail will be decided by the preliminary discussions, the results of PHA and the complexity of the process. It is worthwhile to perform a first-pass hazard study to identify specific areas for deeper study later. A useful first-pass hazard study method is as follows:
  - Select the first activity on the activity diagram, or the first step in the operating sequence.
  - State the intention of the activity. This must identify the materials, equipment, process parameters, and controls. The connections and inter-actions with the total system including the operator and operating sequence must also be identified by reference to engineering line drawings, the batch sheet and the operating procedures.
  - Apply the HAZOP guidewords to the activity selected. For the first-pass study, these are applied to the activity transformation verb, object and subject alone. For example, apply the guidewords to 'Fill vessel'; 'Dry the batch'; 'Load clean ampoules'; 'React A with B'; 'Operator starts pump'; 'Computer regulates flow', etc. Use the guidewords in the widest sense to generate deviations from the intention. The stated intention relates the causes and effects to the drawings and procedures. Several of the deviations generated at the start will be re-generated many times over when applying guidewords to activities later in the study. The first activity studied always generates the most deviations, and, as the study of other activities proceeds, fewer new deviations are generated, as most will have been identified already.
  - For each guideword, the HSL controls the discussion and recording of causes, consequences and safeguards for each deviation to suit the creativity and enthusiasm of the team. When ideas are flowing freely it is best to record only the deviations and their causes. The effects, safeguards and actions can then be discussed when the idea flow ebbs. The discussion of the effects and safeguards will then usually set the ideas flowing again, and so on.
  - Repeat the above steps for the rest of the activities on the activity diagram.
  - Once all the activities have been studied, make a final overview of the whole system. It is useful to use the PHA checklist for this purpose, particularly to identify any conditions that could have an effect on the whole system.
  - The team decide whether to study any activities or equipment items in more detail using the detailed HAZOP batch process method described as follows.

The detailed hazard study examines every step of the batch process sequence. For each step, each item of equipment used is studied element-by-element for each equipment state ('Active', 'Inactive', and any other state in which it may exist). The parameters for each equipment state are then studied using the guidewords. A simplified logic diagram of the process is shown in Figure 7.8.

To perform a study of the whole batch process as thoroughly as this would be excessively time-consuming, so it is important to restrict this degree of detail to the process steps that have been identified from the first-pass study. The Pareto principle that about 80% of the risk lies in 20% of the system can be used as a guide to deciding what to include. The decisions on how to perform HAZOP of a batch process will be governed by the experienced judgment of the HSL.

7.5.4 Consequences estimation

A single hazardous event may have many consequences, some of which may develop over a significant time period. The final outcomes are, thus, difficult to predict with confidence. The Sandoz warehouse fire is a good example of this phenomenon. A fire started in a warehouse containing chemicals that were potential pollutants. The fire developed extremely rapidly and the local population was alerted to close windows and stay indoors to avoid breathing the resultant heavy and foul-smelling smoke. The firemen applied large volumes of water to control the fire as foam alone proved ineffective. The

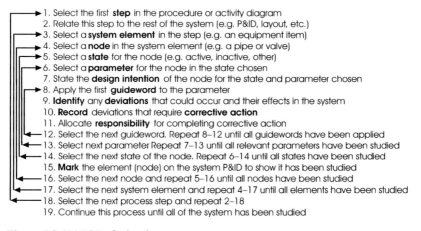

Obtain system operating procedure or activity diagram and all relevant drawings

1. Select the first **step** in the procedure or activity diagram
2. Relate this step to the rest of the system (e.g. P&ID, layout, etc.)
3. Select a **system element** in the step (e.g. an equipment item)
4. Select a **node** in the system element (e.g. a pipe or valve)
5. Select a **state** for the node (e.g. active, inactive, other)
6. Select a **parameter** for the node in the state chosen
7. State the **design intention** of the node for the state and parameter chosen
8. Apply the first **guideword** to the parameter
9. **Identify** any **deviations** that could occur and their effects in the system
10. **Record** deviations that require **corrective action**
11. Allocate **responsibility** for completing corrective action
12. Select the next guideword. Repeat 8-12 until all guidewords have been applied
13. Select next parameter Repeat 7-13 until all relevant parameters have been studied
14. Select the next state of the node. Repeat 6-14 until all states have been studied
15. **Mark** the element (node) on the system P&ID to show it has been studied
16. Select the next node and repeat 5-16 until all nodes have been studied
17. Select the next system element and repeat 4-17 until all elements have been studied
18. Select the next process step and repeat 2-18
19. Continue this process until all of the system has been studied

Figure 7.8 HAZOP of a batch process

firewater dissolved the stored chemicals and eventually flowed off the site and into the nearby Rhine. The Rhine was polluted and suffered severe ecological damage over a length of 250 km. The reparation and litigation costs were enormous. As a result of this incident, legislation was passed to ensure that all warehouses containing potential pollutants were provided with firewater containment to reduce the likelihood of such an event happening again.

The overall consequences of a hazardous event evolve over time in a chain of events triggered by the first event. Although the cause of the event may be determined, the consequences are probabilistic. A typical chain is initiated by an event that causes a loss of containment of energy or hazardous material. Depending on the size of the leak, the efflux will then act as a source for further dispersion in the local atmosphere. The resultant explosion, toxic cloud, fire or combinations of all three may then affect the local population, depending on the weather conditions at the time and the local population distribution. A useful method for evaluating potential outcomes of a hazardous occurrence is to draw an event tree. An example of the event tree for a solvent leak inside a building is shown in Figure 7.9.

The potential consequences arising from many major industrial hazards have been modelled along such chains of events to estimate the effects quantitatively. There are, thus, a great many methods and tools available for estimating the potential consequences of hazardous events that have been developed in the heavy chemical and nuclear industries.

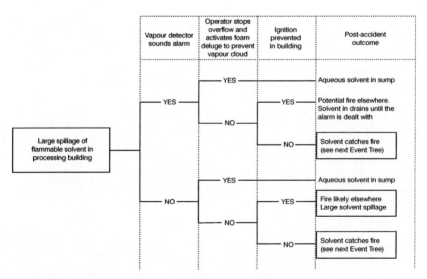

Figure 7.9 Event tree for a solvent leak inside a building

In the pharmaceutical industry, where the inventories of hazardous materials and energy are usually much less than those categorized as major hazards, the immediate consequences of fire, explosion and toxic releases are potentially less severe than in the heavier industries. Nevertheless, the available consequence models can still be used. In addition, there are many pharmaceutical chemicals and intermediates that can present environmental hazards as great as those from the major hazards industries. The consequences of these hazards are best estimated by the models developed and proved for the heavier industries.

Since most pharmaceutical processes are performed inside buildings, even small leaks can generate enclosed flammable atmospheres, which can explode with potentially serious consequences. Suitable models are not yet available for such indoor situations so expert technical advice will usually be required to estimate the consequences of indoor situations. The knock-on effects on adjacent facilities must also be considered.

It is important not to under-estimate the ultimate consequences of fire and explosion in the pharmaceuticals industry. The very high value of pharmaceutical materials, laboratories and markets can cause potentially very large consequential losses in the event of a fire. The chain of consequences that can result is usually quite different from those experienced in the heavy chemical industries as the effects on markets are often greater than on people. The consequential business loss of a pharmaceutical business can be several orders of magnitude higher than that of the low margin high volume industries.

The consequences of hazardous events in the pharmaceutical manufacturing industry can usually be estimated to the nearest order of magnitude by experienced judgment to make a preliminary estimate of severity. The preliminary estimate can then be used to decide whether to use the more powerful consequence models.

The simplest approach to consequence estimation is to consider the 'Worst Case' that can be imagined for each hazardous event identified. The extent of the worst case and the events that must occur to contribute to it can then be determined. Ideas for other scenarios can then be developed by brainstorming around the 'Worst Case'. It is also useful to consider a 'typical' consequence of lower severity as another reference point in the scale of potential consequences. As there are usually several possible outcomes, an event tree approach may be helpful to explore the possibilities, otherwise experienced judgment and risk ranking can be used to select the possible outcomes for the final risk assessment.

When estimating the consequences in this way, it is practical to consider the effect of each identified hazardous event on five key targets:

- people;
- the environment;
- process plant, equipment and buildings;
- the product;
- the business.

By considering separately the potential effects on people, society, the environment, material assets, the product and the business, the severity of the consequences can be estimated fairly consistently. Various yardsticks such as the number of injuries, fatalities, emissions, fires, explosions, or nominal costs in monetary terms can be used to build up a reasonably accurate and quantitative estimate of the overall consequences.

The severity of the consequences can then be ranked in a simple scale of consequences using verbal descriptions such as '*Very Severe*', '*Severe*', '*Moderate*', '*Slight*' and '*Very Slight*' in decreasing order of overall loss to fit a risk ranking matrix, described in Figure 7.12 (see page 232). The consequences ranked as '*Very Severe*' and '*Severe*' may then require quantified risk assessment using more sophisticated models depending on the likelihood of occurrence.

7.5.5 Likelihood estimation

Having identified all the hazardous situations and their consequences, the next step in the risk assessment process is to estimate the likelihood of occurrence. This is very difficult to do consistently without using a logical method and some form of quantification because people are notoriously unreliable at estimating the likelihood of hazardous events. Any human judgments must be explained and recorded so that they can be justified on a logical basis.

The likelihood of occurrence is usually expressed as a frequency (events/unit time) or as a probability (a dimensionless number between 0 and 1). In some situations the likelihood may be expressed as a probability over a specified time interval and for a particular event or individual. Probability theory and the various probability distributions and methods used for reliability estimation are described fully elsewhere and are not covered in this guide.

There are essentially two ways to estimate the likelihood of a hazardous event. The first and most reliable way is to use historical data that matches the event as exactly as possible. The second way is to calculate the likelihood from generic data or from relevant data obtained locally using mathematical models. It is important not to use 'off-the-cuff' opinions to estimate likelihood since these will invariably be misleading.

Estimating the likelihood of hazardous events from historical data

Historical data should always be carefully checked to ensure that it fits the event being studied as closely as possible. Very old data may not be representative of current conditions. The accuracy of the data and the conditions under which it was obtained must also be carefully checked and validated. If possible, confidence limits for the data should be derived using suitable statistical methods.

The stage in the life-cycle of equipment can also affect the validity of the data collected. Typical equipment failure rates follow a 'bath tub' curve through the equipment life-cycle shown in Figure 7.10. The curve predicts high failure rates at start-up, which decrease steeply during the early life, then level out to a constant failure rate for the main life, eventually increasing linearly in the final wear-out stages.

Sparse data should be analyzed using statistical methods to estimate the expected mean and deviation. The negative exponential probability distribution and the Poisson distribution have been used successfully for system or component failure rate estimation in the pharmaceutical industry.

Historical data that matches the event exactly is often very difficult to obtain. This is a particular problem for the events of interest to the pharmaceutical industry. Although there are many databanks containing data of major hazards incidents, fires, explosions, toxic gas releases, etc., there is currently little data that has been derived from the pharmaceutical industry.

The problem of using data that is not exactly applicable when no other data is available is best resolved by adopting a conservative (high) value for the

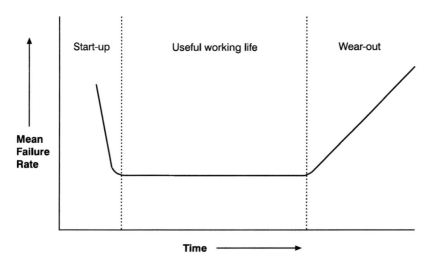

Figure 7.10 'Bath-tub' curve for equipment failure rate

initial likelihood calculation. Once a conservative estimate has been obtained, lower values can then be inserted to assess the sensitivity of the estimate to the data. In many cases, particularly in pharmaceutical manufacturing processes, the equipment data may not have such an impact on the estimate as the human error estimates.

Estimating the likelihood of hazardous events using mathematical models

There are many mathematical modelling methods available for estimating the likelihood of occurrence of hazardous events. Some of the methods suited to the pharmaceutical industry are listed in Table 7.3 and explained briefly in the following paragraphs.

Order of magnitude frequency ranking

A preliminary estimate of likelihood is always useful in deciding whether to use the more time-consuming techniques available. Order-of-magnitude frequency ranking is one of the most effective methods for this purpose. The method uses a combination of verbal and quantitative data to define a frequency band for the event studied. A range of five frequencies can be used as a guideline, stepping up in orders of magnitude to fit the five-by-five risk ranking matrix described later in Figure 7.12 (see page 232). For example, the lowest frequency would typically be one event per ten thousand years (1/10,000 yrs). The highest would then be once a year (1/year) with inter-mediate steps of 1/1000 yrs, 1/100 yrs, and 1/10 yrs.

These could then be described in increasing frequency as '*Very Unlikely*', '*Unlikely*', '*Average*'; '*Likely*'; and '*Very Likely*'. Finer or coarser frequency bands can be used to suit individual system requirements.

Using these broad frequency bands for risk ranking still requires a realistic estimate of the frequency for each identified hazardous event. Realistic, if very approximate, frequency estimates can be based on local records and knowledge or on generic data from the sources previously mentioned.

Table 7.3 Likelihood modelling examples

Basis of modelling method	Description of method
Real events and statistics	Constant failure rates
Expert judgment	Order-of magnitude frequency ranking
Logical algorithms	Fault tree analysis, human reliability analysis
Simulation	Monte Carlo method

Fault tree analysis

A fault tree is a logically constructed diagram used to model the way that combinations of failures cause the event of interest (the top event) to occur. The construction of a fault tree provides valuable insights into the way that hazardous events interact even if no data is inserted for calculations. However, the main use of fault trees is to calculate hazardous event frequencies or probabilities.

The logical arrangement of the 'And' and 'Or' gates of the fault tree is more critical to the overall calculation of the likelihood of the top event than the accuracy of the data inserted. If the logic is incorrect or key elements are omitted, the results will be misleading. It is important to have an independent check of the fault tree logic before accepting the results.

It is advisable to keep the logic as simple as possible. A rule of thumb is that if there are more than twenty elements in the tree then subdivision is worthwhile. In the pharmaceutical industry, if a problem requires a fault tree more complex than this, then a way of avoiding the problem altogether by changing the system is usually sought (Inherent SHE). If a better system cannot be identified and the fault tree cannot be simplified, then experienced safety and reliability engineers should be consulted.

Human reliability estimation

Pharmaceutical production processes rely heavily on human operators in nearly all aspects, ranging from direct intervention in process operation to business decision-making. This can cause problems when attempting to quantify risks accurately as human factors are hard to define precisely.

Although it is relatively straightforward to estimate equipment reliability consistently, human reliability estimation, in spite of many years of research, is still something of an art. It is important to realize that, when estimating the likelihood of a hazardous event, the probability of beneficial action by an operator should not be a critical factor to achieve the target criterion. There should always be adequate protection in place to ensure that the operator action is not critical to the safe operation of the system.

Human tasks can be classified as 'Skill based', 'Rule based' or 'Knowledge based'. Skill based tasks that depend on physical skill and manual dexterity are fairly well understood and can be estimated with some confidence. Tasks where rules or procedures are important are not so well understood. Some guidance is available for formulating clear instructions, but ensuring compliance with rules is governed by human behaviour. It is difficult to estimate the effectiveness of training and management on behaviour. Knowledge based tasks that depend on the knowledge and mental models of the operator cannot be modelled with any confidence at present.

The most effective approach is to make a preliminary estimate of the effects of human reliability to help decide whether a more detailed analysis is warranted. For the best possible circumstances, when an operator is not stressed by the situation or his local environment, is well trained and healthy, a failure probability of 1 in 1000 (0.001) may be assumed. For the worst possible circumstances when the operator is highly stressed, in poor health, in a noisy and uncomfortable environment, and is not trained, it is almost certain that failure will occur (probability of failure 1.0). Values of failure probability 0.1 and 0.01 can be selected between these two extremes to fit the local conditions. For most activities by well-trained staff in the clean and comfortable environments in the pharmaceutical industry, a human failure probability of 0.01 may be assumed as a first estimate. For primary production areas, where the environment is less comfortable and the processes more difficult to operate, a probability of 0.1 may be assumed.

If a more rigorous treatment is indicated then there are several techniques that can be used in consultation with human factors specialists. The 'Technique for Human Error Rate Prediction' (THERP) considers the task in separate stages linked by a fault tree and estimates the probability of failure for each stage. The probabilities are calculated from the likelihood of detection, the chance of recovery or correction, the consequences of failure if it is not corrected, and the 'Performance Shaping Factors' (PSF) governing the task. THERP requires considerable time and specialist expertise to derive the best estimates of human failure probability. Task analysis can be used when a particular task is critical to the business, and the preliminary estimate indicates that more precision is required. Task analysis must be performed by an expert practitioner to be effective and can prove very costly and time consuming.

Monte Carlo method

The Monte Carlo method uses numerical simulation to generate an estimate of event probabilities for complex systems. Although the method is very powerful, it can be very time-consuming if the system failure rate is low. Fortunately there are several computer software packages available to ease this burden and the method has become widely used throughout the industry.

7.5.6 Risk assessment criteria

Risk acceptability criteria govern the management of SHE, quality and business performance. If the criteria are set too high, the costs become exorbitant, but if set too low, the consequential losses become excessive. Risk criteria must be set to give the correct balance between the cost of prevention and protection and the cost of a potential loss. Since obtaining this balance is hampered by uncertainty, risk criteria definition is usually an iterative process with frequent reviews and adjustments. In the pharmaceutical industry, risk acceptability criteria are usually expressed qualitatively to comply with legislation, codes of practice or approved standards. The use of quantitative criteria is still evolving in the industry to meet the requirements of tighter budgets and stricter legislation.

Acceptability

A particular problem that is often encountered is how to decide whether risk criteria are acceptable. Acceptable to whom? Risk acceptability criteria can only be acceptable to the people who will be affected. Sometimes, when the benefits seem to outweigh the perceived risk, people will tolerate a risk until it can be made acceptable. In the pharmaceutical industry, risk acceptability criteria are dominated by product security and quality as these govern the potential consequences to the people who use the industry products. The risks from pharmaceutical manufacturing operations, however, are subject to the same acceptability criteria as the rest of industry. Risks must be managed in such a way that they are tolerable to employees and to the general public.

Risk acceptability criteria range and precision

The range of risk acceptability criteria is very large. Many people seek 'Zero Risk' at the unattainable bottom end of the range. The concept of 'Zero Risk' is often mentioned when the potential consequences of a particular risk are extremely severe yet extremely unlikely. There are some risks that could harm future generations to such an extent that society would never agree to take them. This is the basis of the 'precautionary principle', which is often quoted to stop particular risks from being taken.

There are many practical and achievable risk criteria that society will accept. The industrial regulators have used upper and lower boundaries of risk with risks in between these levels controlled to be 'As low as reasonably practicable' (ALARP). The ALARP principle has been widely and effectively interpreted over many years in the law courts as a practical criterion of risk acceptability.

Recent environmental legislation uses the phrase 'Best available technology not entailing excessive cost' (BATNEEC) in a similar manner. There are many other qualitative definitions of risk acceptability criteria such as these. Unfortunately, qualitative risk criteria, which are not very precise, may be interpreted in many different ways. Comparative risk criteria such as 'Better than' or 'Not worse than' some clearly specified example, are more precise and simpler to interpret.

Approved codes of practice and standards set by bodies such as the American Society of Mechanical Engineers (ASME) and the British Standards Institution (BSI) provide another way of defining risk acceptability criteria. The relevant ASME or BS codes can be specified for particular systems to define an acceptable level of safety assurance. For example, a specified requirement that a pressure vessel is designed to BS 5500 or ASME VIII; Div. 1 is a well-known criterion of acceptability.

Simple risk acceptability criteria

A simple and very useful method for setting risk acceptability criteria, which is easy to explain and apply within the pharmaceutical industry, is 'risk ranking'. Risk ranking is based on the intuitive idea that the events with the worst consequences should have the least chance of occurrence to have an acceptable risk.

By plotting consequence severity against event likelihood, a borderline of acceptability may be drawn between areas of acceptable and unacceptable risks as shown in Figure 7.11 (see page 231). This principle was first described and used in the nuclear power industry. If the curve is represented as a matrix, semi-quantitative risk ranking becomes possible as shown in Figure 7.12 (see page 232). A range of consequence severities is designated along the vertical axis and a range of likelihoods along the horizontal axis. The number of sub-divisions on each axis can be decided to suit individual requirements for precision. A three by three matrix is often used for coarse screening risks, but a five by five matrix is more discriminating. The risk of a specific hazardous event can then be located in the matrix by its severity and likelihood coordinates.

Each square in the matrix is allocated a number to represent the level of risk. The convention used is that the higher the number in the matrix, the higher the risk. For a five by five matrix as shown in Figure 7.12 (see page 232), the top right-hand square is numbered 9 and the bottom left-hand square numbered 1. A diagonal band of 5s might then be defined across the matrix to discriminate between 'Acceptable' and 'Unacceptable' risks. Hazardous events with coordinates above the diagonal band are unacceptable, while events with co-ordinates below the band are judged acceptable. Events with co-ordinates in the diagonal band need further study, as this is an area of uncertainty where the apparent clarity of the method should not be allowed to cloud experienced

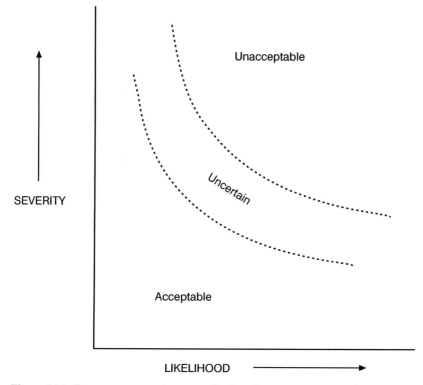

Figure 7.11 Consequence severity versus likelihood curve

judgment. Risk ranking is only a coarse filter of the unacceptable risks from the trivial.

The Risk Ranking Matrix, thus, provides a coarse risk acceptability criterion that can be tailored to suit particular situations. The allocation of the numbers can be skewed to make the criterion as strict or as lenient as required. For example the 5s could be classed as unacceptable. Alternatively different numbers could be placed in the matrix. To reduce the amount of judgmental bias on likelihood, guide frequencies can also be provided along the horizontal axis.

7.5.7 Quantitative risk assessment

The most well defined risk criteria for process design and management are quantitative. Even so, absolute values for risk acceptability criteria are often difficult to justify because quantitative risk assessment (QRA) is not a precise tool and usually involves idealized assumptions and the use of unvalidated data. In addition, QRA calculations, although logical and mathematically exact, often depend on human judgment. This usually means that QRA is mostly used

**Severity of
Consequences**

	Very low	Low	Normal	High	Very high
Very severe	5	6	7	8	9
Severe	4	5	6	7	8
Moderate	3	4	5	6	7
Slight	2	3	4	5	6
Very slight	1	2	3	4	5

Likelihood ⟶

	Very low	Low	Normal	High	Very high
Guide frequency:	1/100,000 yr.	1/10,000 yr.	1/1,000 yr.	1/100 yr.	1/10 yr.

Figure 7.12 Risk ranking matrix

for comparisons or for sensitivity analysis. (Sensitivity analysis is the process of testing the effects of different values of the data or assumptions made on the predictions from QRA models). Sensitivity studies are important for checking QRA models and for pinpointing key risk areas for improvement. The main advantage of QRA is that it enables the final risk decisions to be explained logically and quantitatively against quantified risk acceptability criteria.

Acceptability criteria for risks to people and the environment from fire, explosion, toxic gases and pollution have been developed and agreed in many industrial areas. Some of the most widely used quantitative risk acceptability criteria in the chemical industry are those for fatalities, but there has been considerable debate about using them for regulation because the risks to the public attract much controversy.

The resultant data, experience and techniques give useful guidance for setting risk criteria for potential fatalities or pollution in the pharmaceutical industry. Risk acceptability criteria for product quality and business risks are still under development and are the subject of considerable debate.

Risks to the public

When the problem of controlling major industrial hazards was first being studied, the Advisory Committee on Major Hazards suggested that a 'serious accident' frequency of once in 10,000 years might just be regarded as the borderline of acceptability. This frequency was subsequently used as a basis for arguments about the acceptability of major risks from process plant in many countries. The estimated effects on process personnel and the public from such accidents was also used as a guide to the acceptability of risks to individuals.

One practical acceptability criterion often used is that the risk to a member of the public from a major industrial accident should not be significantly worse than that from the pre-existing natural risks. Using this principle and an analysis of natural fatality statistics, this equates, on average, to a chance death of less than one in a million (1.0×10^{-6}) per year per person exposed. Recent legislation in the Netherlands uses 1.0×10^{-6} per person per year as the maximum tolerable risk for new major hazard plants. For a specific industrial hazard that could kill a member of the public, a target value of 1.0×10^{-7} per person per year has been suggested.

Although it is difficult to agree quantitative risk acceptability criteria, it is necessary to do so in order to be able to do QRA. On this basis, it is suggested that the risk acceptability criterion for pharmaceutical industry manufacturing plant accidents that could cause public fatalities should be less than 1×10^{-6} per person per year shown in Table 7.4.

Risks to process operators

Quantitative risk acceptability criteria based on event frequencies have been widely used for ranking process risks in order of priority for action. A criterion that has often been used for assessing process hazards is that the risk of death for a plant operator should not exceed the risk of death for a fit adult staying at

Table 7.4 Guidelines for QRA in the pharmaceutical industry

Hazardous event	Risk acceptability guideline
Public fatality from a specific plant hazard	$<0.1 \times 10^{-6}$ per person per year
Public fatality from all process hazards	$<1.0 \times 10^{-6}$ per person per year
Process operator fatality from a specific plant hazard	$<7.0 \times 10^{-6}$ per person per year
Process operator fatality from all process hazards	$<35.0 \times 10^{-6}$ per person per year

home. On this basis, the chemical industry for many years has aimed that the risk of death from all process hazards should have a probability of occurrence of less than 35.0×10^{-6} per year per person exposed. It was considered that the risk of death from a specific process hazard should be a fifth of the total and targeted at 7.0×10^{-6} per person per year.

It has also been suggested that the risks to the public should be an order of magnitude less than that for process personnel. This suggestion, taken with the public risk guideline described previously, implies that the risks to plant operators should be less than 1×10^{-6} per person per year. This is of the same order of magnitude as the criterion derived by the chemical industry. Risk criteria for process operators in the pharmaceutical industry can be developed on a similar basis (see Table 7.4 on page 233).

7.5.8 Risk assessment and validation

Risk assessment by hazard study and process validation have had different histories during their evolution (see Figure 7.13). During the last decade, however, the two methodologies have drawn closer together in the pharmaceutical industry so that they overlap in several areas. Figure 7.14 shows these

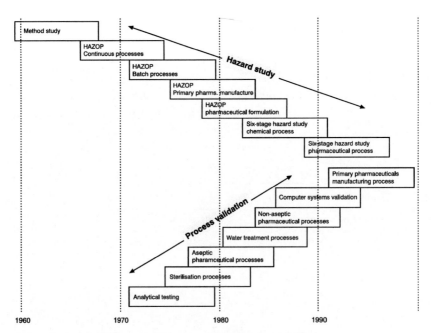

Figure 7.13 A brief history of hazard study and process validation

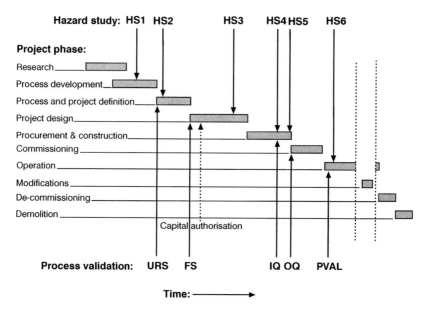

Figure 7.14 The six-stage hazard study methodology and process validation for a typical pharmaceutical product

areas of overlap diagrammatically. The diagram represents a six-stage hazard study applied to a typical pharmaceutical project life-cycle with the associated validation activities included.

As mentioned earlier in this chapter, the six-stage hazard study consists of Hazard Study 1 (HS1) to get the facts, Hazard Study 2 (HS2) to identify significant hazards, Hazard Study 3 (HS3/HAZOP) to perform a hazard and operability study of the final design, Hazard Study 4 (HS4) and Hazard Study 5 (HS5) to check that the hazards identified have been controlled to acceptable standards, and Hazard Study 6 (HS6) to review the project and lessons learned.

Although Chapter 4 provided a full explanation of validation, it is useful to re-state the activities that overlap with the six-stage hazard study process. Process validation starts with the preparation of a User Requirements Specification (URS) followed by a Functional Specification (FS) for engineering design and procurement. Installation Qualification (IQ) and Operation Qualification (OQ) are performed to prove that the URS and FS have been met prior to the final process qualification or process validation.

A quantitative analysis of several hazard studies showed that about 50% of the hazards identified by HAZOP were related to quality and validation issues. The use of the existing guidewords, thus, appeared to be effective from the quality viewpoint. It was further improved by having validation experts in the hazard study teams. Unfortunately, any quality hazards identified as late as HS3 by HAZOP could be costly in time and effort to prevent or protect against. The most important thing to do is to increase the emphasis on quality earlier in the life-cycle at HS1 and 2.

The hazard study of computers has always been difficult to perform with complete confidence that all the main hazards could be identified. The lack of confidence is due to the complexity and volume of the interactions between the hardware and the software. It is impossible to analyze all the computer codes in a reasonable time-scale, in even the simplest systems. Computer Hazard and Operability Study or CHAZOP was developed in an attempt to identify the significant hazards with reasonable confidence. CHAZOP has been successfully used with computer applications data flow and logic diagrams treating the computer operating systems and watchdogs as 'Black Boxes'. CHAZOP and similar techniques are still being improved to provide more confidence that the significant hazards can be identified.

As explained in Chapter 4, the validation of computer and critical automated systems has advanced considerably over the last few years, building on the work of systems analysts, CHAZOP and process validation methods. Computer validation has concentrated on a life-cycle approach, building quality into computer systems from their conception. Computer validation is currently the most effective means of ensuring that computer systems hazards are controlled acceptably.

The synergy between hazard study and computer validation in the pharmaceutical industry is now well established. Hazard study and computer validation operate together and share techniques and information produced by the function that is the most effective.

7.6 Pharmaceutical industry SHE hazards

The pharmaceutical industry has similar SHE hazards to those of the chemical industry, but to different degrees of severity. Chemical reaction, fire, explosion, toxic, environmental, occupational health, mechanical energy and radiation hazards are well described in the literature together with methods of assessing and controlling them. The chapters on primary and secondary production, process utilities and services, laboratory design, and process development and pilot plants also cover these hazards where relevant. This chapter will only briefly consider the particular aspects of these hazards that apply to the pharmaceutical industry. The hazards arising in specific pharmaceutical processes, which are not encountered elsewhere, will also be discussed briefly.

7.6.1 Chemical reaction hazards

Chemical reaction hazards assessment

As explained in Chapter 5, the primary production processes to produce active drugs involve a wide variety of complex reactions and reaction sequences. Many of these reactions may be exothermic or may evolve gases at high rates, and could cause reactor over-pressure. It is, thus, essential to establish the basis for safe operation in the laboratory before scaling up such reactions. It is good practice to perform a methodical assessment (described by Barton and Rogers in the bibliography) summarized as follows:

- define the process chemistry and operating conditions and the process equipment to be used;
- evaluate the chemical reaction hazards of the process, including potential maloperation;
- select and specify safety measures;
- implement and maintain the selected safety measures.

There are many published procedures for evaluating chemical reaction hazards. Whatever procedure is used, it is essential that tests are performed and interpreted by qualified people. This is because there are many factors that may affect the test data such as sample size, container material, heating rate, thermal inertia and endothermic effects.

Control of runaway reactions

Runaway reactions are thermally unstable reactions where the heat of reaction can raise the temperature of the reactants sufficiently to accelerate the reaction rate out of control. The temperature at which the runaway starts is often termed the onset temperature. Such reactions are normally controlled by cooling the reactor, or by controlling the addition of the reactants. Loss of reactor cooling or agitation during the course of an exothermic reaction are two of the commonest causes of runaway reactions. A runaway reaction can cause the reactor contents to boil, generate vapour or explode, and over-pressurize the reactor.

There are several protective measures that can be used to mitigate the effects of a runaway reaction. The most common protection is emergency venting, but containment, crash cooling, drown-out and reaction inhibition provide other options.

Reactor venting

Reactor over-pressurization can occur by overcharging with compressed gases or liquids, by excessive vapour generation due to overheating, or by a runaway reaction. Such events are normally avoided by adopting suitable operating

237

procedures and control systems. When control is lost, the most effective way to prevent damage to the reactor is to relieve the pressure through an emergency relief system. The design of reactor pressure relief systems is well described in the literature and will not be explained here. However, some key questions to ask are as follows:

- what is the maximum pressure that the vessel can contain?
- what pressure will activate the relief system?
- will the relieved material be a liquid, a vapour or a two-phase mixture?
- what is the maximum expected relief rate to avoid over-pressurization?
- is the area of the relief device sufficient to handle the maximum expected relief rate?
- is the pressure drop in the relief system low enough to prevent over-pressurization during venting?
- will the relief device survive in normal reactor operations (for example, bursting disk under vacuum)?
- will the relief device re-seal after depressurization?
- is the material ejected from the reactor toxic or environmentally harmful?
- does the relief system exhaust to atmosphere in a safe place?

7.6.2 Fire and explosion hazards

In the pharmaceutical industry, fire and explosion hazards arise most frequently when handling flammable solvents or finely divided organic powders. Flammable materials or mixtures are frequently used for the reactions such as hydrogenation, nitration, Grignard reaction, and oxidation in primary production processes. Occasionally chemical intermediates or by-products in primary production processes may be pyrophoric or explosive. Flammable solvents and finely divided solids are also encountered in purification and secondary production processes. It is, thus, essential to obtain information about the fire and explosion properties of all materials that occur in the manufacturing processes in order to establish a basis for safe operation.

Material fire and explosion properties

All materials used must be tested for fire and explosion properties. In the pharmaceutical industry it is very important to test dusts and finely divided powder, as almost all of these can form explosive mixtures with air. The test methods and procedures are well described in the literature and will not be described here. It is essential to obtain specialist advice to interpret the test results to achieve a safe process design, although the key parameters that influence safe process design are as follows:

- gases and vapours:
  - lower explosive limit in air;
  - upper explosive limit in air;
  - critical oxygen content;
  - density;
  - minimum ignition energy;
  - auto-ignition temperature;
  - minimum flame diameter.
- flammable and highly flammable liquids:
  - flash point;
  - boiling point;
  - lower explosive limit in air;
  - upper explosive limit in air;
  - auto-ignition temperature;
  - vapour density.
- finely divided powders and dusts:
  - dust classification;
  - maximum dust explosion pressure;
  - critical oxygen content;
  - St rating (maximum rate of pressure rise during explosion);
  - minimum ignition energy;
  - train firing.

Area classification of plants handling flammable gases and liquids

The handling of flammable gases in the pharmaceutical industry is usually restricted to hydrogenation processes and to fuel gases supplied for process utilities and services. The inventories are usually small and leaks can be well controlled, so that the probability of an uncontained gas cloud explosion in the open air is very low. The main hazards occur inside buildings, where even small leaks of flammable gas can form explosive mixtures in air. Risk management of flammable gases in buildings relies on leak prevention, containment, ventilation, and control of ignition sources.

The inventories of flammable liquids in pharmaceutical processes can often be substantial, so fire and vapour cloud explosions are significant hazards. These hazards are exacerbated inside buildings, particularly when solvents are handled at temperatures above their flash point. Risk management relies on similar controls to those used for flammable gases with the additional possibility of vapour knock-down and foam systems to control leaks or spillages.

239

The hazards of handling flammable gases and liquids in plant areas are identified and risks assessed by a team of suitably qualified people to provide suitable controls. This activity is called Area Classification (British Standard 5345) and is performed as follows:

- list all flammable and combustible materials used in the area to be studied, with quantities;
- obtain all relevant fire and explosion properties for the materials listed;
- obtain an engineering drawing of the area to be studied and identify and list the possible sources of flammable atmospheres;
- study the area using the 'Source of Hazard' method described in BS 5345;
- estimate the extent of the following zones around each source using standard procedures:
  - zone 0: A zone in which a flammable atmosphere is continuously present for long periods;
  - zone 1: A zone in which a flammable atmosphere is likely to occur in normal operation;
  - zone 2: A zone in which a flammable atmosphere is not likely to occur in normal operation and, if it occurs, will only exist for a short time;
  - non-hazardous: A zone in which a flammable atmosphere is not likely to occur at all.
- record the decisions on an Area Classification drawing;
- decide the review frequency.

Dust explosion hazards

It is worth re-iterating that most finely divided powders handled in pharmaceutical production processes can form explosive mixtures in air. Dust explosion properties are determined in specialized laboratories by qualified staff that use approved test equipment and procedures. The tests and their interpretation are well described in the literature, but are beyond the scope of this chapter. However, a few rules-of-thumb may be useful for preliminary process design and risk assessments as follows:

- most organic materials with a particle size less than 75 microns will form explosive mixtures in air;
- the lower explosive limit in air for most organic dust clouds is between 15–60 gm m^{-3} depending on the temperature but independent of ignition energy. (*These are very dense clouds that would obscure a 100 W light at about two metres*);
- the upper explosive limit is generally very high at 2–6 kg m^{-3}. Most dust explosions will generate a final contained pressure that is about ten times the

start pressure. (*This means that atmospheric pressure systems designed to withstand 10 Barg should contain a typical dust explosion*);
- most explosive dusts can be inerted by limiting the atmospheric oxygen concentration to less than 8% v/v;
- the most severe consequences arise from secondary dust explosions that are caused by the ignition of very large dust clouds generated by the primary explosion dislodging dust held on ledges, etc. in the vicinity.

There are several methods of protection against dust explosion hazards. The first step is to eliminate potential ignition sources. The possibility for incendive electrostatic sparks must be removed by adequate earthing of metal conductors and electrostatic charges. The next step is to provide protection against dust explosion. The most well known methods are explosion venting, inerting, suppression and containment. The protection most frequently used for dryers, storage vessels and conveying systems is to vent the explosion to the atmosphere via rupture disks or panels. Venting must be to a safe place and must not cause environmental hazards. Inerting is often used when venting to a safe area is not possible or if the vented material can cause environmental hazards. Containment is frequently used for milling and dust separation processes where the equipment can be made to withstand the dust explosion pressure with reasonable economy. Suppression can present quality problems and is usually only used for systems where there are hybrid mixtures of dusts and flammable vapours or gases.

7.6.3 Occupational health hazards

Occupational health hazards arise in the workplace when uncontrolled harmful substances or conditions exist that can adversely affect the health of the workers. The exposure of staff to external hazards from the environment and from their life outside work is also important as it can affect their response to exposure at work. This chapter will only consider the effects of workplace hazards.

To achieve good occupational health in the workplace, hazard identification, risk assessment and the selection of suitable controls against hazardous exposure are essential. Engineering and procedural controls must also take account of the additional controls provided by occupational hygiene. For example, in certain circumstances, it may be necessary to monitor workplace emissions or to provide health surveillance of the operating staff.

Occupational exposure limits

Toxicologists, epidemiologists, physicians, occupational hygienists and research workers provide the essential information for defining the Occupational Exposure Limits (OELs) that are used to define and maintain healthy working conditions. The information for setting these criteria is either obtained by direct experiment or by modelling data from experiments performed in similar systems. The complexity of some of these issues is outside the scope of this brief review.

Occupational health risks arise from operator exposure to materials and physical conditions that occur in the working environment. The materials can be chemicals, biologically active substances or ancillary materials used in the workplace. Exposure to these materials can affect the health of the person exposed by inhalation, skin contact and absorption, or ingestion. The immediate effects are termed acute effects. If exposure is over a long period of time and the effects persist, these are termed chronic effects.

(a) Materials

The OELs for materials that cause chronic effects are usually based on an 8-hour time weighted average exposure. Highly active materials are allocated shorter times such as the 15-minute time weighted average exposure. Some materials may be allocated both long and short-term exposure limits.

The dose-response relationship for a toxic substance is the relationship between the concentration at the site of ingress and the intensity of the effect on the recipient. It is difficult to interpret dose-response relationships for a particular individual, so the assessment of occupational health risks from toxic materials requires considerable knowledge and experience.

Pharmaceutical research and development of biologically active compounds generates occupational health hazards for which the exposure limits are often unknown. New compounds are thus tested for toxic effects as well as therapeutic efficacy as a key part of the research and development programme. In the early research and development stages it is essential to assess substances for occupational health risks even though reliable data may not be available. This is done by defining in-house OELs on the basis of experience and available models assuming that there is a threshold below which there are no adverse effects.

These in-house OELs or preliminary standards may then be altered to match the experimental data obtained as research progresses. From the process design viewpoint, the in-house exposure limits are used as the best information available, but it is important to record the fact in the process documentation. Subsequent changes to OELs will require a re-examination of those system elements that are affected.

A particular problem encountered in pharmaceutical research involving animals or biotechnology is allergic reactions. Allergy depends very much on the individual exposed. Susceptible individuals may respond to minute amounts of allergen that are too small to measure. In these cases it is impossible to define a reliable OEL for control purposes because a threshold cannot be determined. In these circumstances, it is normal to work to approved codes of practice for known allergens, to provide personal protection, and to perform health surveillance of operators exposed.

Great care is needed to interpret occupational health data. As a simple example, the OEL for a nuisance dust is often loosely quoted as $10\,\mathrm{mg\,m}^{-3}$; 8 hr TWA (Time Weighted Average). However, this value is strictly for total inhalable dust concentration: the OEL for the respirable fraction is $5\,\mathrm{mg\,m}^{-3}$; 8 hr TWA. Table 7.5 provides some idea of the range of OELs that may be encountered in the pharmaceutical industry for inhaled substances. These simple examples are only intended to be used for discussing process design issues with occupational health practitioners and are not provided as standards.

(b) Physical conditions

The assessment of the effects of physical conditions such as temperature, humidity, noise, vibration, and electromagnetic radiation is more straightforward than for materials because they have been well researched and the dosage and effects can be monitored more reliably. Physical effects that are not dose-related such as the stresses and strains arising from manual operations are more difficult to assess. Back problems, repetitive strain injury and eye strain are usually controlled by ergonomic workplace and equipment design backed up by education and training based on the findings of medical research and ergonomics. Recent legislation requires that such risks should be assessed at the design stage of new manual systems.

Most of the physical hazards that can occur in the workplace can be controlled by following recognized codes of practice to control dose-related exposure. The number and change rate of physical hazards is much less than for

Table 7.5 A typical range of occupational exposure limits encountered in the pharmaceutical industry

Description of inhaled substance	Range of occupational exposure limits	Typical example
'Nuisance dusts'	$1\text{--}10\,\mathrm{mg\,m}^{-3}$	Starch dust
Toxic substances	$0.1\text{--}1\,\mathrm{mg\,m}^{-3}$	Solvents, Common chemicals
Highly toxic substances	$0.01\text{--}0.1\,\mathrm{mg\,m}^{-3}$	Cytotoxins
Extremely toxic substances	$<0.01\,\mathrm{mg\,m}^{-3}$	Carcinogens

chemical and biological hazards which makes physical hazards relatively simpler to study. The main physical hazards to consider are heat, humidity, air quality, noise, vibration, ionizing radiation, non-ionizing radiation, and electricity. Typical occupational health criteria for physical hazards are given in Table 7.6. These values are solely for discussion purposes with the relevant experts. A qualified occupational hygienist should always decide the relevant criteria for a pharmaceutical project.

Occupational health legislation

The regulations governing occupational health management now established throughout the western world all require risk assessment of occupational health

Table 7.6 Typical physical hazard occupational health criteria

Workplace physical hazard	Typical occupational health criteria	Comments
Temperature	30.0 deg Centigrade (Wet bulb)	Continuous light work
	26.7	Continuous moderate work
	25.0	Continuous heavy work
Humidity	40%–60% R.H.	Guidance for comfort only
Air change rate	>10 changes of air/hour	Rule-of-thumb guide only
Noise	≥90 dB(A) ($L_{EP,d}$)	Ear protection required at or above this level for 8 hr TWA exposure
Vibration	Magnitude: $2.8\,m\,s^{-2}$ rms Frequency: *Whole body*: 0.5–4.0 Hz. *Hand — arm*: 8–1000 Hz.	8 hr TWA level for taking preventive action
Non-ionizing radiation	<50 mW/cm^2 @ 5 cms	Microwaves (2450 MHz)
	<10 mW/cm^2 in workplace Depends on laser classification	Microwaves Laser light
Ionizing radiation	50 mSv (5 rem/year)	Total exposure to radiation (ICRP) for workers whole body
	5 mSv (0.5 rem/year)	Any other person; whole body

hazards. In the UK, the Control Of Substances Hazardous to Health Regulations 1994 (COSHH) requires the employer to assess the workplace risks from handling substances hazardous to health, to identify any control or personal protection measures needed, to maintain these measures and where necessary monitor workplace exposure and/or provide health surveillance. COSHH also requires the employer to provide information, instruction and training about the hazards, the risks and the controls required and also to keep auditable records.

Legislation will often define specific occupational exposure limits for substances or physical conditions that are known to present health risks. The limits for toxic substances under the COSHH legislation, for example, are expressed as Maximum Exposure Limits (MELs) and Occupational Exposure Standards (OESs). MELs are allocated to substances such as carcinogens that have known serious health effects but for which no threshold of effect can be identified. OESs are allocated to substances that could cause serious health effects above a specific and clearly definable threshold exposure.

Occupational health systems description

Occupational health hazard identification and risk assessment can only be performed effectively with a clearly defined system model. The minimum requirement is for a simple process block diagram and a brief description of the activities that can give rise to occupational health hazards. A list of process operations and operator tasks is essential to determine the extent of exposure. The list can be used to prepare an activity diagram of the operator actions and movements suitable for hazard study. The activity diagram information can then be used to plot operator movements on the workplace layout drawing. The model can be improved considerably by indicating the harmful emissions on the same drawing to identify the interactions between the operator, process and emissions.

Occupational health controls

Occupational health hazards are identified by a team of knowledgeable people studying the system model and activity diagram. It is helpful to include an occupational hygienist in the team to interpret the applicable exposure limits and advise on the best controls for emissions that cannot be eliminated. Typical controls are based on containment, ventilated enclosures, local exhaust ventilation, dilution ventilation, personal protection or combinations of these main types. If possible, personal protection should be avoided as it hampers operator activities and is costly to implement and maintain.

245

Occupational health impact assessment

For a typical pharmaceutical project, it is important to write a formal 'occupational health impact statement' that describes the occupational health hazards identified and the principles of the control regime needed to comply with legislation and in-house standards. In the six-stage hazard study methodology this is done as part of hazard studies 1 and 2. The activities necessary to complete this assessment are as follows:

- identify the occupational health hazards present and list them. For chemical and biological materials identify the amounts used in the process and other hazards that they may present (Materials Hazard Checklist);
- obtain the Material Safety Data Sheets (MSDS) for each hazardous substance identified. If a MSDS is not available, consult an occupational health specialist for guidance, particularly if there is no information about OELs or hazard categories for specific materials;
- for each hazard, identify the potential routes of entry into the bodies of the operators or staff exposed to the hazards;
- state the control principles to be used to meet the OELs or other occupational health criteria for each hazard. The control principles for maintenance, cleaning activities, emergencies and abnormal operation are particularly important;
- specify the control measures that will be used and state the test and maintenance procedures to ensure that they remain effective. The exact details may not be known at the early stages, so the aim here is to provide engineering guidance;
- state whether health surveillance or exposure monitoring will be required;
- specify any personal protective equipment (PPE) that may be required;
- state whether any specific training will be necessary for hazard awareness, use of PPE, etc.;
- define the actions and responsibilities for further occupational health assessments that may be required, such as COSHH assessments that will be needed during construction, commissioning and start-up;
- record all the findings and necessary actions in a formal report.

7.6.4 Environmental hazards

The protection of the environment is a major concern of modern society, but opinions about the best way forward vary considerably. In the context of the environmental risks to a pharmaceutical project, the whole life-cycle must be assessed as far into the future as can be reasonably predicted. The following paragraphs provide a brief overview of environmental risk assessment and current environmental legislation.

Environmental hazards in the pharmaceutical industry

In the pharmaceutical industry the main environmental hazards associated with routine operations are solvent emissions to air and emissions to the aquatic environment. Releases due to loss of containment in an accident or during a fire or other emergency can also cause pollution of the aquatic and ground environments.

(a) Routine solvent emissions to air

The pharmaceutical industry emits relatively small amounts of volatile organic compounds (VOCs) but is, nevertheless, under pressure to reduce existing releases. The abatement of routine batch process releases at source is difficult as VOC emissions are usually of short duration and high concentration. The best available technology not entailing excessive costs (BATNEEC) for such emissions is usually 'end of pipe abatement' technologies such as adsorption, absorption, condensation, etc. Unfortunately such measures increasingly require the use of manifolds and catchpots that can cause additional problems from cross-contamination of the product or fire and explosion hazards.

The prevention of cross-contamination is a particular problem in purification and formulation processes where systems to remove solvent vapours are needed to protect the environment. In such systems, the containment of potentially explosive atmospheres may generate an explosion hazard that will require additional protection measures. One solution to this problem is to use inert atmospheres to minimize the explosion risks, but this then adds the risk of asphyxiation of operators and will require suitable controls in enclosed areas.

(b) Routine emissions to the aquatic environment

Aqueous discharges from pharmaceutical processes are usually collected and pretreated to reduce the environmental impact before release off-site. The relatively small volumes involved rarely make biological treatment on-site economical and so this is usually performed at the local sewage works. Solvent discharges are recovered if possible either on-site or off-site. If recovery is not possible it may be possible to use waste solvents as a fuel source during incineration.

It is important to be able to monitor routine discharges to drain from processes that involve polluting chemicals. Process drains should not be buried and should have suitable access for regular inspection. Surface water and process drains should be segregated and studied to identify any potential interconnections during storms or emergencies. Any bunds, catchment basins or effluent pits should be leak proof and regularly checked for integrity to prevent accidental leakages.

(c) Loss of containment

Emergency relief discharges of volatile materials or dusts can contaminate both the aquatic and ground environments. This is a major concern in primary production as the chemicals and intermediates used to prepare crude bulk drugs are all potential pollutants and some may be severe pollutants. The release of such chemicals to atmosphere as a result of a runaway reaction or major spillage, for example, could be potentially damaging to the environment. Catchment or 'dump' systems to collect any emergency emissions may be essential to comply with legislation. Unfortunately, if manifolds or inter-connections are used for this purpose they may cause explosion, over-pressure, or fire hazards that must be controlled by additional protective measures.

Solids handling and particulates can cause risk to the environment at all stages of pharmaceutical production. As previously explained, most of the dry solids handled in pharmaceutical processes can cause a dust explosion hazard. Dust explosions can be contained in equipment designed to withstand >10 Bar g, pressure, and vented, inerted, or suppressed in weaker equipment. If dust explosion venting is used, it may cause serious pollution and more costly alternatives of containment and suppression will be needed to protect the environment. The cost of cleaning up soil contamination from emergency releases of biologically active dusts or solids can be prohibitive.

A large fire on a primary production process or warehouse can lead to environmental pollution. Apart from the environmental damage arising from smoke and soot, fire-fighting water containing dissolved chemicals can cause pollution of local watercourses and damage to water treatment works. Firewater retention systems may be needed to prevent the contamination of local water-courses or ground waters. Fortunately, formulated products present fewer pollution problems as they are usually hermetically contained in small quantities.

(d) Early identification of environmental hazards

The environmental, safety and health risks must always be considered together rather than individually, as there is considerable interaction between them. Environmental protection is usually very costly, so it is important to attempt to avoid environmental hazards by eliminating them at the outset. Since pharmaceutical processes are usually registered before a capital project is started, it is thus important to consider environmental hazards at the research and development stages. At the very least, researchers should perform a rudimentary 'What If?' or Preliminary Hazard Analysis (PHA) to assess chemical routes or process alternatives for environmental hazards.

Environmental legislation

In Europe the Directive 85/337/EEC 'The assessment of the effects of certain public and private projects on the environment' came into effect in 1988. The Directive requires an environmental impact assessment for all projects that could have significant environmental impact before consent to proceed is given. It has been incorporated into the legislation throughout the European Union, and in the UK by The Environmental Protection Act 1990 that is now implemented by the Environment Agency. Established under the Environment Act 1995, the Environment Agency took over the functions of Her Majesty's Inspectorate of Pollution, the National Rivers Authority, Waste Regulatory Authorities, and some parts of the Department of the Environment (internet web-site: *http://www.environment-agency.gov.uk*).

The UK Environmental Protection Act 1990 requires that certain prescribed processes may only be operated with an authorization. The Act defines two systems of pollution control, Integrated Pollution Control (IPC) and Local Authority Air Pollution Control (LAAPC). The Environment Agency regulates IPC and also authorizes prescribed processes. The local authorities and metropolitan boroughs enforce and authorize LAAPC, which covers air pollution only. The local authorities also administer the Town and Country Planning (Assessment of Environmental Effects) Regulations 1988 for which there is a guide to performing environmental assessment procedures (HMSO 1992). Pharmaceutical production processes require environmental assessment under Schedule 2 of these regulations only if they have significant effects on the environment. The industry also has a 'Duty of Care' under Part 2 of the UK Environmental Protection Act 1990 for assessing and disposing of its wastes, even when they are handled by contractors. To decide the level of compliance required by the regulations it is necessary to assess the environmental hazards for all projects.

Environmental protection systems description

Environmental protection systems are usually an integral part of pharmaceutical process systems and appear on the same engineering drawings as other systems. To clarify the interactions of environmental protection and process systems it is advisable to prepare a separate block diagram that shows all the environmental contact points with the process systems. All gaseous, liquid and solid emissions should be clearly identified together with estimates of the emission rates. The procedures for normal operation, cleaning and maintenance should also be studied to identify how process interactions could generate normal and abnormal emissions. Any emergency procedures or provisions such as explosion relief must also be included in the systems description.

Environmental hazards identification

There is much quantitative information available to identify how substances can pollute water. Regulations make use of this information by categorizing substances for their pollution effects. The European Directive 76/464/EEC defined the 'Black' and 'Grey' lists to categorize substances for control purposes. Substances on the 'Black' list are considered to be the most harmful and pollution from these must be eliminated. Substances on the 'Grey' list are considered to be less harmful and pollution levels are controlled at national level. The German Chemical Industries Association (VCI) has developed a system for rating substances for their water endangering potential, and have published tables for a wide range of materials.

Environmental risk assessment

An environmental risk assessment is required internationally by law for most projects that could have significant effects on the environment. The format of the environmental risk assessment may be prescribed by some regulations. The reader is recommended to read 'A Guide to Risk Assessment and Risk Management for Environmental Protection' (HMSO 1995) for an informative description of environmental risk assessment. Although simple risk ranking can be used within a project to make decisions about alternative courses of action, formal approval from the relevant authority may require more quantitative assessment to prove compliance with their criteria.

The aim of most assessments is to ensure that the project management consider the environmental issues at the earliest possible stages of the project. Suitable action can then be taken to prevent environmental damage if necessary.

Environmental risk acceptability criteria

Environmental risk acceptability criteria have become more stringent due to research on the environment that has revealed many previously unsuspected sources of damage, and that has raised levels of public concern for the environment. General principles such as the 'Precautionary Principle', 'As Low as Reasonably Practical' (ALARP), 'Best Available Techniques Not Entailing Excessive Cost' (BATNEEC), and 'Best Practicable Environmental Option' (BPEO) have been discussed as bases for setting criteria, and some have been developed within legal frameworks.

Environmental risk acceptability criteria are defined separately for gaseous, aqueous and solids emissions to atmosphere, water courses, ground water and soil. The limits imposed by the authority that governs a project will vary considerably and it is essential to define these at the project outset. An environmental impact assessment must be made so that the project design complies with these limits.

Environmental impact assessment

Although some pharmaceutical projects may not require a formal environmental impact assessment by law, it is essential to perform the assessment for project design purposes and to meet SHE management criteria. A typical environmental impact assessment should include the following headings:

- site selection;
- visual impact;
- building and construction;
- normal emissions;
- abnormal emissions;
- site remediation.

7.6.5 Specific pharmaceutical process hazards

Laboratories and pilot plants

(a) *Laboratories*
As explained in Chapter 9, research, development, production, analytical and quality control laboratories are designed and engineered to high standards, and are typically operated under Good Laboratory Practice (GLP) by experienced and well trained staff. Laboratory risk assessments are performed to comply with legislation, such as the UK COSHH regulations, during the design and engineering of new laboratory projects. Laboratories are extremely important business assets.

The main risks in laboratories arise from uncontrolled changes to the original design and operating systems. For example, when new equipment is installed it will usually contain integrated circuits and computer controls. The ease with which the software can be modified may allow in-built safeguards to be inactivated or to generate unexpected hazards. The new owner of such equipment may lack the knowledge to assess its hazards and inadvertently cause an accident.

The use of automated equipment or robotics to perform potentially violent chemical reactions can also lead to accidents in laboratories. It is essential in

251

these circumstances to perform a rigorous HAZOP and CHAZOP to define safe operating procedures, to enable validation, and to implement adequate change controls to avoid unacceptable risks.

Some laboratory equipment may incorporate hazardous materials in a way that the purchaser may not be aware of. An example of this is the use of Nuclear Magnetic Resonance (NMR) equipment. NMR equipment uses super-conducting magnets that are cooled by liquid nitrogen and helium. The cooling systems are provided with emergency pressure relief to prevent hazardous over-pressurization in the event of overheating. Unless suitable ducting to atmosphere is provided, the pressure relief may discharge gases directly into the working area where anyone present could be asphyxiated.

Scaling up the use of liquid nitrogen for storing tissues, etc. in closed laboratories or confined spaces is another hazard that may not be recognized without a hazard study. Laboratory workers can become very accustomed to using small quantities of liquid nitrogen but may forget the asphyxiation hazard if the scale of use increases. Whenever significant amounts of liquid nitrogen are to be used it is essential to perform a risk assessment beforehand to design safe handling and control systems.

The hazards of using fume cupboards on a temporary basis without suitable fire and explosion protection are well known. This problem can be encountered in laboratories where there is a high rate of change and fume cupboard space is limited and can be exacerbated when potentially exothermic reactions, or reactions involving flammable liquids, are run automatically outside normal working hours. It is essential to implement a strict change control system for such circumstances.

(b) Pilot plants

The design of pilot plants is described in Chapter 10. However, effective risk assessment of new pilot plants is often difficult because it is not possible to define exactly what the plant will be used for. This problem is usually addressed by specifying 'Worst Case' and 'Typical' process conditions and materials to define a reasonably realistic model suitable for risk assessment.

The main hazard in pilot plants is uncontrolled change. Once a pilot plant has been built and is in operation, strict change control procedures must be enforced. Comparison of proposed changes with the original system design can help to decide whether further risk assessment is necessary.

A six-stage hazard study and risk assessment for new pilot plant projects will ensure that the users and engineers can agree the user requirements. The added advantage is that the methodology may generate new ideas and eliminate significant hazards before any capital is spent.

Crude bulk drug production

The production of pharmaceutical intermediates and crude bulk drugs involving fine chemical or biotechnological batch processes involves many hazards such as fire, explosion, toxicity, pollution, product contamination, health hazards and energy release that are well known both inside and outside the industry. Most of the processes that contain such hazards are designed using codes of practice, hazard study and risk assessment to minimize the risks.

The following list of problems that have been encountered and successfully resolved by using hazard study and risk assessment indicates the range of application:

- the design, operation and maintenance of safe systems for handling toxic materials;
- control of potentially exothermic reactions;
- effluent control and environmental hazard control systems design, operation and maintenance;
- nitrogen inerting systems design, operation and maintenance;
- safe systems of operation using batch process control computers;
- dust explosion prevention and control systems design, operation and maintenance;
- electrical earthing systems design, operation and maintenance;
- fire protection and prevention systems design, operation and maintenance;
- sampling systems design, operation and maintenance;
- fermenter 'Off gas' filtration;
- fermenter downstream processing;
- cleaning and maintenance systems and procedures;
- designing process systems to cope with the increasing activity and cost of bulk drugs.

Purification

Bulk drug purification is the final stage of primary production and produces the purest material in the product supply chain (see Chapter 5). For many years effective hazard study and risk assessment of the production processes has enabled this purity to be achieved safely, securely and with minimal environmental impact.

Purification processes involve mainly physical changes to the crude drug. The process hazards involved may be less severe than those encountered in crudes production and the main concern is product quality. The typical purification operations of dissolution, carbon adsorption, filtration, chromatographic processes, ion exchange, drying, milling and so on, are all amenable to

253

conventional hazard study and risk assessment. The list of known hazards would include dust explosions, solvent fires, environmental pollution and many of the process hazards associated with cleaning, sampling and maintenance that were listed for the crudes processes. However, it is the hazards to product quality that require particular attention. Hazard study, particularly HAZOP, can contribute to improved operability and quality of purification processes. Risk assessment may also be used to balance quality criteria and SHE criteria.

Quality assurance may sometimes compete with SHE criteria. One example is the routine testing of fire-fighting systems in bulk crude and drug purification facilities. Reliable fire prevention and protection is essential to protect the business from potentially serious interruption. The problem of testing sprinklers, water deluge systems and foam pourers, without causing product quality problems has raised many arguments between the quality assurance staff and the fire engineers in the past.

Secondary production
The design of second production processes has been described in Chapter 6, so only specific hazards and risk assessment topics are considered here.

(a) Formulation
The cleanliness and product security of formulation processes is obtained by removing ancillary equipment from the processing area to 'Plant Rooms'. The design of the plant rooms is often less demanding than for processing areas. Designers sometimes regard plant rooms as peripheral and only give design priority to such rooms when they are critical to GMP, such as for the provision of demineralized water or water for injection. Even then, the room layout is rarely optimized. Plant rooms are often congested, difficult to access, and difficult to work in. Valves and controls are often badly positioned for manual operation or maintenance. Plant rooms located in the process area ceiling space or in basements may have low headroom and rarely have natural lighting, so require reliable emergency lighting during electrical power cuts or fires. Safe systems of work for lone working in plant rooms are essential. In addition to these hazards, plant rooms may sometimes be used for unauthorized storage of equipment and materials. Plant rooms are essential targets for hazard study and any pharmaceutical project for a new facility should include the hazard study of plant rooms in a six-stage hazard study programme.

The major problem of granulation and tabletting processes is the control of biologically active and combustible dust clouds. As was the case with bulk drug purification processes, a key requirement of the process design is the control of such dusts by containment to minimize operator exposure and to comply with

GMPs. Containment may generate potential harm to the operators and to the environment from dust explosions in equipment such as granulators, dryers, mills and conveying systems. The balance of risk between toxic and combustible dust hazards will govern the basic process design and is best achieved as part of a six-stage hazard study. (If flammable solvents are used, the risks are increased considerably).

Alternatively, for a new formulation project, an inherently dust free process may be sought. Direct compression, microwave drying, mixer-granulators, and other such developments aimed at eliminating dust exposure and explosion problems may bring their own particular hazards. The selection of the process must be done as early in the project as possible to allow time to evaluate such options satisfactorily.

Tablet or spheroid film coating with solutions in flammable solvents involves the hazards of fire and environmental pollution. These hazards can be eliminated if aqueous coating can be used instead, although very powerful incentives may be needed to develop aqueous film coating for existing solvent-coated products because of re-registration problems. A comprehensive hazard study together with a combination of QRA and cost benefit analysis can help to decide the most effective alternative.

A typical formulation project will include many items of equipment that are purchased and installed as modular packages 'off the shelf' such as autoclaves, sterilizers, freeze-dryers, chillers, Water for Injection (WFI) units, demineralized water units, centrifugation units, fluid bed dryers. The hazards that can arise will vary depending on the materials processed and the type of process performed. It is very important to determine the level of hazard study and risk assessment that has been performed by the supplier and to check that it meets SHE and quality criteria. Many suppliers perform FMEA, HAZOP and risk assessments as part of their equipment design process, but integrating their equipment into a pharmaceutical project may generate unforeseen hazards. In many project situations, it may be necessary to perform a risk assessment of each module before it is installed in the pharmaceutical system.

(b) Packaging

New packaging facility design and operation can be improved considerably by six-stage hazard study. Although the safety, health and environmental hazards involved may not be as severe as in other pharmaceutical processing activities, the potential quality improvements, the minimization of minor accidents and the improvements in layout and operability that can be achieved are very worthwhile. Hazard study and risk assessment are particularly beneficial if the project is to accommodate aseptic filling or new packaging technology. The increasing use of computerized control systems for packing lines may require FMEA and CHAZOP to complement HAZOP during a six-stage hazard study and as part of the validation exercise.

(c) Warehousing and distribution

Warehouses containing expensive pharmaceuticals are always scrutinized closely by accountants as major centres of working capital. However, the high stock value may not be as important as the potential business interruption arising if it were lost. The hazard study and risk assessment of warehouses and their contents is thus very important to pharmaceutical business activity.

Fire is the main warehouse hazard, so risk assessment is essential to decide the best combination of fire prevention and protection to be provided. As prevention is better than protection, the 'Inherent SHE' principle suggests that the fire load and potential business loss should be minimized by suitable compartmentation or stock separation. However, this principle may conflict with productivity improvements such as high-rise automated warehousing. If fire prevention is not possible, passive or active fire protection must be used. The quantitative risk assessment of fire protection systems, however, may prove to be difficult as reliability data is often unavailable. The consequences of a fire may also be difficult to estimate. Insurers often use the 'worst case' complete destruction scenario, but a very small fire can still generate enough smoke to contaminate all the stock held. Depending on the type of stock held, firewater retention may also be required to comply with environmental regulations.

In countries where earthquakes occur, the location and construction of warehouses require specialized risk assessment and design. Similarly the risks of flooding in some locations require risk assessment.

Archives

The value of pharmaceutical archives in business terms is generally very high — a fact which is often overlooked when designing new facilities. The archived documents, samples of product, new chemical entities, tissues and other materials must be stored securely to meet legislative requirements. A hazard study of existing archives and sample stores will often reveal that significant risks have been taken inadvertently; for example, it would not be unusual to find documents stored in basement areas with no special fire precautions or that storage is under fragile pipes or service drains. Archive areas may be visited infrequently and rarely audited for fire safety.

Most pharmaceutical projects will review archive requirements during HS1 and HS2 study of business risks and Quality Assurance. The PHA guideword

'Other Threats', interpreted by an experienced hazard study team, may also prompt a study of archiving.

7.7 Safety, health and environment legislation

The pharmaceutical industry must comply with both SHE legislation and the pharmaceutical product regulations explained in Chapter 2. This section only considers the SHE legislation.

7.7.1 Overview of SHE legislation worldwide

All engineers and designers need to have an understanding of the law and its relevance to risk issues in their sphere of operations. In most pharmaceutical companies, it is recognized that the legal SHE requirements provide a minimum standard for risk management and assessment. Most organizations operate to more stringent standards in the interest of product security and business risk management. Since SHE legislation is being updated and augmented continuously, it is essential to keep abreast of changes in the law by using commercially available legal databases, preferably electronic and accessible through e-mail, such as those by OSHA and EPA in the USA.

7.7.2 Overview of UK SHE legislation

In the UK, most SHE legislation has been, and still is being, updated and amended to comply with the requirements of recent EU Directives. The Health and Safety Executive (HSE) have powers and duties under the Health and Safety at Work etc. Act 1974 to ensure that UK industry complies with the regulations passed under this and subsequent acts and regulations. The HSE provides useful guidance booklets that are published for all the safety and health regulations in force in the UK. Environmental legislation is implemented by the Environment Agency, established by the Environment Act 1995. A list of some of the main UK regulations that govern SHE in the pharmaceutical industry is given below as an overview, although readers should always check with HSE and the Environment Agency for up-to-date legislative requirements:

- Health and Safety at Work Etc. Act 1974;
  - Management of Health and Safety at Work Regulations 1992;
  - Manual Handling Operations Regulations 1992;
  - Provision and Use of Work Equipment Regulations 1992;
  - Workplace (Health, Safety and Welfare) Regulations 1992;
  - Personal Protective Equipment at Work (PPE) Regulations 1992;
  - Health and Safety Display Screen Equipment Regulations 1992;

- o Control of Substances Hazardous to Health Regulations 1994 (COSHH);
- o Genetic Manipulation Regulations 1989;
- o Genetically Modified Organisms (Contained Use) Regulations 1992;
- o Control of Asbestos at Work Regulations 1987;
- o Supply of Machinery (Safety) Regulations 1992;
- o The Ionizing Radiation Regulations 1985;
- o Noise at Work Regulations 1989;
- o Pressure Systems and Transportable Gas Containers Regulations 1989;
- o Electricity at Work Regulations 1989;
- o Chemicals (Hazard Information and Packaging for Supply) Regulations 1996 (CHIPS);
- o Carriage of Dangerous Goods by Road and Rail (Classification, Packaging and Labelling) Regulations 1994;
- o Carriage of Dangerous Goods by Road Regulations 1984;
- o Control of Industrial Major Accident Hazard Regulations 1984, 1988, 1990 (CIMAH);
- o Control of Major Accident Hazards (COMAH) 1998;
- o Construction (Design and Management) Regulations 1994 (CDM);
- o The Construction (Health, Safety and Welfare) Regulations 1996;
- o Health and Safety (Safety Signs) Regulations 1996;
- o Reporting of Injuries, Diseases and Dangerous Occurrences Regulations 1995 (RIDDOR);
- o The Health and Safety (Consultation with Employees) Regulations 1996;
- • Fire Precautions Act 1971;
  - o Fire Safety and Safety of Places of Sport Act 1987;
  - o Fire Precautions (Workplace) Regulations 1997;
- • Building Act 1984;
  - o Buildings Regulations 1991;
- • Environmental Protection Act 1990;
- • Factories Act 1961;
  - o Highly Flammable Liquids and Liquefied Petroleum Gas Regulations 1972.

7.7.3 Litigation

The foregoing legislation in the UK comes under Criminal Law. However, individuals can seek redress through the Civil Law by the process of litigation. Lawyers, particularly in the USA, have been actively increasing their business in this area. Several successful lawsuits against large organizations have led to extremely large financial awards and it is now very common for individuals to sue for redress.

Engineers, process designers, managers, and risk assessors may often be exposed to litigation, or have to act as expert witnesses on behalf of their organizations. It is essential in these cases to have the best legal representation and advice available. The process of the law is complex and upheld by the lawyers. Technical or moral quality is of no use without a thorough knowledge and understanding of the law.

Bibliography

Gillett, J.E., 1997, *Hazard Study and Risk Assessment in the Pharmaceutical Industry*, ISBN 1-57491-029-9, Interpharm Press Inc.

Barton J. and Rogers R., 1993, *Chemical Reaction Hazards*, ISBN 0-85295284-8, Institution of Chemical Engineers.

Pitblado R. and Turney R., 1996, *Risk Assessment in the Process Industries*, 2nd Edition, ISBN 0-85295-323-2, Institution of Chemical Engineers.

Kletz T.A., Chung P., Broomfield E. and Shen-Orr C., 1995, *Computer Control and Human Error*, ISBN 0-85295-362-3, Institution of Chemical Engineers.

Waring A., 1996, *Practical Systems Thinking*, ISBN 0-412-71750-6, International Thomson Business Press.

HS(G)51, 1990, *The Storage of Flammable Liquids in Containers*, ISBN 0-11-885533-6 HMSO.

HS(G)50, 1990, *The Storage of Flammable Liquids in Fixed Tanks (Up to $10,000\,m^3$ total capacity)*, ISBN 0-11-88-55-32-8 HMSO.

Dept. of the Environment, 1995, *A Guide to Risk Assessment and Risk Management for Environmental Protection*, ISBN 0-11-753091-3 HMSO.

Design of utilities and services

8

JACKIE MORAN, NICK JARDINE and CHRIS DAVIES

8.1 Introduction

The design engineer may ask why this book covers the design of utilities and services and their maintenance, as these are common throughout industry. However, these systems have become important parts of asset management and should no longer be an afterthought following the completion of the 'pharmaceutical' part of the design. Consideration throughout the design makes the validation stage so much easier.

The impact on the design of engineering workshops for maintenance and servicing of production and the utilities is outlined in this chapter and aspects particularly relevant to the pharmaceutical industry are emphasized.

Regulatory inspectors spend a lot of time looking at the design of water supplies, air conditioning systems, their operation and cleaning, and how they impact on pharmaceutical processes. They also want to know how the business is run and organized and who is responsible.

The ideal pharmaceutical facility (using the USA convention to mean the entire building, services plant, services distribution and production equipment) is:

- simple;
- has accessible plant and services;
- reliable;
- does not breakdown, go out of adjustment or wear out;
- fully documented.

The engineer wants the information on the plant in a form that his people can understand, to enable them to maintain it easily and find a quick solution to a problem. The production department wants a flexible plant available at all times, while the quality assurance department wants a plant which performs to design, with written procedures that are always followed and documented and where all changes are recorded and validated. The company wants all this at minimal cost.

Engineering has moved from being a service to becoming an essential part of overall profitability and is now spoken of in terms of asset management. Asset management is the consideration of the activity as the ownership of a major company resource, i.e. the plant and equipment rather than as the 'fixer' connotation normally appended to maintenance.

Maintenance is now using fault analysis, more sophisticated monitoring of the equipment and methods to assess performance to concentrate effort where it is needed. Less maintenance, correctly performed, can be shown to give increased up time.

There are two aspects to achieving trouble free operations:

- management and organization;
- engineering design and specification.

Management requires a clear understanding of the objectives of the engineering function to enable organization and planning and to ensure people are available when required. Clear objectives enable the choices to be evaluated and selected.

Organization requires a clear statement of responsibilities and functional relationships of staff and contractors, selection and training of people, setting up external contracts, followed by a system to measure the performance of the engineering department and make improvements.

Planning ensures the information is fed into the design at the right time, the facility is designed, built and tested to the design, the people and systems are in place when the plant is in use and the facility is maintained.

The design of the engineering space and content of workshops and offices is a subsidiary design exercise based on all the above.

Engineering design requires use of all available engineering knowledge, analytical skills and design experience, by a systematic questioning of the design for operability and maintenance throughout the design and construction.

8.2 Objectives

The engineering function in a pharmaceutical facility is a cost centre (it has a direct impact on the costs and profitability of the company) and, therefore, must be justified. Engineering costs, as with all costs in the industry, are constantly being reviewed. 'Downsizing', 'internal customers', 'process re-engineering', 'delighting the customer' and 'core activities' are terms in common use in the pharmaceutical industry. The emphasis is on trouble free operation for the customers and they expect no breakdowns, the lowest cost and to be able to plan production without concern over availability of the equipment.

No longer is a new facility designed with a clean slate to set up the maintenance department. A greenfield site does not automatically have an engineering complex with all the essential functions of machine shop, welding and fabrication, instruments, design office and a full set of satellite workshops throughout the production areas. The objectives and measures for the engineering function, therefore, need to be determined with the customers. These will depend on company policy, the location and the type of operation in the facility. For example, the following may have different objectives:

- 'Over the Counter' (OTC) facility;
- 24-hour freeze drying operation;
- handling cytotoxic products;
- a local packaging operation.

8.3 Current good manufacturing practice

The attention of the inspecting bodies is moving from the process and production operations to the research and development activities before production and to the services plant and maintenance during production. Increased importance is being placed on validation of the plant and equipment and maintenance of the validated state. For a new product, the pre-approval inspection will require a fully validated plant. Subsequent inspections will examine production records and follow these through the maintenance routines.

FDA guidelines cover these principles (see Chapters 2, 3 and 4). They require:

- appropriate design of facilities;
- equipment history and records or database;
- written procedures and evidence that procedures are followed;
- a maintenance programme.

This enables engineers to set up systems to ensure control of their activities. It places a requirement to know the plant and equipment and to be able to show that it is receiving the correct maintenance. It requires method statements of the maintenance and the description of the tasks.

To do this requires planning, systems and records. There is only one good time to start this — at the inception of the project.

8.4 Design

During design, decisions are made which affect the maintenance and operation of plant and equipment. Maintenance considerations are as important as the process, the production capacity of the facility or the tests to be performed by a quality assurance laboratory. Access and routes for maintenance are as important as those for production and quality assurance staff and for supplying materials to the facility.

Maintenance requirements must be considered during the design stage, as the cost at this stage is minimal compared to the costs after completion and the consequential costs of poor performance.

The maintenance strategy should be part of the initial design study and will determine action during design and installation. The maintenance staff should be part of the project team. The engineer responsible for maintenance should be appointed and be responsible for design decisions and acceptance of the plant and facility. The validation master plan will have been formulated and an essential part of validation is the clear trail from design intent to finished facility.

Co-ordination of the services and the structure are critical. The question to ask of every service line and connection is 'Why do I need access and how do I check it?'

8.4.1 Building

The materials of construction and general size and shape of the building are important. Heights of floors and size and location of plant rooms are part of the design process. The service loads should be calculated in the front-end design to size the main elements of the plant and an allowance made for the inevitable increase in these during design development. This determines the area for services and the location of main plant areas.

Separate engineering floors can be justified. Floor to floor heights should be generous. The increase is in structural cost of floors and envelope. The floors are needed to support the plant, so are not extra and the increase in envelope cost is minimal. The cost of services plant and its controls can represent up to 60% of the total project and the civil structure up to 10%.

All ducting requiring inspection should be on the plant floor and not hidden in false ceilings. This leads to structural slab ceilings in parenteral areas. Where services are run above a false ceiling (such as an office suite) there are beams supporting the floor and, if it is a reinforced concrete structure, there are caps on the columns, which will reduce the space locally.

The structure must allow for access for services. The increasing electrical power and controls require co-ordination and affect the structural design. A reinforced concrete structure can become complex when many conduits pass through an area.

Thought must be given to future service requirements, for example, in an analytical laboratory additional services may be required or the bench layout may be changed to suit new methods. The floor must be designed to permit these changes without affecting the strength and a grid of soft spots may be required.

The trend is to locate the drives and services of production plant in service areas. These should be designed with good access and enough space for maintenance.

Inlet and exhaust should be located to suit the prevailing wind and may require a special study on a multi-building site.

Details such as the design of windows or atria for cleaning are important. A glass stair tower may look good but will be costly to clean. It may need specialist contract equipment, which will require steel reinforced concrete pads. Building expansion joints should not run through critical areas and should be kept away from heavy traffic routes. Parenteral production areas should be on a good slab to minimize floor cracks.

Wet services should not run over critical areas. If this is unavoidable there should be no joints and all items requiring maintenance should be located away from the area. Inspection points and clean-outs for drains should be located in service or plant areas. Particular attention should be paid to the design and construction of service penetrations to process areas.

8.4.2 Maintenance access routes

Movement of engineering personnel should be part of the overall people and materials movement study. Separate engineering floors allow separate access routes for staff, which reduces contamination of the production space; fire escape routes or separate external entrances can provide access, for example. A WFI plant may require a specific changing area and decontamination for parts to be fitted to the plant.

8.4.3 Plant access

The structure and the openings to the plant areas must be designed to allow removal of the largest maintainable item without affecting the integrity of the production facility.

Adequate access for maintenance of the plant and services should be provided. Any valves requiring maintenance must be accessible even if this means locating them away from the heater batteries. Test points should be accessible. With conventional design tools basic decisions such as location of pumps, motors, valves, traps, filters, etc. can be made.

The mechanical, electrical and control services in a modern pharmaceutical plant area need co-ordination to ensure that there are no clashes and that a normal-sized person can reach all areas of the plant requiring access. Drawings to 1:20 scale in plan and elevation of plant areas are required to check this. Alternatively, modern design software using 3D could be used.

8.4.4 Storage of consumables

As part of the strategy, a decision is required on the storage of spare filters and other consumables that are used infrequently. If they are held on-site then dry, safe storage is needed.

Solvents

In this context solvents are considered as organic liquids that provide a vehicle for bringing reactants together, moderating reaction conditions, preferentially extracting one component from another, or cleaning equipment, but are not themselves reactants.

Most solvents are flammable, often with low flash points, usually of low reactivity and generally non-corrosive.

Bulk, drum or IBC storage and distribution

In any design, one of the early decisions must be the choice between bulk, drum or IBC storage and distribution. In the absence of other factors, bulk storage is the preferred option since it usually provides advantages in terms of economics, minimized labour involvement and more effective integration in automatically controlled processes. Despite the greater inventory, bulk storage also has the better safety record.

In practice, this early decision will be made primarily on the basis of the individual batch quantities combined with estimated campaign or annual consumptions. The choice may be influenced to a lesser extent by the existence or otherwise of a tank farm, site space considerations, capital versus operating costs, or occasionally the package availability of the solvent involved.

Bulk storage is the method of choice for much primary production but for pilot plants with reactors of say 0.2 to 1 m^3 capacity, drummed solvents often provide an appropriate solution where one-off batches or very short runs are common.

In secondary production, solvents are frequently needed only for equipment cleaning and in such relatively small quantities that supplies in drums or even smaller containers often suffice.

Bulk storage siting

One of the initial decisions relates to the location of storage tanks; whether in a dedicated tank farm serving a number of buildings or by placing alongside the production units they supply.

In laying out a greenfield site, space could be set aside for a tank farm specifically able to meet the initial site needs but capable of expansion to cater for bulk solvent demands as the site develops. However, the benefits of centralized control, minimized space and facilities for tanker unloading and sampling and reduced vehicle movements on-site must be weighed against higher first costs for set-up and piping distribution to production buildings. Once established the marginal costs of adding further solvents or destinations are likely to be small in comparison with the alternative approach of siting storage tanks adjacent and dedicated to individual production units as and when the need arises.

The majority of new designs will, however, be applied to existing plants where choices between centralized versus local storage are not applicable and location will be dictated by the site philosophy and space availability.

When decisions are taken to locate tanks adjacent to the buildings they serve, conscious recognition must be given to the additional restrictions imposed, particularly in ensuring the safety of the facilities in the event of fire. Such limitations can affect the total quantities stored against the proximity of building walls, the nature and fire resistance of their structure, location of doorways, windows and fire escape routes.

Wherever tanks are located a prime requirement is good road tanker access, not only to allow safe docking for unloading but also to facilitate rapid vehicle exit if required by a serious incident. Siting should minimize or avoid obstructing site roads during unloading which, with quality control checks can occupy several hours per visit.

For the most part storage tanks should be located above ground. Although below ground installations provide some advantage in terms of fire protection, environmental concerns and the costs of providing satisfactory protection and leakage detection often prove prohibitive.

General

Having located the storage area and associated tanker bay, facilities are needed to allow safe sampling of the cargo before discharge, often in the form of a height adjustable overhead gantry. Occasionally, weather protection is provided by canopies, usually without sidewalls, which inhibit ventilation and dispersion of flammable vapours.

Over recent years pumping has become the preferred method of emptying road vehicles, in contrast to the increasingly rare use of compressed air discharge with its attendant drawbacks of formation of flammable atmospheres, potential for solvent contamination and unnecessary emission of vapours. Though some road tankers are equipped with pumps, one forming part of the storage facility itself will give greater assurance of cleanliness especially if dedicated to one material.

Provision of static earthing, safety showers, self-sealing hose couplings and vapour balance connections (between tank and tanker head spaces) are safety or environmental features of an almost mandatory nature.

In sizing storage vessels the main factors will be the annual consumption together with individual batch and campaign requirements. Consideration should, however, also be given to ensuring that tanks are sized to contain a full tanker load plus a margin to allow for order lead times as well as unexpected late deliveries (caused by inclement weather, for example). Typically 10% ullage is applied once the actual storage volume has been determined.

The normal practice is for tanks to be installed within bunds, mainly to protect the environment against leakage. More than one tank can be located in a single bund provided its capacity is adequate to accommodate the capacity of the largest one plus 10%. Good bund design should allow adequate access between bund and tank walls for maintenance and to ensure ease of escape in an emergency. For similar reasons, wall heights need to be limited. Low walling has the additional benefit of promoting vapour dispersion. Since bunds collect rainwater, arrangements are needed for its periodical removal.

Where several tanks are located together or single tanks are located close to buildings, drench systems can provide cooling in the event of fire in adjacent vessels or buildings. The need or otherwise of such protective devices is determined in conjunction with insurers, the Health and Safety Executive or similar authorities.

Most storage tanks for highly flammable solvents (flash point below 32°C) are blanketed with an inert gas as a safety precaution. For some materials, excluding oxygen and moisture helps to maintain solvent quality and for this reason it is also applied to less flammable situations. Nitrogen is the most common inerting gas, but carbon dioxide is an occasional alternative with typical blanketing pressures of 10 to 20 mBar.

Distribution

From the storage tanks, solvent will be distributed via a system comprising pumps, distribution pipework and usually metering devices and filters.

An appropriate control will be built into the scheme. Distribution pumps are generally located outside the tank bunds on plinths arranged to drain away any leaked fluid. The pumps (duty plus standby for critical situations) and distribution main are sized to support the number of vessels that need to be filled simultaneously. Branches to individual users will be based on the desired filling time for that vessel. With non-water miscible and hydrocarbon solvents, particular emphasis should be placed on reducing fluid velocities to minimize static build up. This applies especially where the presence of moisture may create two phases.

Pump differential heads are determined using standard calculations accommodating pipeline, filter and instrument losses and static heads including pressure in receiving equipment. Calculations should cover the full operating envelope of the system. Centrifugal pumps provide low cost, reliable service with packed glands or single mechanical seals suitable for the majority of installations. Magnetic drive pumps eliminate the leak potential of rotating seals.

To minimize leakage and avoid establishing unnecessary zoned areas, welded lines are preferred with the minimum of joints for maintenance purposes. Solvents do not usually need to be distributed through ring mains — continuous circulation wastes energy and can cause unwanted heat and static. Long pipe runs especially where subject to temperature variations (for example exposure to direct sunlight) require protection against hydrostatic overpressures, most commonly in the form of a small relief valve returning to the source vessel.

Protection of pumps with inlet strainers is good practice as is end-of-line filtration, largely to remove rust scale and similar particulates. Where higher standards are demanded, micron filters can be fitted usually alongside a downstream piping specification change to stainless steel to avoid potential recontamination from lower grade materials.

Batching meters are a common and economical means of metering solvents into receiving vessels with satisfactory levels of volumetric accuracy for most purposes. Versions are available with output signals suitable for integration into computer and other control arrangements. Load cells, level gauges and transmitters and other devices on either source or receiving vessels provide alternative means of metering with varying degrees of applicability, accuracy and cost.

Most bulk distribution systems are connected to many destination vessels so that the final shut-off valve will be exposed to its internal conditions. This final valve must, therefore, provide positive shut-off to ensure no back contamination; where circumstances are more critical, the final solvent valve can be mounted on a manifold with other services, with double protection being provided by another valve between the manifold and vessel.

To eliminate risks from static, it is vital that all metal components of flammable solvent systems are checked for earth continuity both before bringing into use and after modification. Arrangements to allow entering solvents to run down vessel walls helps to eliminate 'free fall' static generation.

Materials of construction

Most solvents are produced in plant fabricated from carbon steel. Hence, this material is adequate for many pharmaceutical grade solvent storage and distribution systems. For certain solvents or where absence of colour is important, stainless steel is an alternative constructional material.

Carbon steel is similarly suitable for the majority of distribution pipework although it may well be upgraded to stainless steel downstream of final filtration. Such upgrade avoids pick-up of particulate downstream of the filtration, minimizes internal corrosion where the tail end of the solvent pipe may be exposed to reactor contents and provides cleaner, maintenance-free piping within the process areas. The latter point is particularly important for ensuring GMP in secondary manufacturing plant.

Plastic and glass-fibre are rarely, if at all, employed for flammable solvent handling. Difficulties of static elimination, potentially inadequate chemical resistance and above all their lack of fire resistance make them unsuitable. Solvent suppliers are always willing to offer advice on suitable materials of construction and their advice should be sought if there is any doubt over the suitability of one material over another.

Recovery

Most plants have some form of solvent recovery plant to reduce the costs of purchasing new solvent and disposing of contaminated solvent waste. Steam stripping is usually used for this application, so non-polar solvents with low boiling points are preferred.

Recovered solvent is usually stored separately from new solvent, with the facility to top up with new solvent as required. It is common to use the recovered solvent in the initial stages of production with the new solvent being used for the final filter washes. New solvent is always used for cleaning.

8.5 Utility and service system design

There is a temptation to specify spare capacity and duplicates of plant for run and standby. Care should be taken with this approach, as over-sizing fans and pumps can lead to control problems.

Run and standby may require more control. For example, do you alternate between the two or have 'run' installed and 'standby' unbelted or as a non-installed spare? On WFI a simple system with no dead leg is required. Duplicate pumps require more valves and give dead areas unless complex controls are provided. Standardized flange spacing and a non-installed spare can replace duplicate steam reduction sets.

A risk analysis or Failure Mode Effect Analysis (FMEA) may need to be carried out to decide the strategy.

Shut off valves should only be used sparingly. Breaking a complex service into many sub-sections with shut off, in the hope of being able to carry out selective shutdown, is expensive and you will have to prove that the flows in the part plant are still within design limits. It may be better to shut down the whole system for repair work.

Multiple-use HVAC plants should be avoided. They are difficult to keep in balance and prevent cross-contamination.

Table 8.1 shows the type of system categories that may be required and the areas of utilization.

Table 8.1 Utility system categories

Utility category	Type of system	Possible area of utilization	
		Plantroom	Packing
Compressed	Service comp air	✓	✓
Gas and Vacuum	Process/instrument comp air	✓	✓
	Breathing air	✓	
	Special gases		
	Vacuum-cleaning		✓
	Vacuum-service		✓
	Vacuum-process		
Water	Domestic H&C	✓	✓
	Purified		✓
	WFI		
	LTHW	✓	✓
	Condensate	✓	
	Chilled water	✓	✓
	Cooling water	✓	
Steam	Service	✓	
	Clean		

Some examples are discussed in this section. The intention is not to provide prescriptive solutions, but to indicate factors that will influence the design and to suggest sources of information that may be useful to the designer. It is important to achieve a clear understanding of the requirements of the system under consideration, in terms of quantity and quality, at the outset of the design process, as this will allow a proper assessment to be made of the best methods available for meeting the requirements of the system. Of particular importance when specifying the quality will be the likelihood of contact with the product, i.e.:

- part of final product, e.g. water;
- direct contact, e.g. solvents;
- indirect contact, e.g. Clean In Place;
- no contact, e.g. thermal fluids.

For fluids with no contact with the final product there are many similarities with standard chemical manufacturing facilities, but these areas will also be discussed for the sake of completeness. This chapter will also discuss the effects of forthcoming regulatory requirements, allowing for any future expansion and systems to prevent cross-contamination of utilities with process uses. Table 8.1 gives a checklist for determining possible requirements for utility systems in various types of pharmaceutical facilities.

Possible area of utilization

Laboratory	Creams liquids	Tablets oral	Aerosols	Sterile bio
✓	✓	✓		
✓	✓	✓	✓	✓
		✓		✓
✓		✓	✓	✓
		✓		✓
✓	✓			
	✓			✓
✓	✓	✓		
✓	✓	✓		✓
				✓
✓				
	✓	✓		✓
✓	✓		✓	
✓	✓		✓	✓
✓	✓	✓		✓
✓	✓			✓

8.5.1 HVAC

The types of HVAC systems commonly found in secondary pharmaceutical facilities are extremely diverse and are selected mainly on the basis of the required environmental conditions and the specified level of product contain-

Table 8.2 HVAC system designs

Description of HVAC system objectives	Possible applicable areas
1. Natural ventilation only	Plant rooms, warehouse
2. Mechanical ventilation	Plant rooms, warehouse, changing
3. Mechanical ventilation with heating and/or cooling	Warehouses, receipt and despatch, changing, bin floor, dry products, creams/ointments, packing hall, corridors, offices.
4. Air conditioning i.e. heating and cooling and humidity control to meet a specified band of temperature and humidity	Warehouses, receipt and despatch, changing, bin floor, computer rooms, dry products, creams/ointments, packing hall, corridors, offices
5. Full air conditioning i.e.: heating and cooling and humidity control to meet a specified condition of temperature and humidity	Offices, stability rooms special stores, computer rooms, dry products, creams/ointments, packing hall, offices.
6. As 4 or 5 below but including a low humidity set point (i.e. below approx. 50% RH)	Dry filling, capsule manufacturing, aerosols, dry products.
7. As 4, 5 or 6 below with specified clean room conditions	Creams, dry products, aerosols, steriles.
8. As 7 below but with Class 100 laminar flow distribution	Steriles, dry products (for dust control)
9. As above but recirculation in lieu of Total Loss	As above
10. Separate systems for each work centre and total loss systems to minimize risk of cross-contamination. Terminal HEPA filters on supply and extract. Sterile (positive) or containment (negative) pressure cascades. Low humidity. Dust extract. Specified classification of clean room	Sterile, dry products, aerosols, cytotoxics, vaccines, clinical trials, bio pharms.

ment. As the degree of control associated with these factors increases, the complexity, and therefore, cost of the HVAC system increases proportionately. Table 8.2 details the main types of HVAC systems commonly used in secondary pharmaceutical facilities.

Associated plant	Temp and humidity control	Filter standard	Clean room class
As 6 below + terminal HEPA Filters + Dust extract. *Note: Total loss demands the highest possible plant loads*	As 6 below	HEPA	100–100,000
As above but reduced plant loads	As 6 below	HEPA	100–100,000
As 7 below	As 4, 5 or 6 below	HEPA	100–100,000
As 4, 5 or 6 and HEPA filtration	As 4, 5 or 6 below	HEPA	100–100,000
As 5 below and dehumidification	Any manufacturing conditions with low RH i.e.: 19°C 30% RH or 18°–22°C 30% RH Max.	EU3 to EU9	NIL
Input/extract fans, heating and cooling, filters and humidification. *Note: Requires greater plant capacity than 4 below*	Specified manufacturing conditions (not lower than 50% RH) i.e.: 21°C 50% RH summer and winter	EU3 to EU9	NIL
Input/extract fans, heating and cooling + filters and humidification	Comfort conditions usually a specified band for summer and winter i.e.: 20°C–24°C 30%–60% RH	EU3 to EU9	NIL
Input/extract fans + heating and/or cooling + filters	Max or min temperature control only i.e.: Max 25°C	EU3 to EU9	NIL
Input/extract fans + filters	5°C–10°C above external temperature in summer	EU3	NIL
High and low level louvres	10°C–20°C above external temperature in summer	NIL	NIL

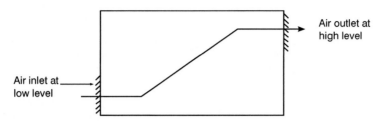

Figure 8.1 Natural ventilation

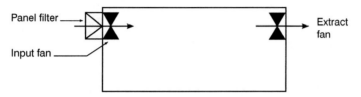

Figure 8.2 Mechanical ventilation

These systems are shown in schematic form in Figures 8.1 to 8.10.

8.5.2 Air

Compressed air is used in pharmaceutical applications for driving pumps and back flushing bag filters. Atmospheric air is passed through a 50 μm or smaller aperture filter to remove insects, dust and pollen before it enters the compressor. Care should be taken to ensure that the air intake is not immediately adjacent to sources of solvent vapour or combustion fumes.

The air is compressed to an appropriate pressure for the system, taking into account the maximum required design pressure and distribution system pressure drop.

The air is then filtered again using a 0.1–0.5 μm filter and dried to remove any compressor oil and condensed water. The pipework is usually carbon steel or galvanized carbon steel.

A general specification for air for these duties is:

- particulate filtration to 0.1 micron;
- pressure dew point at 7 Bar g + 3°C;
- oil filtration to 0.01 ppm;
- normal operating pressure 7 Bar g.

Instrument air is used for actuating valves. Compressed air is filtered to remove dirt and oil mist, which can clog the actuator. The pipework is usually carbon steel or galvanized carbon steel. The specification of the air varies according to

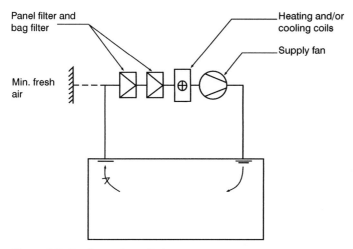

Figure 8.3 Heating and ventilation

user requirements and guidance should be sought from valve suppliers. A general specification for instrument air is:

- particulate filtration to 0.01 micron;
- pressure dew point at 7 Bar g $-40°C$;
- oil filtration to 0.003 ppm;
- normal operating pressure 7 Bar g.

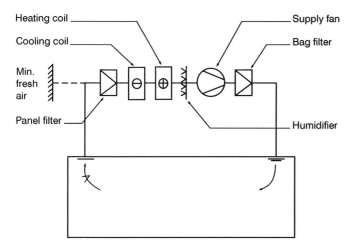

Figure 8.4 Air conditioning

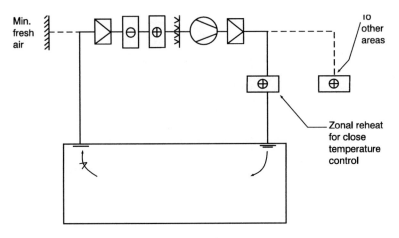

Figure 8.5 Air conditioning with zone reheat

Breathing air is used to protect personnel from dust and toxic fumes by supplying air to hoods or full suits. British Standard BS4275 covers the design of distribution systems for breathing air.

The breathing air system is usually supplied from the compressed air system. The air is then filtered, purified and dried before distribution to the end users. The use of compressed air for breathing means that the location of the compressor air inlet is especially important to prevent toxic fumes from entering the breathing air system.

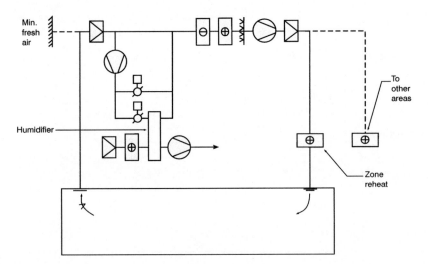

Figure 8.6 Low humidity air conditioning

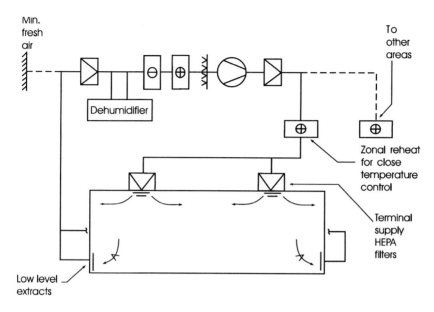

Figure 8.7 Low humidity clean room air conditioning

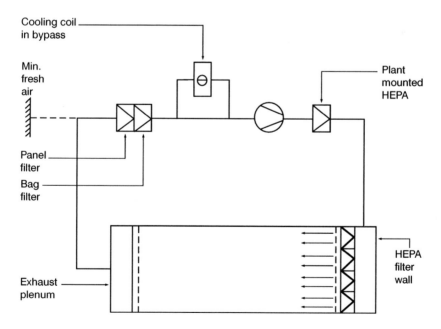

Figure 8.8 Laminar flow clean room

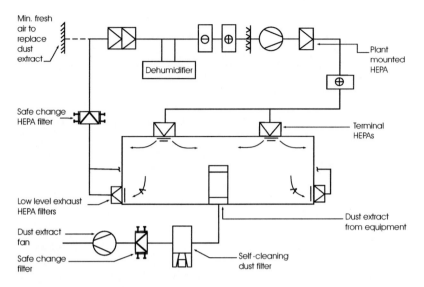

Figure 8.9 Low humidity containment clean room

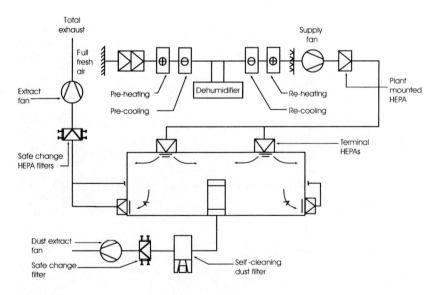

Figure 8.10 Low humidity total loss containment clean room

Air purification units may contain the following equipment:

- 0.01 micron pre-filter to remove solids;
- activated carbon adsorption bed to remove hydrocarbons;
- desiccant drier to remove water;
- catalytic element to remove carbon monoxide;
- final filter;
- carbon monoxide monitor alarm;
- flow meter;
- low pressure alarm.

BS4275 states that provision must be made to warn operators if the system fails. An emergency supply facility is usually provided in the form of a storage tank or cylinder.

There should be a minimum number of manual isolation valves in the distribution system due to the possibility of these valves being mistakenly closed whilst the system is in use. The materials of construction for pipework can be galvanized carbon steel or degreased copper. The distribution system ends in self-locking fittings that feed directly into the PE air hoods or suits.

Process air is used for feeding to fermenters or for processing equipment for parenterals. Process air is sterile, i.e. filtered to 0.2 micron. For fermenters, the air may have other gases added such as carbon dioxide; the gas used being dependant upon the cell culture being grown. Materials of construction are usually stainless steel and the pipework and fittings must be suitable for occasional steam sterilization. As a guideline, the general specification for instrument air (see page 275) is also applicable as it is the basic source of air for this purpose.

8.5.3 Vacuum

General vacuum systems are normally connected to a number of process vessels through a common pipeline and are used for evacuating process equipment prior to nitrogen blanketing, filling head tanks from drums and transferring from one vessel to another. The actual vacuum achieved is not critical, but is of the order of 200 mBar g.

For filtration, a vacuum pump is normally connected to a single filter via a receiver. The vacuum is connected to the liquid outlet of the filter and used for transferring filtrate from the filter to the receiver. The vacuum is applied to the receiver and the receiver is usually fitted with a vent condenser to prevent the vapours from reaching the vacuum pump. The pipework is commonly stainless steel as a minimum, as the filtrate is often reused either directly or after distillation.

For drying, the vacuum may be used to dry the solid on the filter by applying to the top of the filter or dryer. There will be a vent filter on the dryer to prevent the solids from entering the vacuum system. The solvent vapours will be condensed using a condenser supplied with refrigerant and collected in a receiver. The vacuum used for drying will depend upon the maximum temperature which can be applied to the product balanced against the likelihood of pulling solids into the vent filter causing a blockage.

The use of vacuum in distillation systems on pharmaceutical facilities is common, in order to depress the boiling point of distillation mixtures where some component of the mixture is sensitive to heat. Since depression of boiling point is inversely proportional to the system pressure, this duty gives the greatest demand for high vacuum with requirements for system pressures of 1–2 mbar g being commonplace.

There are two main types of vacuum pump:

- liquid seal;
- dry running.

Liquid seal pumps use fluid to provide a liquid seal between the pump casing and the central impellor. As the maximum achievable absolute vacuum is the vapour pressure of the seal fluid at the operating temperature, the choice of sealing fluid is important.

The seal fluid can be run on a single pass or on recirculation. A single pass type is the most appropriate choice for vapour streams containing solids, condensed solvent vapours or corrosive gases. This is due to the flushing action of the sealing fluid preventing the build-up of contaminants to corrosive concentrations leading to pump damage. The downside to this, however, is the increased amount of effluent produced, which is costly in terms of sealing fluid. Recirculating seal fluid systems require additional equipment such as a cooler (to remove heat from the condensing process vapours and the power of the pump motor) and a pot that can be topped up with fresh sealing fluid and which has an overflow to drain. The recirculating system produces less effluent but if not correctly maintained or cleaned can become blocked with solids or the seal fluid can be completely displaced by solvent. A further downside is that if the cooler is not effective, the exhaust gases may also contain a greater amount of solvent and the pump may produce a poor vacuum due to the increase in vapour pressure of the seal fluid at the higher operating temperature.

Dry running pumps are similar in operation to liquid ring pumps but use oil for the lubricating fluid. The tolerances within the pump are much smaller and, therefore, much less oil is required. The choice of lubricating oil is important as this can react with the process vapours and choke the pump.

Dry running pumps are also intolerant to some corrosive gases but, unlike single pass liquid ring pumps, they do not have the protection of the flushing action. These pumps are capable of very high vacuums and in clean process conditions, are superior to liquid ring pumps, with less effluent produced.

Multistage units can produce very high vacuums required for purification of primary product from close isomers by distillation.

Vacuum pumps are usually fitted with an inlet condenser or small vessel to receive any liquid carryover or condensate. All pipework should fall towards the catch pot to prevent back flow of condensed vapour to the equipment item. If the vacuum pump is used for more than one vessel, care should be taken that vapours and condensate cannot reach the other vessels. The pipework should be arranged to minimize pressure drops and pipelines should have long radius elbows or pulled bends to prevent erosion due to solids carryover. There should be the lowest possible number of in-line devices to avoid blockages.

A condenser that uses a refrigerant can be used, but care should be taken if water is being removed from the vapour and gas stream. The discharge of the pump is fitted with a device to remove entrained liquid prior to discharge to atmosphere. Care should be taken to ensure that discharge pipework has a low pressure drop as this will control the absolute vacuum the pump is capable of achieving.

The pipework is suitable for the process but care should be taken, in the case of a reduced specification at the receiving vessel, that no dirt or corrosion products could back flow to the vessel.

Care must be taken when cleaning, especially in the case of filter failure on a dry vacuum line, that any change in pipework specification occurs after the high point, in order to ensure that no corrosion products can back flow in the condensed vapour. If the condensed solvent is to be recycled, the use of stainless steel pipework throughout is recommended to ensure cleanliness.

8.5.4 Clean steam

Clean steam is used in pharmaceutical applications where steam or its condensate is in direct contact with the product. The end use of steam demands that it is supplied dry, saturated and free of entrained air. The requirement for chemical purity is primarily what differentiates clean steam from plant steam. The prohibition of corrosion inhibitors and anti-scaling additives influences generator design and materials of construction. Clean steam and plant steam systems should be completely separate.

The requirement to use clean or pure steam is governed by the cGMP to avoid contamination of the product.

The major use for clean steam is in the sterilization of process and specialist water systems. Clean steam is also used in autoclaves and sometimes for the humidification of clean rooms. Pure steam is used in processes producing parenterals, which demand the use of WFI and here the steam must not be contaminated with micro-organisms or endotoxins (pyrogens). The steam must be of the same specification as the WFI (to BP or USP standards for WFI) and is also used for the sterilization of WFI systems.

The uses of clean steam in pharmaceutical plant are fundamentally different from the uses of pure grades of water, as steam is rarely used as part of the product and only traces come into contact with the final product. It could be argued that the steam need not be to such a high specification, but it is generally used in the final stages of production where precautions against contamination are most stringent.

Clean steam and pure steam are usually produced in a dedicated steam generator. The generator is heated using plant steam. The heat exchanger is double tubesheet with an air gap between plant and clean sides which prevents contamination.

The generator is fitted with a device to remove entrained liquid droplets that may contain bacteria or endotoxins from the vapour stream. This may take the form of a demister pad or some sort of baffle arrangement.

The generator is usually manufactured in stainless steel 316 L or possibly titanium due to the corrosive nature of pure water. It is important not to let too much non-condensable gas (0.5% by volume) into the steam distribution system, as this will form a coating on the vessel surface and prevent efficient heat transfer. There is normally an aseptic sampling device before and after the generator to allow for sampling for endotoxins. The feed water to the generator is purified and free of volatile additives such as amines or hydrazines. As generators will only usually reduce the endotoxin concentration by a factor of 1000 whatever the quality of the feed material, it is important to control endotoxins in the inlet water to minimize the chance of spikes of high endotoxins in the pure steam system.

Steam is a sterilizing agent so although the materials of construction are required to be 316 or 304 stainless steel for reasons of corrosion resistance, the pipelines do not require special internal finishes and can be connected using flanges. The main consideration for distribution systems is their ability to remove condensate. Condensate poses the risk of micro-organism growth and reduces the effectiveness of sterilization. To ensure effective removal of condensate there should be steam traps at all low points and at 30 m intervals of pipework. The pipelines should incline towards the point of use by 1:100 and be properly supported to prevent sagging. Any in-line fittings should be designed to prevent condensate collection. Any lines not used continuously should be fitted with their own steam trap arrangement to prevent the build up of condensate above the isolation valve.

Condensate should not be recovered for use as clean steam. It could be returned to the plant steam boiler if not heavily contaminated, although the small quantities of condensate involved make this impractical and it is therefore usually sent to drain. There should be an air break between the condensate lines and the drains to prevent back flow of condensate. The drains should be suitable for dealing with hot corrosive water. The steam traps should be 316 stainless steel, free draining, with the minimum number of internal crevices i.e. thermostatic type. Condensate quality for clean steam systems should comply with the USP or BP specification for WFI.

For fermenter systems growing recombinant or pathogenic organisms, where there is a possibility of contamination, the condensate should be fed to the kill tanks (see Section 8.9).

8.5.5 Inert gases

Nitrogen is used to blanket vessels, for liquid transfers, filtration, cleaning bag filters, and for blowing process lines clear. It is also used for inerting explosive atmospheres in solids handling equipment and for pressure testing vessels.

Nitrogen can be produced in pressure swing absorption systems from air, by other means from air, or from liquefied nitrogen in storage tanks and cylinders. Pressure swing absorption can produce nitrogen at a reduced specification if the unit is undersized and, therefore, should not be used for critical applications such as inerting of mills. Liquid nitrogen can be produced in many different grades and, therefore, it is important to select the correct grade for the application. It must be remembered that the grade must be for the highest requirement if the system is for site wide nitrogen supply. Some grades of nitrogen contain hydrocarbons (dependant upon the manufacturing route) and these would be unacceptable for flammable environments. cGMP requirements normally specify nitrogen to be filtered to 0.1 micron when in contact with the primary product, i.e. once the bulk pharmaceutical chemical has been produced.

The material of construction for pipework is usually carbon steel. The highest pressure required and the maximum line pressure drops set the pressure of the main. The back flushing of filters is usually the highest pressure and is of the order of 6 Bar g. Normal maximum operating pressures for systems of this type are of the order of 10 Bar g.

Hydro fluoro alkanes (HFAs) are a group of gases that have been developed to take the place of the old CFC refrigerant gases. They are used as propellant for pharmaceutical aerosols and their main property is their degreasing effect, which means that diaphragm pumps are usually used for transfer. They are expensive and, therefore, leakages in the system should be kept to a minimum. In the interest of cleanliness the materials of construction are stainless steel for HFA systems.

8.5.6 Specialist water supplies

This section offers an overview of the main aspects of water and steam production and use in pharmaceutical facilities. This area is covered in far more detail in the ISPE 'Baseline Pharmaceutical Engineering Guide Volume 4: Water and Steam Guide'.

There are many types of water to be found in pharmaceutical facilities. A few of the main types are as follows:

- *towns water* is usually straight from the mains and may vary in quality throughout the year. The specification can be obtained from the local water company and is usually given as a yearly average. There may be two or more water sources for a given plant, and the characteristics of water from these different sources may vary widely;
- *process water* is normally towns water that has passed through a site break tank;
- *de-ionized/demineralized and softened water* has passed through some form of water softening process to remove calcium and magnesium ions that can cause scale on heat exchanger surfaces and in reactors;
- *purified water* has usually been softened and passed through a UV source to remove bacteria. There are various specifications for this as discussed later in the section. The most suitable of these depends upon the market for the final product but generally the water is soft and contains a reduced number of bacteria;
- *water for injection/pyrogen-free water* has been softened and has a low bacterial count and a reduced endotoxin loading. There are a number of different specifications for this type of water. The USP and BP specifications are the most commonly used for WFI.

Towns and process water is treated to give all the other types of water by using the following processes (amongst others):

- organic scavenger — removes organics (may be naturally occurring);
- duplex water softeners — removes calcium and magnesium salts on a continuous basis;
- coarse filtration — removes dirt and debris;

- break tank — protects water supply and protects against short-term failure of supply. Often a mandatory requirement under water bye-laws;
- reverse osmosis unit — removes solids, salts and bacteria;
- electrical deionization — removes the ions present, effectively softening the water;
- UV sterilization — kills a significant number of the remaining live bacteria.

Potable water is used widely in the pharmaceutical industry as a solvent, a reagent and a cleaning medium.

Purified water is used in the preparation of compendial dosages. While Water for Injection is generally used for sterile products, it is also used for cleaning equipment used to make such products.

Specifications for specialist waters are laid down by British Pharmacopoeia (BP), European Pharmacopoeia (Ph.Eur.) and United States Pharmacopoeia USP. These documents also describe the tests that must be carried out to prove the water is to specification.

Historically these specifications were much the same. Recently however, there have been moves to harmonize the BP and Ph.Eur. specifications but the USP specification has changed. This change has lead to a drastic reduction in the number of tests required and specifies only Total Organic Carbon (TOC) and conductivity, both of which can be measured continuously using online monitoring equipment. It would also appear that the specification of the water has been tightened by change.

At present there is some confusion about the specification of WFI and purified water mainly because of the wide differences in requirements between the USP and BP/Ph.Eur. water specifications. The main problem is that WFI must be produced by distillation in the BP and Ph.Eur. specifications, but can be produced by reverse osmosis in the USP specification. Although it would appear that the BP and Ph.Eur. will probably follow the USP at some point in the future, it has left manufacturers who market their products in both the Europe and America with something of a dilemma. With this in mind, it is important to be clear of the desired final product specification when initially specifying a new water system.

After treatment to produce purified water or WFI, the water is collected in a receiver, which is either jacketed or has an in-line heater. The vessel is normally a cylindrical dished end vessel designed to withstand the vacuum that may occur during steam sterilization. The vessel is 316 stainless steel to prevent corrosive attack by the hot purified water.

The tank is fitted with a relief device and possibly some sort of relief monitoring device. The vent is fitted with a HEPA filter to prevent the ingress of microorganisms and is normally heated, to prevent blockage of the hydrophilic filter packing with water. The vent is fitted with a drain via a steam trap to allow any condensate in the vent line to be drained off.

The water is pumped from the vessel through a heat exchanger and then to the distribution system. The heat exchanger can be a shell and tube or plate variety but must be of the double wall type. The water is maintained at 80–90°C by heating with steam. The distribution system can be a ring main or closed line. Ring mains are favoured as the hot purified water continuously cleans them, but single lines are acceptable to the Food and Drug Administration (FDA) as long as they are regularly cleaned and validated. The pump must be fitted with a casing drain to allow drainage after sterilization.

8.5.7 Heat transfer fluids

Hot oil is used for reaction temperatures greater than about 180°C and is dedicated to a small number of reactors.

The system consists of an electrically heated element, pumped loop, distribution pipework and expansion tank. The tank may be vented to atmosphere or nitrogen blanketed. The latter increases the life span of the oil by reducing oxidization of the hot oil at the surface. The system will need periodic draining and cleaning to prevent build up of carbon on the heat transfer surfaces.

The type of oil specified is dependant upon the desired operating range, but the oils are normally silicone based and, therefore, have high boiling points and are highly stable at sustained high temperatures.

Heat transfer oils may also be used where it is critical to prevent water reaching the reagents, for example, if this produces an explosive reaction. The vessel will then be heated using a pumped loop with the normal services (steam, cooling water, refrigerant) on a heat exchanger in the loop. This system will also need an expansion tank.

8.5.8 Refrigeration systems

'Fridge' systems are used to cool reactors, in batch crystallization or as vent condensers on volatile solvent tanks. Glycol is usually used as the heat transfer medium with ethylene glycol being used for nonfood use and propylene glycol for food use.

There are usually two tanks, with one to hold the chilled glycol supply and the other to receive the refrigerated glycol return. The glycol in the return tank is then passed to the supply tank via the chiller or may overflow to the supply side via a weir system.

The concentration of glycol is specified by the desired minimum operating temperature of the process vessels, so care must be taken to ensure that the glycol concentration remains at the required level. Low glycol concentrations may cause freezing of the line's contents, whilst excessive concentrations of glycol may cause problems in the pump due to its viscosity exceeding the pump specification.

The heat removed from the glycol in the chiller is either discharged to the cooling water or to the air via forced draft coolers.

8.6 Sizing of systems for batch production

The sizing of utilities requires a good knowledge of all the operations in the plant including the other utility operations and HVAC requirements. A large amount of information is required and the processing part of the plant needs to be designed before the utilities are designed.

Information required includes:

- mass balance;
- energy balance;
- batch times;
- mode of operation i.e. 24 hr, 5 day etc.

The first step is to produce lists of users for each utility with some assessment of the mode of operation, i.e. continuous/intermittent. The next stage is to attempt to assign a quantity to the users for each operation. Some trivial requirements can be ignored.

Electricity

A motor list is usually made which details power requirements and whether the power requirement is intermittent or continuous. Depending upon the electrical zoning of the plant, it may be necessary to construct a switch room for housing the MCC panels and control equipment.

Cooling water

Using the mass balance and batch times it is possible to calculate the cooling requirements of the process. The summertime cooling water temperature should be used to give a worst case. The cooling requirements of utility systems, for example HVAC and refrigeration equipment, need to be included here. If the process has more than one stage running concurrently a Gantt chart needs to be constructed and the heat loads for a day/week should be considered. From the data, a graph of duty versus time can be produced from which the peak requirements can be ascertained. The designer should also look at the worst possible case and at situations which are not part of standard operation i.e. start-up and shutdown. Future expansion requirements should also be considered. It should be noted that cooling towers come in a limited range of sizes, which vary between suppliers. The final choice of actual size is, therefore, constrained by the supplier chosen. As with all design sizing there is a balance between capital cost and flexibility of operation.

Steam

The method for sizing steam-raising systems is as described above, but an additional consideration is the required pressure. This can either be standard site steam pressures or an individual consideration of the desired final temperature within the process vessel. The flow rate of steam at the desired pressure can be calculated for all the duties and from the above the overall heat duty for the system can be ascertained. Allowance should be made for heat losses in the distribution system and for future expansion of the system.

Nitrogen

The flow rate for purging can be calculated but care must be taken in designing these systems for plant including filtration operations, as these are batch operations. Using the batch cycle time (or an estimate of this), the volumetric flow rate for this duty can be found. Some flow rates will be specified by suppliers, such as backflushing of bag filters. From the volumetric flow rate at the user pressure, the volumetric flow rate at the distribution pressure can be calculated. Again using a graph of duty versus time for the process the overall flow rate at the supply pressure can be found. The supply pressure will depend upon the users' maximum requirements.

The system will normally have an accumulator depending upon the critical nature of the uses to which it is put and the method of producing nitrogen. The supply pressure will be reduced within the plant to give the variety of pressures required. There is usually a relief valve after the pressure-reducing valve to protect downstream equipment from an overpressure within the nitrogen system. The main criteria are:

- pressure required at end user and supply;
- quality;
- quantity;
- temperature at user;
- application e.g. tank blanketing, reaction control.

Compressed air

The ratio of the maximum to minimum capacity of the utility is known as the turn down ratio. All systems should have the capacity to be turned down if part

288

of the plant is under maintenance or if the process is changed for any reason. To allow for future expansion, new systems should not be designed to be operating at their peak loading for 24 hours a day.

If the ratio of maximum to minimum load is greater than about 10, consideration should be given to the use of two or more smaller units, which increases the flexibility of the utility. This would increase initial capital cost but would, if properly controlled, reduce the running costs of the plant. Multiple units may also reduce down time, as the plant may be able to operate on a single unit when not under peak loading.

Duty/standby

Critical systems should have a duty standby facility such that some of the equipment is not run continuously. This allows time for maintenance without the necessity for shutdown periods.

If there is a single duty of short duration with high flow rate, capital costs can be reduced by having some sort of accumulation system to allow a smaller unit to be installed.

8.7 Solids transfer

For charging biologically active solid materials into reactors, it is important to determine:

- the quantity to be added;
- the sizes of kegs to be used;
- the Occupational Exposure Limit (OEL);
- whether the material is explosive;
- whether contact with air is acceptable;
- whether waste bags and filters can be removed safely?

Glove-boxes are used for solids input and kegging of primary product. The requirements in primary production are usually controlled by the characteristics of the product, i.e. the particle size range, the explosive characteristic of the material and whether it is necessary to exclude air or moisture.

8.8 Cleaning systems

All reactor systems require cleaning if a batch has failed or for period maintenance. Some items of plant are also used for different processes and cleaning between these is required, and often this must be validated.

In batch reactor systems, cleaning can be carried out by boiling either water or solvent in the vessel to give the degree of cleaning required. Validation of the cleaning procedure will be necessary.

8.8.1 Clean in Place (CIP)

The first thing to consider for CIP is what is to be achieved by this process and what is to be removed. The systems themselves are very simple, consisting of a tank filled with the correct concentration of cleaning medium, heated by recirculation to the required cleaning temperature and then introduced in the pipework or vessel. This is pumped through the lines and back to the tank or to a drain. The lines are then flushed with water and may be blown with nitrogen before the system goes back to production. The important consideration here is the superficial velocity of the cleaning medium.

8.8.2 Steam in Place/Sterilize in Place (SIP)

Cleaning of lines and vessels using steam can be broken into two main types — Steam in Place or Sterilize in Place, with the main difference being that Steam in Place does not have a quantitative check on the microbial content of the lines after cleaning and that the procedure is not validated. If the requirement is to minimize the biological loading of the system without the total removal of the biological population then Steam in Place is the most appropriate choice. Sterilize in Place is used in biotechnological processes to clean the vessel between batches and for periodic cleaning of Water for Injection (or purified water) storage and distribution systems. This process requires validation to ensure that the cleaning process can be repeated with confidence.

Steam to be used for cleaning must be pure steam (see Section 8.1.4) and is usually reduced down to 1.2 Bar g at the point of use, corresponding to the usual sterilization temperature of 121°C, which is the temperature at which *Bacillus Stereothermophilis* spores are destroyed. The vessel is normally cleaned by CIP first, as the steam will only sterilize the surface, and the vessel internals are checked to ensure cleanliness. Steam is injected into the highest point and collected at the lowest. The time taken for sterilization is determined by the initial bacterial loading and the final bacterial loading required and is governed by the exponential equation:

$$N = N_0 e^{-Kt}$$

where: N is the number of colony forming units (cfu/ml) at the end of the sterilization;

N_0 is the number of colony forming units (cfu/ml) at the start of the sterilization;

k is an empirical constant for the organism in question at the sterilizing temperature;

t is time in seconds.

Clearly, the same percentage level of reduction in biological loading can be achieved by sterilizing for longer at a lower temperature.

The actual time is usually determined during commissioning by covering the vessel or pipework with thermocouples, and timing from when the coldest spot reaches the required sterilization temperature and then relating back to the vessel temperature probe reading.

Vessel requirements

The vessel must be capable of withstanding any vacuum produced by the sudden condensation of the steam. Care must be taken in the design of any vessel that is to be cleaned in this manner to minimize crevices in the vessel and any connecting pipework. The vessels are normally dished end design.

Care must be taken that the condensate produced by the cleaning process can drain away as pockets of warm condensate will not adequately be sterilized. The process is validated by swabbing or by strips impregnated with a substance that changes colour when exposed to a given time/ temperature combination.

8.9 Effluent treatment and waste minimization

The following section is a brief overview of a broad area of knowledge. More detail can be found in the standard texts on the subject. All chemical manufacturing processes produce waste streams and, as all treatment and disposal costs money, it is sensible to reduce waste wherever possible. Waste minimization can save money but all effluent treatments have costs. The chosen waste disposal strategy is based on economics, regulatory compliance and commercial secrecy. Health and Safety has a part to play in any decision, as legislation requires pollution control to follow an integrated approach. It is unacceptable simply to move pollution from one form to another, for example, air stripping of ammonia from a liquid effluent to produce a gaseous discharge.

8.9.1 Types of effluent produced by process

Pharmaceutical processes do not tend to produce large amounts of solids but produce large amounts of waste water contaminated with solvent, reaction products and inorganic salts, some waste solvents, tars from solvent recovery, scrubber liquors, and contaminated gaseous waste streams.

This tends to produce small amounts of high Chemical Oxygen Demand (COD) waste broth, large amounts of wash waters and some gaseous effluents, all of which may be contaminated with microorganisms. There may be commercial reasons as well as environmental to prevent the organisms leaving the site, such as if the organism is novel or genetically engineered.

In general, most waste streams pass to a jacketed vessel known as a kill tank. Periodically the vessel contents are heated to the temperature required to kill the organism. Here the costs of any treatment process (capital, operating, maintenance, disposal) must be weighed against the present cost of disposal.

8.9.2 Options for effluent treatment (in order of expense)

- direct recycling;
- sell to waste processor, for example, waste IPA is used in car screen washes and waste aluminium hydroxide (from Friedel Craft's reactions) is used in antacid tablets;
- recovery and reuse with some form of clean up, such as solvent recovery;
- to the foul sewer with simple gravity separation and pH modification;
- incineration, although some materials such as iodine based contaminants cannot be incinerated because they form acid flue gases which corrode the incinerator;
- landfill is becoming increasingly expensive due to the reduced number of suitable sites, pressure by local populations and the substantially increased Landfill Tax.

8.9.3 Regulatory requirements
There are a number of regulations that relate to waste, including the following:

- Control of Substances Hazardous to Health (COSHH) (1994);
- Environmental Protection Act (1990) — the main aspects being that a producer of waste is responsible for knowing where that waste ends up;
- Water Industries Act (1991) — controls operation of water treatment companies, as well as companies delivering waste to them;
- Trade Effluent Prescribed Substances Regulations (1991) — Red List — this determines which chemicals cannot be released to air or atmosphere.

292

8.9.4 Licensing and regulatory bodies

Water company

The local water company grants consents for discharge of chemical waste to the foul sewer. Here industrial effluent is mixed with sewage and eventually ends up at the sewage treatment works where it is treated by various physical means, before being fed to bacteria and other organisms. If it is an existing site, a consent limit will already be set detailing flowrates and levels of contaminants.

The water company may require information on the toxicity of the effluent to bacteria that break down the sewage and can ask for further information until they are satisfied that the effluent is not a danger to the works.

For discharges from a process the amount, concentration of major contaminants and likely disposal method for each stream are required. The COD load of the process can be calculated and any Red List chemicals identified.

Environmental Protection Agency (EPA)

The Environmental Protection Agency (EPA) grants consents for discharge to the river system. The limits for discharge to rivers are much stricter than to the sewage treatment works, but it is very unusual for a pharmaceutical plant of any appreciable size to be discharging to rivers and not to the sewage treatment works.

EPA regulates Integrated Pollution Control reports for all notifiable processes. A report must be submitted to the EPA which details equipment, process, effluent produced, control strategies.

The EPA has also taken over the duties of the old Her Majesty's Inspectorate of Air Pollution (HMAIP) and consequently grants consents and regulates releases to atmosphere.

8.9.5 Gaseous effluents

The release of gaseous effluents is always controlled by regulation. There are no cost savings other than a reduction in raw materials costs to be offset against the cost of installing and operating abatement equipment.

Characterizing gaseous waste streams

- contaminant characteristics;
- gas stream characteristics;
- design and performance characteristics.

Commonly used treatment processes

- particulate:
  o hydrocyclone;
  o fabric filters;
- vapours:
  o wet scrubbing;
  o biological scrubbing;
  o absorption, adsorption;
  o combustion;
  o condensers.

8.9.6 Liquid effluents

For discharges to the foul sewer, the local water company usually asks for the following information on any aqueous effluent:

- Chemical Oxygen Demand (COD);
- Biochemical Oxygen Demand (BOD);
- Suspended Solids (SS);
- flow rate;
- pH;
- heavy metals;
- contaminants such as cumulative or persistent materials, which will not be broken down at the works, may build up in the water supply system. Phenols are also a problem as they may taint the taste of the final drinking water if water for potable use is abstracted downstream of the sewage works outfall. Many phenols also have a bactericidal effect, and may therefore compromise the operation of biological treatment plants.

The water company treats each effluent on a case by case basis but will give a consent limit for the whole site.

Pre treatment

(1) Equalization

For batch processes, a useful method of reducing loading on the pre-treatment system is to allow streams to mix to a more standard effluent. This optimizes the treatment process and reduces the amount of chemicals added, as some neutralization takes place within the buffer storage. This is normally achieved by a system of sumps or receiving tanks to smooth out the differing streams from a batch process. The pH is then modified to neutrality and suspended solids removed.

Primary treatment processes

(1) Removal of suspended solids

This may be achieved by a number of techniques, including flocculation and skimming or addition of aluminium/iron salts and gravity separation. This process may also remove colour and polar molecules. Turbidity, pH and flow are usually measured at the exit to the foul sewer and there should be some means of sampling the waste stream.

(2) Removal of liquids

Many effluents are contaminated with organic solvents, greases, and the like. These may be removed by means of a simple interceptor, where liquids are separated by means of one floating on the other, or by one of the more complicated systems for enhancing liquid/liquid separation. Lamella plates may be introduced into the interceptor, as in the American Petroleum Institute separator; fine bubbles may carry lighter substances to the surface for skimming, as in Dissolved Air Flotation (DAF); or hydrocyclones may be used to enhance gravity separation. All these techniques tend to decrease the required plan area of plant at additional capital and/or running cost.

Secondary treatment processes

(1) Biological treatment

This uses a number of processes, which are conventionally split into two main groups, based upon whether they are carried out in the presence or absence of air.

Anaerobic processes are carried out in the absence of air — the organisms carrying out the process are actually poisoned by oxygen. These processes carry an advantage over aerobic processes, in that the end products of fermentation include hydrogen, methane, and other flammable substances. These substances can be burned to produce heat, or used in modified diesel engines to generate electricity. The process can, therefore, be a net energy producer if carried out at sufficient scale. The plant required for conventional anaerobic treatment can be very large, but newer techniques are reducing the size of unit operations. The higher the COD of the effluent, the more likely it will be that anaerobic treatment will prove suitable. Far stronger effluents can be treated anaerobically than aerobically, and the total containment of the system that is required to exclude air means that highly odorous effluents can be treated without causing a public nuisance.

Aerobic processes may use passive air, active air, passive pure oxygen or active pure oxygen to provide suitable living conditions for bacteria that degrade organic (and some inorganic) substances, mostly to carbon dioxide, water, and oxidized inorganic salts. There are a great number of techniques for aerobic treatment, differing in how the oxygen is brought into contact with the organisms, whether the organisms are free in suspension, or attached to some media, and whether the process is continuous or batch. There are many other small differences between the generic and proprietary systems on offer, but those preceding have the greatest effect on the important system characteristics, such as resistance to shock loading, running costs, capital costs and unit sizes.

(2) Sludge treatment

All flocculative and biological treatment processes produce quantities of sludge, irrespective of what some manufacturers may claim. Biological treatment sludge is produced in quantities proportional to the total COD put to treatment. There are two main problems with these sludges: their 'instability' (their likelihood to rapidly commence to rot, releasing noxious gases) and their bulk (since most biological sludges are greater than 95% water).

Sludges may be stabilized by means of an additional biological treatment stage, for example aerobic digestion, or by chemical means, such as lime addition. This is another area with a wide range of competing solutions. Having consulted with specialists and decided upon the stabilization strategy, some means of reducing the volume of sludge is usually found desirable, especially if it is to be transported off-site.

The main strategies for volume reduction are analogous to standard dewatering and drying techniques. Not only do they often start in a non-Newtonian state, their characteristics may change with feed conditions to the treatment process, and as a result of continuing biological activity.

The resultant stabilized, concentrated sludges may be in the form of slurries, cakes, pellets, etc. These may be incinerated, landfilled, or sold for soil treatment.

Physical/chemical treatments

As well as conventional biological secondary treatment systems, there are several physical and chemical treatments, removing either specific contaminants, or groups of contaminants with similar properties.

Ozone, peroxides, pure oxygen, air, and a number of other agents may be used. Although these processes tend to take up less space than biological methods, they can be very expensive in terms of running costs, especially with respect to the power costs of ozone systems.

Tertiary treatment

In order to allow recycle or reuse of effluent treated by means of the preceding processes, or in the case of discharge direct to watercourse, it may be necessary to give the cleaned effluent a final polish or moderate its properties in some other way. There are again a number of different techniques for this, with ultrafiltration being common as a good final barrier method to prevent recirculation of undesirable substances.

8.9.7 Solid effluents

The solid effluent such as bags, filter cartridges, etc., are incinerated or landfilled and sludges from primary and secondary treatment processes are treated as previously described. There may be additional constraints on some solid waste, for example laboratory sharps, clinical waste, or waste contaminated with specific biological or chemical agents. These often require separation, marking of containers, and final disposal route.

8.10 General engineering practice requirements

8.10.1 Production area workshops

Space is required in the production areas for:

- storage of change parts for product changes. These should be in purpose built units with clear identification;
- tools for changeover adjacent to the equipment. In the pharmaceutical industry, there are many short runs on packaging equipment and change over time can be lengthy particularly with blister packs;
- diagnostic equipment for fault finding;
- measuring equipment to check the environment and calibrate instruments on the production equipment;
- manuals and records of maintenance, although the latter can consist of a computer terminal. This promotes cleanliness and ensures a single central record is maintained;
- minor repairs and modifications;
- overhauls of equipment.

This can be a combination of local storage units and area workshops and is determined by the working methods agreed in the design brief.

8.10.2 Records

A master plant record — a logical, comprehensive set of information on the facility should be assembled starting at day one of design with the design brief and following design through all stages. Any changes in design intent and design decisions made should be recorded. Engineering change control will ensure this happens and is necessary to show the trail from design concept to completion. It is also a good project cost control tool.

The framework for the record system should be established early. A numbering system for drawings and plant should be agreed. The finance department will want to record the asset value and ideally the same system of numbers should be used. This system will ensure that the required information:

- is available;
- can be found;
- can be updated;
- can be put into systems to monitor, calibrate, and record repairs and use of spares;
- can show that the plant is maintained and performing to design.

8.10.3 Plant numbering

All plant systems, will generally be numbered sequentially from 001. There is a P&I diagram for each system. A system list gives the locations and the areas served, which are shown simply in the P&I diagram. All items on the P&I are numbered sequentially with functions indicated by the symbol and prefix letters e.g., MDM001245 is a motorized damper modulating in system 1 and is item 245 on the P&I diagram. Building management control system outstations sometimes control more than one plant system and this will need to be covered in the BMS system documentation.

A similar method can be devised for electrical panels and distribution boards.

8.10.4 Measurement and calibration

All product significant controls or measuring elements must be calibrated and the calibration traceable to a National Standard. It should be possible to identify the procedures associated with the processes that would detect an instrument problem and, if there are problems, whether the procedures would detect them every time and soon enough. The most serious implications will be associated with critical instruments or instruments in safety-related applications, so a greater margin for error should be used in these cases.

The implications of instrument malfunction are frequently so serious that a cautious estimate of the calibration interval is justified — if an interval is over-cautious, it will soon be revealed as such.

This is an activity that may be desirable to keep in-house. List the types and numbers of instruments to be calibrated and the frequency of calibration to determine the staff and space required. It may be possible to draw on the experiences of calibration from other sites; the same instrument in a similar application may exist, with several years worth of calibration history (e.g. magnehelic gauges) and an optimized calibration interval. This will improve the level of confidence in an estimated calibration interval but must not be used as a substitute for a thorough evaluation of each application; each will be unique in some respect.

Bear in mind that the period between calibrations can be increased if successive calibrations show no deviation. For example, after three successive calibration checks without need for adjustment, it may be possible to double the calibration interval.

8.10.5 Computer systems

Software packages, such as Computer Aided Maintenance Management System (CAMMS) are available but these will only assist with handling the data rather than determining the system. The software package must be validated (see Chapter 4).

The system should be chosen early and records added. The cost and problems of trying to enter the information after the plant has been handed over usually result in incomplete records. If staff who will ultimately use the system enter the data as the work progresses, they will learn the system and the plant. It is essential to manage the quality of this data not only at entry but also throughout its required life. Failure to do this effectively will render the CAMMS system useless and an expensive burden on the operation.

A corresponding reference system for manuals should be set up. Backup copies of all software and records should be made and stored in a secure fire resistant area.

8.11 Installation

8.11.1 Staff duties

The maintenance engineer should be part of the project team.

The technicians should be on-site from the beginning and they should be sent on acceptance trials of major plant. There should be a budget for minor changes, to improve maintenance, and a rigid change control followed.

The technicians should be involved in the IQ/OQ and should ensure that all drawings represent 'as built' and are marked up as the installation progresses.

8.11.2 Training

The core team of technicians may have been selected for their knowledge and experience but they will need further training in analytical skills and fault finding. Involvement in the project is a good training activity and technicians can be trained at suppliers during construction. If it is planned to contract out maintenance, their designated staff should be trained on the equipment. They will also need training on cGMP practices and, if a CAMMs system is in place, they will need training on its use.

8.12 In-house versus contractors

Suppliers of large capital equipment such as refrigeration plant and specialist systems such as fire sprinklers have contract maintenance departments.

In USA and Canada the trend is to contract out facilities management, and consultants and contractors are set up to carry out this service. Major contractors in the UK are now investigating the feasibility of offering this service, as they already have the organization to manage sub-contractors and have established working relationships with preferred suppliers.

An Invitation to Tender or Request for Proposal will be needed, which will specify the requirements and the measures used to compare bids and monitor performance. An Invitation to Tender will typically have the following headings:

- background;
- objectives;
- present situation;
- proposed system;
- company needs;
- nominated staff and qualifications of staff;
- job functions;
- tasks for various job functions;
- reports required;
- confidentiality;
- vendors qualifications;
- timing of proposal;
- format of reply;
- contractors guarantees;
- evaluation of proposal.

The evaluation of proposal lists all the information that is needed to compare bids, such as rates, response times, references, safety record etc.

Partnership is another concept, with agreed performance and profit sharing on improved efficiency.

Contracting out the maintenance is not a simple option. There will still need to be sufficient in-house expertise to effectively control the relevance and quality of the external work.

8.13 Planned and preventive maintenance

8.13.1 Reasons for planned maintenance

- improved equipment reliability;
- reduce lost production time;
- cost avoidance;
- unscheduled repairs and downtime;
- cost control;
- more accurate budgets;
- satisfy FDA and local requirements;
- we deserve a good nights sleep!

8.13.2 Planned maintenance

Planned maintenance, in its simplest form, is applying the manufacturer's routines to the plant at the frequencies they recommend. If done conscientiously and properly, this will reduce breakdowns but it is labour intensive and can result in application with no thought to hours run, duty and environment. Many routines are invasive and can affect the plant if not done correctly. It can result in over-maintenance and rarely can be completed due to pressure to reduce downtime.

Improvements have been made using hours run meters on the starters and BMS systems to log hours run.

8.13.3 Preventive maintenance and reliability centred maintenance

This requires a better understanding of the plant and its use. It involves more extensive examination and review of inspection reports and repair work; an assessment of the potential for failure; emphasis on methods of assessing failure and effort concentrated on those items likely to fail and whose failure has the most significant effect on the facility.

It uses techniques of *condition monitoring*:

- observation and use of analytical skills;
- analysis of oils;
- vibrations analysis;
- Sound/sonic testing;
- Infrared testing.

All the above require a base line of the 'as installed', new condition as a reference.

Good preventive maintenance requires:

- systems (manual or computerized) to track, schedule and record the preventative maintenance;
- system of identifying equipment uniquely;
- good equipment records;
- written procedures;
- following procedures;
- technically competent resources;
- safe working practices and training.

8.14 The future?

More companies will offer contract maintenance and facility management services. The engineering function will reduce in number but increase in engineering and management skill. Plant will be computer monitored and controlled. Confidence and knowledge of computer systems and software will increase and BMS will be used more, removing parallel monitoring and measurement systems. (This is dependent on the BMS software being validatable).

Trouble free operation requires effort. It starts by clearly defining the engineering operating objectives at the beginning of a project and using these to determine the strategy and organization of the engineering department and to prepare a plan to bring this about. Then it requires a lot of detailed effort throughout design and construction on the design and organization.

Then, once this is in place, performance should be measured, reviewed and improved.

Bibliography

1. Haggstrom, M., *New Developments in Aseptic Design Relating to CIP and SIP*, Biotech Forum Europe 3 (92) 164–167.

2. Latham, T., 1995, Clean steam systems, *Pharmaceutical Engineering*, March/April.
3. Smith P.J., 1995, Design of clean steam distribution systems, *Pharmaceutical Engineering*, March/April.
4. *FDA Guide to Inspection of High Purity Water Systems*, July 1993.
5. Honeyman, T., *et al.*, 1998, Pharmaceutical water: In over our heads? *European Pharmaceutical Review*, Aug.
6. *Pharmeuropa*, 1997, (9) 3 Sept.
7. *Clean Steam*, booklet published by Spirax Sarco.
8. US Pharmacopoeia 23 Fifth Supplement, *Water for Pharmaceutical Purposes General Information*, pp. 3547–3555.
9. US Pharmacopoeia 23 Fifth Supplement, *Purified Water*, pg 3443, *Water for Injection*, pp. 3442.
10. Metcalf and Eddy, 1991, *Wastewater Engineering Treatment Disposal and Reuse*, 3rd ed (McGraw Hill, USA).
11. *Baseline Pharmaceutical Engineering Guide Vol 4: 'Water and Steam Guide*, ISPE, 1999.

Acknowledgements

The following persons are thanked for their invaluable help with the writing of this chapter: Roger Freestone, Ken Gutman, Trevor Honeyman and Sean Moran.

Laboratory design

9

DUNCAN LISLE-FENWICK

9.1 Introduction

9.1.1 A need for quality control

Safety has escalated to number one on the agenda of pharmaceutical companies worldwide. Quality Control is the mechanism by which safety is achieved and measured and the Quality Control (QC) laboratory provides a crucial and integral role in achieving the safety objective.

There is a wide spectrum of laboratory types, from schools through to genetic research, undertaking tasks which may take only a few minutes or literally years to complete. The focus of this chapter is on QC laboratories and their purpose, operational requirements and design features, many of which are common to other laboratory types. The QC laboratory has an important place in pharmaceutical production. The activities undertaken in the laboratory rarely contribute directly to the pharmaceutical manufacturing process, but the function of the laboratory remains essential to the final product.

9.1.2 Complex issues require clear procedural guidance

Earlier chapters provide a basic understanding of the complexity of pharmaceutical production. To appreciate how important it is to have a structured and quantifiable approach to any production process it is necessary to examine the process beyond chemistry and biotechnology. Dividing the process into functional categories reveals opportunity for failure error in each. In developing an understanding of how, even within the most highly automated facilities, there is infinite scope for something to go wrong, it is clear that due attention should be paid to the preparation and implementation of safe operating procedures. Consider the implications for controlled functionality in each category:

- facility:
  - construction and materials;
  - maintenance and cleaning;

- o age — wear, corrosion, deformity;
- o control and measure — accuracy, calibration;
- o warning systems — detection and alarm.
- operatives:
  - o skill level — training, experience;
  - o awareness — familiarity, tiredness;
  - o attitude — positive, safety conscious, composed.
- environment:
  - o temperature, humidity, air flow — direction, velocity;
  - o contamination — to the product, to the environment;
  - o hazardous — explosive, flammable, toxic.
- raw materials:
  - o quality — composition, constitution;
  - o storage and transportation — stability, containment, shelf life;
  - o dispensing, handling, containment.

Design, construction and operating codes and standards exist to ensure all factors are given due consideration and a consistent approach. The designer, constructor and operator use knowledge and experience in the endeavour to provide a facility which functions safely and correctly time after time.

In pharmaceutical production scientific accuracy is the major factor contributing to repeatability. Accuracy is the degree to which measurement can be recorded. The principle measurements are: weight, volume, velocity, duration (hence flow rate), temperature and pressure. Measurements apply to solid, liquid or gas states or any combination producing slurries, solutions, vapours etc. Precise measurement is essential to avoid potentially catastrophic reactions and of course it is crucial to the effective product formulation. The process itself introduces stringent specifications for equipment and machinery to attain the high tolerances imperative to the uncompromising quality demanded.

9.1.3 What is the purpose of quality control laboratories?

The pharmaceutical process involves design, material selection, product manufacture and finishing. Each process conforms to codes of practice, regulatory standards and statutory legislation in an effort to produce consistent product quality. A clearly defined, structured and regulated process is the quality assurance demanded by the market for any product to succeed. Quality control establishes the measure of confidence that the market has in any product. All products rely upon consumer confidence.

Manufacturers build their reputations on the quality of their products, reputations that are established by years of faultless products. Reliability can only be achieved through strict quality control. Pharmaceutical production demands strict quality control maintained by thorough checking and inspection, constant monitoring and rigorous testing performed scientifically against exacting specification criteria — enter the quality control laboratory.

9.1.4 What purpose do quality control laboratories serve in pharmaceutical production?

To appreciate how important quality control is to pharmaceutical production, the analogy of a familiar, tangible product, similar, albeit simpler, in process to that operated in pharmaceutical manufacture will be used. Consider the humble cornflake, we know exactly what to expect, a consistent product time after time. Quality control procedures guarantee to deliver the same quality product virtually every time. Confidence that quality is maintained, the product is purchased without hesitation or doubt, yet the level of quality control that produces cornflakes to satisfy the publics' discerning palette is not high enough to meet the demands of pharmaceutical production. On the rare occasion that a burnt cornflake is encountered in the breakfast bowl, it is simply removed without thought. Subconsciously, a quality control inspection has been conducted, as happens every day before anything is bought or consumed. This ultimate quality control inspection is an impossible task when applied to a pharmaceutical product. Typically the product is artificially coloured, artificially flavoured and has an artificial aroma. To further confound the senses, the active ingredient is a fractional component of the dosage form. Consequently, human senses and judgment cannot be relied upon to verify the quality of pharmaceuticals. Fortunately, manufacturers can be relied upon to supply the precise dosage of active drug every time.

9.1.5 How does the quality control laboratory benefit pharmaceutical production?

Regulatory compliance is the subject of Chapter 2 and reference should be made to that chapter for a detailed understanding of regulatory aspects. With regard to QC laboratories, regulatory compliance is concerned with the continuance of the product licence. Laboratory samples and test results must be strictly maintained and catalogued for easy access. Product traceability is essential as it is the essence of validation. Should the burnt cornflake scenario ever occur in a pharmaceutical product, the consequences could be fatal and widespread, and it is vital that the root cause is quickly identified and isolated. The priorities for traceability are:

- prevention of further unnecessary victims arising;
- evaluation and quantification of the problem;
- possible development of an antidote;
- identification of other affected products;
- rectification of the root cause.

Validation is an all-encompassing process; it begins at the design stage and continues through into operation. Each step must satisfy regulatory guidelines and be precisely documented. This approach to pharmaceutical production is known as Good Manufacturing Practice (GMP) (see Chapter 3).

Quality control is one thing, but care must be taken not to confuse validation with quality control. Quality control is an integral part of validation. The onerous procedures pursued in securing a validated product must only be applied to the appropriate steps of the process to avoid unnecessary expense administering the procedures and exhausting effort maintaining the high standards that are a prerequisite of the regulatory authorities.

Perhaps surprisingly, regulatory compliance is complementary to commercial viability. Commercial viability of pharmaceutical products relies on consumer confidence in the product. This confidence is based on the manufacturing companies reputation. The company's reputation is built on their ability to demonstrate repeatability and reliability. Independent regulation provides an ideal vehicle for marketing that ability. There are other commercial benefits to GMP. Pharmaceutical manufacture is an expensive business, whether batch or continuous process. Rigorously structured and controlled production improves efficiency, reduces waste and manages plant shutdown. Large pharmaceutical companies lead in the field in development and improvement of production facilities. Whether inadvertently or planned, developments in manufacturing technique and improvements in equipment and control systems have led regulatory authorities to raise the standards of acceptability.

9.2 Planning a laboratory

9.2.1 Design concept
The most important factor in designing a laboratory is safety.

Aspects of safety that should be considered when evaluating laboratory design should fall into two categories:

- physical space;
- air flow control.

Physical space

The definition of physical space is controlled by a number of criteria, often conflicting and always challenging the skill of the designer to harmonize between regulatory compliance, functional requirement and available space.

Function and operation

Establish the activities undertaken in the laboratory. Determine the space requirements for each activity and any special features associated with the function.

- **bench space**: Typically determined by the laboratory equipment size and any peripheral equipment such as PCs and printers. Depth should be considered as well as length;
- **bench height**: 900 mm is standard for activities undertaken from a standing position with normally transient attendance by the operator. Stools are usually provided for occasional use. 750 mm is standard for activities conducted in a seated position, usually where the operation duration is extensive;
- **bench frame construction**: A variety of frame options are available, generally of steel construction. Each provides a combination of features:
  - o underbench unit: floor standing or suspended;
  - o frame visibility: exposed or concealed;
  - o structural support: floor or wall and/or spine;
  Selection of a frame type will depend on a number of criteria:
  - o flexibility: ease of repositioning/replacing units;
  - o cleaning: access to floor space below and behind unit;
  - o integrity: load capacity depends upon combined structural integrity of structural supports, the frame and the under bench units;
  - o appearance: exposed frames can dominate the overall appearance; carefully considered, they can add feature interest to the laboratory design. Concealed frames reduce the amount of dirt traps providing a more hygienic aesthetic;
  - o cost: flexibility, cleanliness, strength and aesthetics each come with a price tag — specify appropriately to the task duty and responsibly to respect the budget;

- ○ special: special heavy-duty frames with anti-vibration mounts are available where vibration sensitive equipment is to be used such as finely calibrated balances.
- ○ storage space: well-planned and ample storage is essential to safe laboratory operation. Every instrument, container, reagent etc. should have a dedicated and purpose designed home to promote efficiency and safety in the laboratory. For this reason, a diverse range of storage unit types are available; from a simple, eye level, glass reagent shelf to special ventilated cabinets in fireproof construction with automatic door closers.

The range of storage unit styles is too extensive to list here. Each manufacturer has a large selection of modular units to complement their laboratory bench systems. A popular solution to storage problems is storage wall systems, integrating a variety of unit types within a modular frame over the entire length and height of a wall. The generic requirements for each type of storage unit are discussed in this chapter.

Typical types of storage to be considered include:

- **under bench**: Cupboard, drawer or combined units. Internally cupboards may be provided with shelves or may house equipment such as vacuum pumps, waste disposal units, etc. Drawers may be supplied with an array of guides specifically designed to hold equipment, glassware, etc. in an efficient, tidy and safe manner;
- **safety cabinet**:
  - ○ personnel emergency safety equipment;
  - ○ breathing apparatus — gas masks etc.;
  - ○ first aid — medical kit and instruction;
  - ○ fire fighting — hand held extinguishers;
  - ○ hazard spillage — absorbent sand.
- **pull-out storage**: Each with entire pull-out units or individual pull-out shelves. Each designed to provide easier access to otherwise deep storage space where there is a risk of upsetting objects stored close to the front. It is particularly useful for glassware and chemical storage;
- **solvent/flammable storage**: Provided with a system of mechanical extract ventilation discharging to atmosphere, designed to prevent the build up of flammable vapours within the cupboard. Enclosed in a fire resistant casing to contain any fire for a specified period. Fitted with an automatic door closer that is activated on detection of fire. This type of cupboard may be fitted with carousel, rotary shelving to reduce the risk of accidental spillage whilst containing any vapours within the cupboard. Shelves are lipped and a

removable collection tray is housed in the bottom of the cupboard to contain any spillage;

- **acid/alkali/chemical storage**: Provided with a system of mechanical extract ventilation discharging to atmosphere, designed to prevent the build up of toxic fumes within the cupboard. This type of cupboard may be fitted with pull out shelves and is usually lipped to contain spillage. A removable collection tray in the bottom of the cupboard is provided to contain any excess spillage;

  Construction materials used for the storage of chemicals, solvent, acid and alkali must be considered carefully, particularly where spillage is likely to occur. All materials have some inherent weakness that causes it to react with the chemical resulting in corrosion, softening/dissolving, ignition/fire, toxic emission or simple mechanical failure. Common materials used include fiberglass, galvanized steel, stainless steel, polypropylene and glass — each selected for chemical compatibility and physical suitability.

- **controlled temperature**: Often in laboratory operations, it is necessary to store materials at low temperatures. This may be in refrigerator units with storage temperatures a few degrees above zero or freeze units providing sub-zero storage or at the extreme, cryogenic storage systems achieving $-83°C$. Each of these units may require floor space within the laboratory. Typically they are freestanding vertical units with a single door, internally divided into compartments with individual pullout trays. Temperature controls and displays are clearly visible on front of the units. Usually units are designed to suit a 600 mm module.

Operational considerations

(a) Laboratory equipment
The requirements for laboratory equipment will depend upon the procedures to be conducted within the laboratory. Laboratory procedures are generally analytical. The laboratory operator prepares a schedule of equipment with approximate sizes, which will indicate the safety considerations for each piece of equipment, specifying where fume hoods or fume cupboards are required to control emissions. The schedule may include useful information on service utilities for equipment, power, gas, water, air, etc., complete with loads, flowrates, and diversity figures. Typically the equipment includes:

- gas chromatographs (GCs);
- high pressure liquid chromatographs (HPLCs);
- rotary evaporators;

310

- ovens;
- furnaces;
- ultrasonic baths;
- balances.

Armed with this information it is possible to evaluate the basic quantity of furniture items required to satisfy the demands of the laboratory operation.

(b) Ancillary equipment

A host of equipment and storage facilities is required to support any laboratory operation. Guidance is required from the laboratory operator as to the most appropriate and essential items, but generally these will include:

- glassware washers and driers;
- refrigerators;
- freezers;
- safety station — eyewash and safety shower;
- water purifiers;
- vacuum pumps;
- gas generators or cylinders;
- all types of storage.

(c) Personnel and ancillary space

Laboratory operators undertake a number of functions within the laboratory and, whilst they may spend a lot of time at the workbench, they also need an area for report writing and filing. Outside the laboratory, facilities are required for personnel washing and changing, rest and recreation and archive storage of records and samples.

(d) Workflow

The definition of space requirements discussed above provides a quantitative analysis of space requirements for the laboratory. To begin to plan a laboratory into a useful layout requires an understanding of workflow.

The laboratory operator has the best understanding of workflow and work patterns within the laboratory. A simple flow chart or bubble diagram by the laboratory operator will ensure the laboratory design satisfies the demands of the busy schedule of activities in the contemporary laboratory.

Workflow should aim to be in one direction with necessary support facilities provided at each step. Back tracking and cross-over should be avoided as these dramatically increase the risk of accident.

Timing is important — analytical processes may take minutes or hours to complete. The slowest process dictates the throughput of the laboratory.

Workflow is improved by increasing the numbers of critical equipment items (subject to budget). 'Bottlenecks' should be identified and recorded.

Whilst it may be practical to provide utility services to all bench areas, costs aside, it is not always practical to provide additional space for process and utility activities in sufficient number to meet the demand; any limitation must be accepted by the laboratory operator.

Storage space is essential. Storage must be well distributed around the laboratory. Glassware and other implements should be readily available from a number of local storage units. Chemicals should generally be dispensed from a central safe storage location. Trolleys may be used to transport chemicals safely and as a mobile workbench. The laboratory layout must make adequate provision for safe parking of the trolley whilst it is in use as an extension to the work area.

(e) Material flow

Sample receipt, handling and storage feature highly in the work flow requirements for the laboratory.

Once a sample is received into the laboratory it is catalogued before being processed further. The sample is then dispensed into a number of units for different analytical procedures, each catalogued according to the batch requirements. All handling operations must be undertaken with due regard to safety, requiring the use of safe working practice and safety procedure. The use of adequate protective clothing and specialist equipment are essential. The laboratory design must make provision for storage of safety equipment, clean and dirty protective clothing. Changing facilities with showers may be required for some facilities. Clearly identifiable disposal units, segregated according to hazard are as important to safety as safe handling of materials. The laboratory operators must have reasonable access to a safety shower and eyewash facility.

The route for analytical procedures should be planned to be in one direction only, with no crossing of paths or doubling back. Consideration must also be given to the segregation of the different operations — for example, wet chemistry areas are designed to contain spillages and splashing whilst balances are often placed in separate rooms to minimize the effects of adverse room air turbulence and moisture.

(f) Work scope

In large laboratory buildings, different functions are undertaken in separate laboratory rooms, each with appropriate facilities and finishes. In major research complexes, individual laboratory buildings may be designed for

312

different research areas including chemistry, biology, microbiology, biotechnology and animal research (which owing to its political sensitivity is more often referred to as Central Research Support Facility or Biology Support Unit).

There are many support functions which may be undertaken within laboratories such as small-scale production (for clinical trials), kilo labs, instrument and equipment calibration, dispensing and preparation of chemical additives (subject to regulatory restrictions), physical testing.

(g) Personnel flow

Laboratories are hazardous places. The high level of manual handling of dangerous materials, including flammable, toxic, corrosive, radioactive, carcinogenic, bacterial, viral and pathogen, place operators into potentially lethal environments. Whilst laboratories are generally restricted to small quantities of such materials, the consequences of an accident may not be confined to the laboratory, placing the environment and local communities at risk.

Whatever the risk or consequence, strict manual handling policies must be adopted. The laboratory designer must consider the philosophy when establishing the basic design and layout. Personnel need to be able to move around the laboratory freely without cause to disturb colleagues who may be undertaking hazardous operations (albeit with controlled conditions). The operator may also be required to manoeuvre a trolley or cart, carrying hazardous materials, around the laboratory. To ensure these functions are undertaken safely, adequate space must be provided between benches. Fume cupboards need to be positioned where operators have room to manoeuvre freely without being cramped by walls or other fixtures and clear from potential collision with other operators and mobile equipment. The diagrams in Figures 9.1–9.3 (pages 314 to 320) illustrate the general principles of spacing within a laboratory.

(h) Fume cupboards

The use of fume cupboards within a laboratory varies considerably depending upon the nature, frequency and duration of activities undertaken which are either hazardous or are susceptible to contamination. When considering what operations are undertaken within a fume cupboard, it is important to evaluate the viability of multi-function use. Where apparatus can be set up and dismantled in a relatively short time and frequency of use is low, fume cupboards may be utilized for a number of different activities. Keeping the number of fume cupboards low not only saves space and capital costs, it also aids HVAC design. Fume cupboards extract enormous volumes of air from the room. By the nature of a fume cupboard operation, this air must be exhausted to

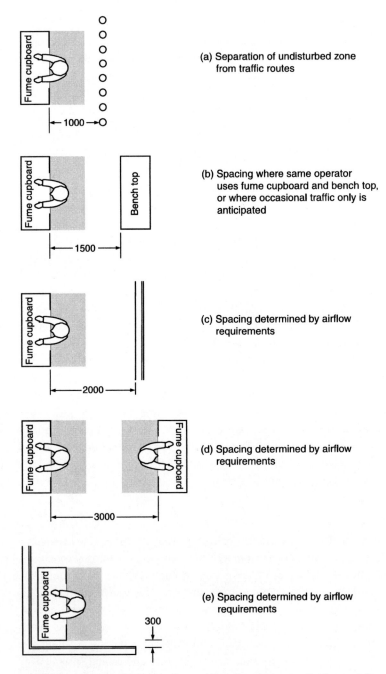

(a) Separation of undisturbed zone from traffic routes

(b) Spacing where same operator uses fume cupboard and bench top, or where occasional traffic only is anticipated

(c) Spacing determined by airflow requirements

(d) Spacing determined by airflow requirements

(e) Spacing determined by airflow requirements

Figure 9.1 Minimum distances for avoiding disturbances to the fume cupboard and its operator

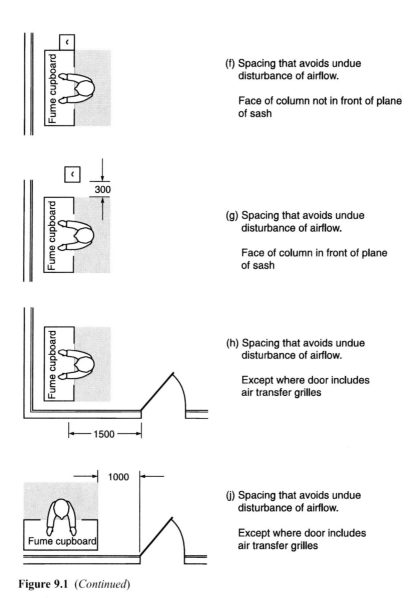

(f) Spacing that avoids undue disturbance of airflow.

Face of column not in front of plane of sash

(g) Spacing that avoids undue disturbance of airflow.

Face of column in front of plane of sash

(h) Spacing that avoids undue disturbance of airflow.

Except where door includes air transfer grilles

(j) Spacing that avoids undue disturbance of airflow.

Except where door includes air transfer grilles

Figure 9.1 (*Continued*)

atmosphere. Detail on the design of air systems is discussed later in this chapter in Section 9.5.

There are a number of different types of fume cupboards available depending on operational requirements. The construction details of each are described

315

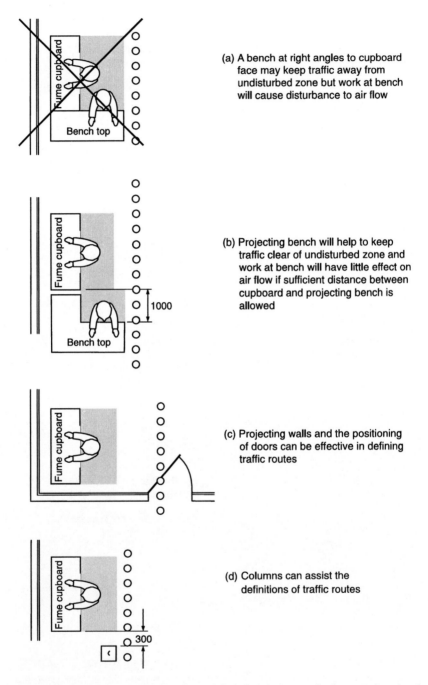

(a) A bench at right angles to cupboard face may keep traffic away from undisturbed zone but work at bench will cause disturbance to air flow

(b) Projecting bench will help to keep traffic clear of undisturbed zone and work at bench will have little effect on air flow if sufficient distance between cupboard and projecting bench is allowed

(c) Projecting walls and the positioning of doors can be effective in defining traffic routes

(d) Columns can assist the definitions of traffic routes

Figure 9.2 Planning arrangements for avoiding disturbances to the fume cupboard and its operator from other personnel

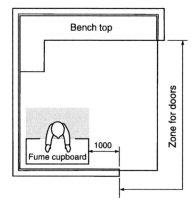

(e) In a small laboratory, the fume cupboard should be clear of personnel entering through doors

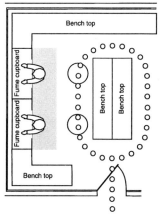

(f) Too much movement in front of fume cupboards should be avoided by providing more than the minimum distances between faces of fume cupboards and bench tops

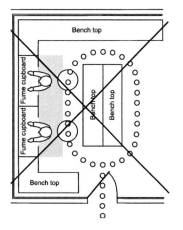

(g) Too much movement in front of fume cupboards should be avoided by providing more than the minimum distances between faces of fume cupboards and bench tops

Figure 9.2 (*Continued*)

317

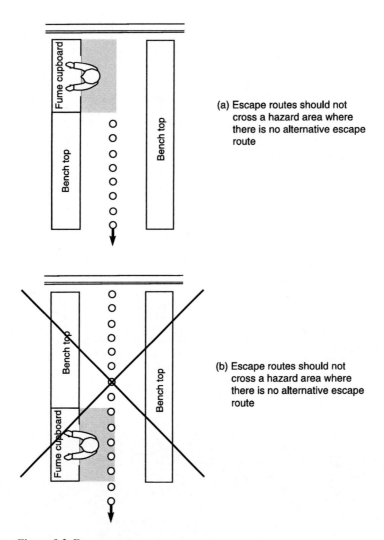

(a) Escape routes should not cross a hazard area where there is no alternative escape route

(b) Escape routes should not cross a hazard area where there is no alternative escape route

Figure 9.3 Escape routes

later in this chapter in Section 9.4. The principle selection criteria are summarized below:

- size: Generally available in modular widths to complement laboratory benches: 1200 mm, 1500 mm, 1800 mm, 2000 mm, 2100 mm are typical;
- sash: Sashes come in a variety of configurations, with vertical sliding being the most common. Horizontal sliding is restrictive but when

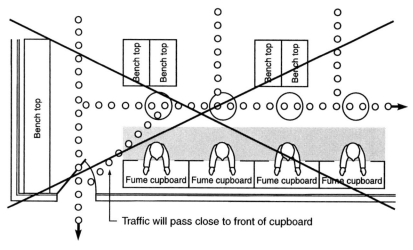

Traffic will pass close to front of cupboard

(c) Principle escape routes should not cross hazard areas

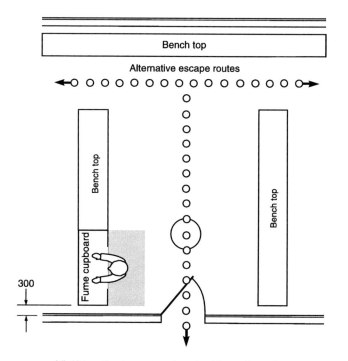

(d) Alternative escape routes should supplement an escape route that crosses a hazard

Figure 9.3 (*Continued*)

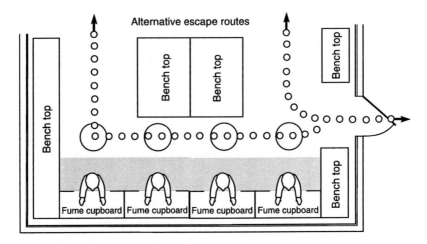

(e) Alternative escape routes should be provided from all hazard areas
in laboratories with more than one fume cupboard

Figure 9.3 (*Continued*)

combined with vertical sliding it provides a more versatile arrangement.
Large sashes are often split horizontally to limit travel and headroom
requirements. Sashes normally start at bench top level. If large equipment
is envisaged then a lower level is appropriate. Some fume cupboards are
'walk in' to accommodate large or heavy apparatus.

○ safety: Most fume cupboards designs are intended to protect the operators
from the hazardous materials being handled. This is achieved by creating a
negative air pressure across the open sash face. The velocity across the
face is usually measured at $0.5\,\mathrm{m\,s^{-1}}$ (termed face velocity). Maintaining
the face velocity for a variety of sash sizes and opening heights is the
fundamental design principle for fume cupboards.

Other safety considerations are that the fume cupboard must offer
protection to the operator from fire and explosion, both of which demand
careful consideration in the selection of suitable construction materials.
Details of construction are discussed later in Section 9.4.

○ facilities: The function of the fume cupboard determines the nature and
number of facilities. These will include utility and laboratory services such
as power, water, air, gas, etc.; equipment frames; sinks and troughs etc.
Details of all available services are included later in Section 9.6.

○ air systems: Although air systems will be covered in depth further on in
this chapter, it is worth mentioning that there are two fundamentally

320

different types of fume cupboard — those which extract all air and those which recirculate air. Recirculatory fume cupboards rely upon local filters to ensure a safe working environment is maintained. This type must only be used for low risk operations. Total extract type remove all air to atmosphere, thus, providing a safe working environment within the laboratory.

9.3 Furniture design

9.3.1 Bench construction systems
Figure 9.4 on page 322 illustrates some of the common bench construction systems.

Pedestal furniture
The pedestal system of benching provides a rigid bench construction by directly supporting the work surface on the underbench units of furniture.

The system is highly cost-effective and commonly features a wide range of modular size units to suit most installations. This pedestal system is an ideal solution in those applications where there is an infrequent requirement for underbench furniture to be interchanged, although should any changes be needed, they can easily be carried out utilizing the services of a maintenance department.

'C' frame bench construction system
This type of bench construction system is ideal for applications where flexibility in the choice of units is a requirement, together with a clear floor space for cleaning.

The system provides a rigid bench construction capable of taking heavy loads. It does not require any floor or wall fixings.

This type of system accepts both suspended and movable types of furniture. Also with both types of unit, the framing allows the units to be placed adjacent to one another without gaps.

Cantilever bench framing
Cantilever bench framing is ideal in installations where flexibility and ease of floor cleaning is required, as there is no horizontal floor leg to cause any obstruction. The design allows for suspended, movable and removable underbench units to be placed anywhere along the length of the benching and repositioned at any time without interference.

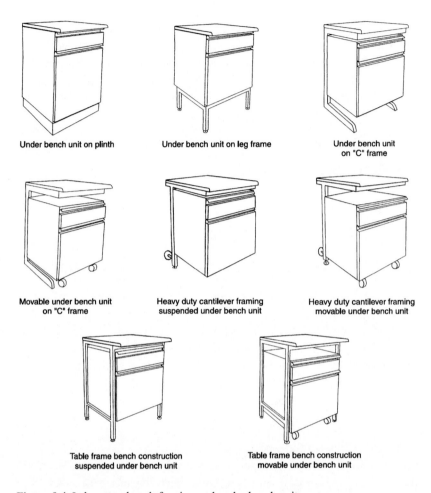

Under bench unit on plinth

Under bench unit on leg frame

Under bench unit on "C" frame

Movable under bench unit on "C" frame

Heavy duty cantilever framing suspended under bench unit

Heavy duty cantilever framing movable under bench unit

Table frame bench construction suspended under bench unit

Table frame bench construction movable under bench unit

Figure 9.4 Laboratory bench framing and under bench units

Available as both standard and heavy-duty cantilever supports, each requires a degree of wall-support for any perimeter benching. Both types are also suitable for island and peninsular benches.

- standard cantilever framing: This system is designed solely for use with removable under bench units; the units themselves provide the necessary additional support to the worktop;
- heavy-duty cantilever framing: The alternative heavy-duty cantilever support system is manufactured from heavier steel sections and, although needing more robust wall and floor fixings, is suitable for movable and suspended underbench units.

Table frame bench construction system

The construction of the table frame is designed to offer both flexibility and economy. It is rigid and can accept heavy loads with minimum deflection. The design will accommodate either suspended or movable furniture units.

When used against a service spine accommodating the mechanical and electrical outlets, further flexibility may be achieved by using modular table units, which obviate the need for long runs of benching. Table frames are generally fitted with adjustable feet for levelling.

Tall storage cupboards

A wide range of tall storage cupboards is available:

- acid/alkali cupboard;
- solvent storage cupboard;
- safety cabinet;
- storage cupboard with pull-out shelves.

Accessories

A large range of integrated accessories is available, such as a comprehensive range of drawer dividers.

9.3.2 Bench top materials

There is a varied range of bench top material available to suit any application. Materials are extensively tested. The most popular materials are detailed below, but this is by no means an exhaustive list. Sizes quoted are typical for the material. The suitability of each material for use with a range of chemicals is summarized in Table 9.1.

Laminate

These bench tops have a thickness of 30 mm. They are covered with laminate with a rolled front edge and bonded to a high-density particle board base. All ends are sealed with a 4 mm thick edging strip of polypropylene.

Epoxy resin

Epoxy resin tops are manufactured from solid epoxy resin and are self-supporting. The tops generally have a thickness of 15 mm, with a dished edging strip 10 mm high, giving a 25 mm thick edge; an alternative is available with a thickness of 19 mm with a 6 mm raised edge.

Table 9.1 Chemical Resistance Chart

	Kambala Iroko	Melamine Laminate	Solid Grade Laminate	Stoneware	Epoxy Resin	Stainless Steel	Tiles	PVDF	PP	PVC	Glass	Slate	Linoleum
ACIDS													
Sulphuric	■	□	□	●	○	○	○	●	●	●	●	○	○
Hydrochloric	○	○	○	●	●	■	○	●	●	●	●	○	○
Fuming Nitric	■	■	■	●	□	■	●	●	●	●	●	□	■
Perchloric	■	□	□	●	●	□	●	●	●	●	●	●	■
Nitric	■	□	□	●	●	□	○	●	●	●	●	□	■
Chlorine	□	○	○	●	●	□	●	●	●	●	●	●	□
Hydrofluoric	■	□	□	○	○	□	●	●	●	○	○	□	■
REAGENTS													
Ammonia	○	●	●	●	●	●	●	●	●	●	●	●	○
Sodium Hydroxide	○	○	○	●	○	●	●	●	●	●	●	○	○
Silver Nitrate	□	□	□	●	○	●	●	●	●	●	●	○	○
Potassium Permanganate	○	○	○	●	○	●	●	●	●	●	●	○	○
Iodine (in 15% Potassium Iodide Soln.)	○	○	○	●	○	●	○	●	●	●	●	○	○
Bromine	□	□	□	●	□	□	□	○	□	□	□	●	□
STAINS													
Malachite Green	○	○	○	●	○	●	○	●	●	●	●	○	○
Crystal Violet	○	○	○	●	○	●	○	○	○	○	●	○	○
Carboxy Fuchsin	○	○	○	●	○	●	○	●	●	○	●	○	○
SOLVENTS													
Acetone	○	●	●	●	○	●	●	●	●	○	●	●	○
Toluene	○	●	●	●	●	●	●	●	●	○	●	●	○
Methyl Alcohol	○	●	●	●	●	●	●	●	●	●	○	●	○
Carbon Tetrachloride	○	●	●	●	○	●	●	●	●	○	●	●	○
Diethyl Ether	○	●	●	●	○	●	●	●	●	●	●	●	○

● = No effect
○ = Slight staining after wiping surface clean
□ = Severe staining and potential corrosion after prolonged use
■ = Not suitable

Solid grade laminate

These tops are normally fabricated from 20 mm thick boards with the edges cut square and polished. Alternatively the front edge can be radiused and polished.

Solid wood

These tops are generally available in Iroko, Kambala or Beech with a thickness of 25 mm or 30 mm. They are constructed from narrow boards jointed with special 's' joint and waterproof glue. They can be linseed oil finished or varnished.

Stoneware

The tops are of solid acid-resistant, glazed stoneware. All tops have a thickness of 30 mm, with a raised front edge 7 mm high. End edging strips of Polybutylene Teraphthalate (PBTB) are available for protection and dishing of ends.

Stainless steel

Two types of stainless steel top are normally available either bonded onto a wood core or self-supporting with suitable reinforcing on the under side. Tops are usually manufactured from Type 316 acid resistant stainless steel. The standard construction is either flat with an overall thickness of 25 or 30 mm, or with a raised edge with a thickness of 32 or 37 mm.

Tiles

Tiled worktops are manufactured utilizing a laminated board base, with all surfaces double-sealed with epoxy resin. First grade chemical-resistant tiles are bonded to the base and jointed with chemically resistant cement to an epoxy grout.

These tops are available as flat worktops or with a raised edge. Flat tops have a thickness of 30 mm. Raised edge tops have a thickness of 37 mm.

Plastic veneered

Plastic veneered tops are available in three types of veneer:

- polyvinylidene fluoride (PVDF);
- polypropylene (PP);
- polyvinyl chloride (PVC).

All tops are covered with one of these materials bonded onto a high density chip board core, with either a flat edge, or with a raised edge all round. In the latter case, the special raised front edge section is welded to the work surface and taken down from the front edge.

Flat tops have a thickness of 30 mm. The raised front edge has a thickness of 37 mm.

Slate
Slate bench tops are used almost exclusively for balance benches. High quality tops are of Welsh Blue Slate with polished edges and thickness of 25 mm or 30 mm.

Glass
These tops are manufactured from a core of block board, covered on both sides with white melamine laminate and veneered with 6 mm thick glass. The toughened glass top surface may be acid etched to give a matt finish. These tops are usually available either flat with a front plastic edging strip in a cumulus green colour or dished with a plastic edging profile. All joints are sealed with silicon rubber sealant.

Linoleum
These tops are manufactured from a core of block board, edged with an insert of heavy-duty linoleum. The tops have a thickness of 25 mm or 30 mm.

Chemical resistance chart
Table 9.1 (see page 324) shows the chemical resistance, at specific concentrations, of the materials used for bench top surfaces and fume cupboard liners. Note that it is only intended to indicate the possible effect of the more commonly used acids, reagents, stains and solvent. It is not intended as a fully comprehensive guide.

9.3.3 Service spine systems
A wide variety of service spines are available, ranging from conventional box spines through to different types of flexible multi-service spines and modules to suit specific applications.

Bench mounted box service spines
Commonly manufactured from melamine faced board with all exposed edges veneered in polypropylene or similar. All electrical outlets are mounted onto the vertical front fascia while mechanical services and drip wastes (if required) are positioned on the top fascia. Where necessary, reagent shelves can also be fitted to these spines.

Floor mounted box service spines

Floor mounted service spines offer the advantage of flexibility; loose benching may be positioned up to them and not necessarily attached. Also, all services can be installed and tested prior to final bench installation.

All spines are supported from an angle iron framework, which accommodates the mechanical service pipework, electrical conduit and cladding panels.

Flexible, multi-service spine system

This is a pre-fabricated self-supporting spine. Consisting of a metal section, with adjustable feet, it can accommodate and support a number of different mechanical services and waste lines.

For maximum flexibility, a capping strip may be fitted at bench level. Alternatively, where flexibility is of minimal concern, the work surface can be taken flush to the spine.

Situated above the work surface is the mechanical service strip, generally made of solid grade laminate, which can either be of a closed type — the strip being taken down to worktop level — or of an open module design which allows a gap above the worktop.

Trunking for electrical outlets is usually above the mechanical services. Over this, trunking may be height adjustable reagent shelves specified in glass, melamine laminate or solid grade laminate. The reagent shelf support may also incorporate scaffold supports suitable for small diameter rods.

Designed in modular lengths to suit most applications, all services are pre-installed in the factory enabling pressure testing to be undertaken before dispatch. Thus, on-site installation time is minimized because it is only necessary to make the joints at the module ends.

Compact, multi-service module

A multi-service module allows for a high-density distribution of mechanical and electrical service outlets.

This type of module is suitable for use as a service bollard with either table frames or mobile trolleys placed against it (an ideal situation for analytical instrumentation) or, alternatively, mounted above a wall bench to provide a high density of outlets in a limited space.

A further use of this module is to site it between two fume cupboards. This enables services to be supplied to both cupboards from a single source and obviates the need for service outlets to be sited in the fume cupboard itself.

Typically, the module is fabricated from moulded sections with a lower section accommodating the mechanical services and the top section housing the electrical outlets. Intermediate sections can be added to accommodate additional mechanical services or outlets for clean instrument gases. The service feed pipes for these modules can either be sited overhead or below. Suitable cladding panels may be used to conceal these service pipes.

Overhead service boom

The use of the overhead service boom, in conjunction with mobile tables, ensures that maximum flexibility is achieved in laboratory benching layout. When used with standard benching, all services are supplied from the boom, leaving the work surface completely free for apparatus and instrumentation.

Booms are available single-sided for wall benches and double-sided for island/peninsular benches.

Boom frames are constructed from metal sections to accommodate the mechanical service outlets with electrical trunking above for 13 amp electrical outlets. Solids grade laminate panels are fitted as a closure to the bottom of the boom. Double-sided booms may be fitted with guardrails at the bottom.

The units are suspended from the soffit on uprights fitted with mounting plates. Services are supplied to the boom from overhead and may be enclosed in a dropper box.

9.3.4 Balance and instrument benches

Balance benches

These benches are specially designed to support analytical balances and other sensitive instruments.

Benches are usually constructed from heavy-gauge steel sections and fitted with adjustable feet. The framing supports, via anti-vibration pads, an anti-vibration work surface consisting of a heavy, thick terrazzo plate. The whole metal structure is often clad in a separate melamine veneered enclosure to give additional protection.

Instrument benches

These are compact benches specifically designed to house analytical instruments together with associated computer and printer equipment.

Benches are based on mobile trolleys fitted with two fixed and two lockable castors. Uprights are fitted to the back of the bench to accommodate the cable store, removable cladding panels, electrical and mechanical services, shelf and swivel monitor stand.

Typically, a melamine laminate worktop is included, under which may be housed additional units, fitted with either cupboards or drawers or a pull-out writing flap or pull-out shelves.

Electrical and mechanical services (such as instrument gases) are connected to the bench from socket and service outlets on adjacent benches via flexible cables and service pipes.

9.3.5 Tables and trolleys

Tables

Two types of table frame are available: the 'C' frame support and the 'H' frame support. Both types are available in various lengths, depths and heights, or in continuous runs to suit specific applications.

The 'C' frame table support

This is normally manufactured from rectangular steel with connecting rails. The cantilever support is fitted with adjustable feet for levelling. Tables are usually fitted with melamine laminate worktops or with other materials. According to availability 'C' frame support tables are designed to carry a limited load.

The 'H' frame table support

The leg frames are usually manufactured from rectangular steel sections, are of welded construction and are fitted with levelling feet. Longitudinal rails are also steel section. Tables are commonly supplied with melamine laminate tops but any other materials may be specified.

Trolleys

These trolleys are typically manufactured as for 'H' frame tables but are fitted with double-wheel castors equipped with rubber tyres, one diagonally opposed pair of castors being lockable. These trolleys are fitted with melamine laminate worktops and shelf, and have a good load carrying capacity. Other worktop materials are always an option.

9.4 Fume cupboards

When considering the layout of a laboratory, the design and positioning of fume cupboards is of critical importance. Poor design or bad positioning of a fume cupboard is not only a safety hazard, but it can detract from the working environment (see Section 9.2 on planning a laboratory).

9.4.1 Typical fume cupboard construction

Support system

Fume cupboards can be supported on pedestal unit furniture, cantilever 'C' frames or table frames with suspended or movable units of furniture. Frames are usually of epoxy powder coated rolled hollow section (RHS) mild steel.

Carcass materials

Mild steel frame sections are commonly used to support external panels of epoxy powder coated steel or compensated laminate-faced medium-density fibre board. (In compensate laminate a balancing laminate is applied to the hidden inside face to prevent exposed facing laminate distorting the board).

Top cover access panels

Designed to be easily demountable, top cover panels may be either epoxy-coated steel or laminate finished board to match fume cupboard outer panels.

Basic internal construction

The back panel is constructed from solid grade laminate, whereas the side and top panels are melamine-veneered boards. Generally the top panel has a cut-out fitted with laminated safety glass, complete with a removable light cowl and light tube. Explosion flaps may also be fitted in the top panel.

A back baffle of solid grade laminate is specifically designed to give an even face velocity. It should include slots to ensure good scavenging at the sides and at the back corners of the cupboard. Scaffold points may be fitted to the back baffle.

Sash design

The vertically sliding sash is commonly made of toughened or laminated safety glass in a metal frame with profiles finger pull to improve airflow characteristics at the lower edge. Suspension is usually by stainless steel cables and lead counter balance weight, the cables running over ball raced nylon pulleys, all arranged on a fail-safe principle in the event of cable failure. Sashes may include horizontal sliding side sashes within the vertical sash frame or horizontally split sashes used where a limited room height restricts normal sash operation.

Airflow

Either a by-pass is fitted above the sash to reduce the face velocity at the lower sash openings and to give a constant extract volume, or a microswitch is fitted to signal the extract system to reduce the extract volume by way of an actuated damper or variable speed fan motor. A profiled metal sill fitted at the front of the work surface ensures good low-level extraction.

The top of the cupboard should be fitted with an aerodynamically designed take-off manifold of fire resistant polypropylene, or similar, ready for connection to the extract system. The manifold should include a condensate collar and, if necessary, a condensate drain.

Utilities

Service outlets are fitted on the centre back wall or the side-walls of the cupboard with control valves fitted into a front fascia rail which also accommodates the electrical outlets. Alternatively, controls may be located on each side of the fume cupboard. Refer to Section 9.6 for details of the services available and distribution systems.

9.4.2 Fume cupboard liner and baffle materials

There is no single, practical construction material for fume cupboard liners that is suitable for all reagents. A comprehensive range of construction materials is available, with each suited to the specific use to which the cupboard is to be put. See Table 9.1 on page 324 for material selection guide.

Liners and back baffle materials

- melamine: veneered high-density board;
- duraline: modified resin and fibreglass filled sheet;
- solid grade laminate;
- polypropylene;
- PVC;
- stainless steel — Grade 316, natural finish;
- toughened glass with backing.

Melamine veneered high-density board

This is highly suitable for use as a construction material for side and top panel. Careful consideration must be given to the detail design and construction of the cupboard to ensure that exposed sides or ends do not come into contact with fumes.

331

This material is only suitable for general-purpose fume cupboards. It is not suitable for use with perchloric acid, radio-isotopes or cupboards which have heavy duty acid use, i.e., metallurgical digestion cupboards or those fitted with a water wash facility.

Duraline
A cost-effective, modified resin and fibreglass filled sheet designed to have good flame retardance, mechanical strength and chemical resistance.

Solid grade laminate
This can be used either for the construction panels of the cupboard, utilizing a thick board, or for the lining panels and back baffles, requiring a reduced thickness board.

This material is very suitable for general-purpose fume cupboards and for cupboards used in low-level radio-isotope applications. It is not suitable for perchloric or heavy acid use.

Plastic
Polypropylene or PVC liners and back baffles are typically fabricated from 16 mm thick material. The plastic liners are excellent for fume cupboards used predominantly for heavy acid applications. Some solvents will cause the plastic to soften. However, once the solvent has evaporated, the plastic will usually appear unaffected. The disadvantage of these liners is their relatively low temperature tolerance. PVC softens at 60°C and polypropylene at 90°C.

If electric hot plates are used in fume cupboards with these types of liner, the power supply should only be energized once the extract fan is switched on. If gas hot plates are used, a solenoid should be fitted in the supply line to inhibit the use of these hot plates when the extract fan is switched off. The fan minimizes the effect of radiant heat on the plastic liners.

Stainless steel
Stainless steel liners are manufactured from acid resistant (Grade 316) stainless steel. They are normally available as either fabricated sectional liners with joints sealed with silicon rubber or one-piece liners and worktop with all corners radiused for ease of cleaning. Care should be taken in selecting this material for specific applications as stainless steel is, to some degree, affected by acids (see Table 9.1 on page 324).

When used for acid applications i.e. perchloric acid including Kjedahl digestion, these fume cupboards should be fitted with water washing jets to enable washing away of any condensed acids after a series of experiments.

Stainless steel fabricated liners are suitable for use in low-level radio-isotope applications. For higher-level use, one-piece liners should be specified.

Epoxy resin liners

Solid epoxy resin liners are generally fabricated from 6 mm thick epoxy resin sheets. All joints are sealed using epoxy resin grout. These liners are suitable for general-purpose fume cupboard use and for high acid use. Some staining may occur when they are used for concentrated acid applications, although the base material normally remains unaffected. Some solvents may also affect this material (see Table 9.1 on page 324).

9.4.3 Fume cupboard work surface materials

Fume cupboard work surfaces may be selected from the higher specification range of bench top materials where chemical resistance and the ability to provide an integral raised rim are important selection criteria.

Work surfaces

- solid epoxy resin;
- solid grade laminate;
- stainless steel — either heavy gauge with reinforcing on underside or light gauge with all edges turned over and under and bonded to a WBP plywood base;
- quarry tiles — on WBP plywood base, bedded and pointed with acid resistant cement;
- polypropylene — bonded to a WBP plywood base.

It is advisable to incorporate raised edges to work surfaces to contain spillage.

9.4.4 Fume cupboards for specific purposes

Fume cupboards for use with some specific reagents or for certain types of analysis require special consideration. Detailed below are cupboards designed to meet some of the more common of these applications.

Fume cupboards used for Kjeldahl digestion

Due to the problems of both heat and condensed acid, either stainless steel or polypropylene liners should be used. Ideally, the necks of the Kjedahl digestion flasks should be manifolded together to enable the majority of the acid fumes to be extracted via a water vacuum pump — the fume cupboard only being used as a secondary containment device. Alternatively, a proprietary digestion apparatus, incorporating its own heater and local extraction may be used.

Polypropylene liners give the best chemical resistance and are quite acceptable if electric heating mantles are used. However, if Bunsen burners are used, care must be taken not to overheat or burn these liners. It is good practice to have a solenoid valve in the gas supply line energized by the extract fan motor. This inhibits the use of the gas burners without the extractor fan switched on.

For both liner materials, it is desirable to fit a water wash device in the cupboard to facilitate washing down after a series of digestions.

Fume cupboards for use with perchloric acid

Fume cupboards designed for this use should be fitted with either stainless steel or polypropylene liners. When stainless steel liners are used, there can be a certain amount of acid attack on this material; however, the by-products of this corrosion are safe and their presence can be minimized by the frequent use of the water wash system. The disadvantage of propylene liners is that when perchloric acid is used, it is normally heated and the heat generated can cause distortion of the plastic liner. Therefore, care must be taken to ensure that the heat source is not placed too near the sidewalls or back baffle.

Due to the possibility of explosive perchlorates being formed by the condensed acids, the fume cupboards and associated duct work should be fitted with water wash jets to enable the system to be washed down after a series of experiments.

Consideration should also be given to fitting a fume scrubber immediately adjacent to the cupboards before the main fume extraction ductwork so that any condensed acid can be washed out. If this is done, then the ductwork after the scrubber will not need to be fitted with the water wash jets.

Fume cupboards for use with hydrofluoric acid

If significant quantities of hydrofluoric acid are to be used (and evaporated), the fume cupboard should be fitted with polypropylene liners. The cupboard should also be fitted with either a water wash system to enable washing down of any condensed acids after a set of experiments or with easily removable baffles to enable manual washing of the inside of the cupboard.

Additionally, the extract system should be fitted with a fume scrubber, either adjacent to the cupboard or on the roof before the extract fan, to inhibit fluorides being emitted into the atmosphere. It should also be remembered that because of etching, the sash should be made of plastic i.e., clear PVC or polycarbonate, rather than glass.

Fume cupboards for use with radio-isotopes

When considering fume cupboards for radio-isotope work, several factors which affect design need to be taken into account. These include the isotope's level of activity, its half-life, the need for filtration and the suitability of the cupboard's face velocity.

If the cupboard is only to be used for tracer work, standard solid grade liners with a face velocity of $0.5\,\mathrm{m\,s^{-1}}$ may be suitable. For dilution work or high levels of activity, the fume cupboard may need a one piece welded liner of stainless steel together with an extract system fitted with high efficiency particulate air (HEPA) filters. Carbon filters may be required for some work.

9.4.5 Special design fume cupboards

Low-level fume cupboards

Low-level (distillation) fume cupboards allow work requiring tall items of equipment to be carried out. The sash opens to the full height of around 1800 mm.

Normally two proportionally opening sashes are fitted. Both are interconnected and operate on a fail-safe principle. Services are supplied to the cupboard from either an adjacent multi-service module or from a service fascia strip built into the underbench unit.

Walk-in fume cupboards

Walk-in fume cupboards provide an especially large workspace with a clear inside height of around 2100 mm and cupboards that are usually fitted with two independently movable front sashes. Sashes are steel-framed with the upper one often being fitted with two horizontal sliding sashes.

Frequently the standard cupboard sides are fitted with access ports with top hung flaps to allow cables and hoses to be passed through from adjacent multi-service modules. Alternatively, front fascia panels are fitted to house the mechanical and electrical controls with the mechanical outlets fitted to the sidewalls of the cupboard.

Special application fume cupboards

This fume cupboard is specifically designed for heavy duty, aggressive chemical use, such as for acid digestions where the significant amounts of condensed acids produced could affect the life of conventional cupboards.

Ideally, the cupboard is fitted with a two-piece, angled back baffle that is easily removable to allow decontamination and cleaning of the whole interior of the cupboard. The baffle is designed to give one third of the total extract volume extracted from the lower baffle opening and two thirds of the extract volume extracted through the top baffle opening. The internal configuration of the cupboard combined with the baffle openings ensures that fumes generated within the cupboard are first directed towards the lower baffle opening, then the fumes migrate up and adjacent to the back baffle and are extracted via the top baffle opening. The baffle is fitted with a condensate trough at the bottom with connection to drain.

While the carcass and sash construction of this cupboard is generally the same as basic models, the special application cupboard and its back baffle should be lined with approximately 5 mm thick ceramic, the top panel of solid grade laminate and the sash of laminated safety glass. For hydrofluoric acid use, the cupboards need to be lined with polypropylene and fitted with an aluminium back baffle that is polyamide-coated. Polycarbonate is recommended for the sash.

The special application fume cupboard should be specifically designed to accommodate a scrubber/demister unit for the removal of contamination from the extract air system before discharge into the atmosphere, especially important where perchloric or hydrofluoric acids are used.

9.5 Extraction hoods

Local bench extraction hoods

For many types of operation, where only small amounts of noxious fumes (smoke, vapour, gases) or occasional high temperatures are generated, local extraction at source is ideal.

A local bench extraction hood uses laboratory supply air to produce a cone-shaped vortex within its confines to capture any noxious substances and extract them efficiently and quickly.

Hoods may be fabricated from PVC or epoxy coated steel or stainless steel. It is important that the maximum height of a hood, above the source of emissions, should not be greater than its diameter.

A variable speed axial flow fan for supplying air from the laboratory may be mounted at the back of the casing, or the hood may be ducted to a central extract system.

Drop front steel extract hoods

Hoods are generally fabricated from steel and finished in epoxy powder coated paint. They normally have a vertically adjustable front cowl and are suitable for extracting radiant heat from ovens, muffle furnaces etc.

Chromatography spray hoods

These hoods are usually fabricated in PVC and are specifically designed for the spraying of chromatography plates. They are fitted with a louvred back baffle to give good extraction, and also with chromatography plate holders.

The hood is suitable for wall mounting or can be fitted at the rear corner of a fume cupboard. It is advisable for the extract duct to be flexible enabling the hood to be lowered during use or pushed up out of the working area of the fume cupboard when not in use.

Fume hoods

Fume hoods are available fabricated from epoxy powder coated steel, aluminium, PVC or polypropylene. They are available in a wide variety of styles to suit individual requirements. These hoods may be fitted with internal baffles to produce a high velocity peripheral extraction in order to improve containment. They may also be fitted with side and back panels.

9.6 Utility services

Services may range from simple installations, requiring just hot and cold water, drainage and possibly natural gas, to more sophisticated installations which use high-quality instrument gases.

Service pipework may be carried out both in-factory, using pre-plumbed service spines or by traditional plumbing methods with the pipework being battened to the wall or clipped to the furniture units.

9.7 Fume extraction

One of the most important areas of laboratory design is in the design and engineering of fume cupboard extract systems. No matter how good the design of the fume cupboard itself, safe containment remains critically reliant on the performance of the extract system. Not only must the system achieve the correct volume flow required for a particular cupboard or cupboards, consideration must also be given to noise, condensate drainage and to ensuring that ductwork does not contravene fire regulations.

Design criteria

Extract systems should be designed to provide a maximum duct velocity of $5–6\,\mathrm{m\,s^{-1}}$. This velocity is sufficient to ensure good scavenging of the duct in order to inhibit any build up of contamination within the duct, whilst not being high enough to generate undue air noise within the ducting system. Generous radius moulded bends are recommended in all systems up to 600 mm diameter. Rectangular ductwork, and circular ductwork above 600 mm diameter may have fabricated bends.

Careful consideration should be given to the routing of all ductwork, so that it is taken outside the building, or to a firebreak service void, by the most direct route. Horizontal ductwork is to be minimized; where long runs are necessary, they are to be laid to a fall with a condensate drain at the lowest point. All extract systems, whether they serve a single cupboard or several cupboards (in which case a manifolded system may be used subject to safety criteria), require volume control dampers (butterfly type) to be fitted for system balancing. Normally all joints in ductwork are solvent welded socket and spigot type. If required, flanged ductwork, with Neoprene gaskets may also be specified for particular applications.

Materials of construction

The most commonly used ductwork material is UPVC, which is suitable for most applications. Where necessary, due to fire regulations, this ductwork can be GRP-coated to give 30–60 minutes' fire resistance, negating the requirement for fire dampers that introduce an additional safety hazard. For very specific applications stainless steel or galvanized steel ductwork is available.

Fume extraction fans

Fume extraction fans are fabricated from either UPVC or polypropylene. Fans should be generously sized to enable the impeller speed to be kept to a minimum for quiet operation. Flexible sleeves are recommended to isolate the fan for connection to ductwork.

Multi-vane forward curved blade type impellers provide maximum efficiency. They may be either directly driven or with indirect drive via 'V' belts and pulleys.

Motors with either single or three phase supply are available depending mainly on the load. Motors should be suitable for external use, as most installations find fans mounted on the roof.

Typically a fan unit is mounted on a galvanized steel angle frame complete with anti-vibration mounts.

9.7.1 Specialized ancillary equipment

Fire dampers

In those situations where it is sometimes necessary for ductwork to pass through firebreak walls or into general purpose building service ducts, it may be necessary to fit fire dampers. Fire dampers must provide the same corrosion resistance as the ductwork. Consequently fire dampers are usually fabricated from a stainless steel outer casing fitted with a stainless steel folding curtain shutter.

The shutter is fitted with stainless steel constant tension closure springs and is held open by a fusible link which releases the shutter in the case of fire. The fire damper is fitted into the partition wall and access hatches are provided in the ductwork for maintenance and testing. Owing to complex routing requirements or simply the sheer quantity of individual ducts, the configuration of fire dampers often makes it impossible to provide accessible access hatches. In these cases, motorized dampers provide an acceptable alternative.

Water wash systems

For some applications, such as extract systems handling perchloric acid, it is necessary to fit a water wash system. Spray jets are fitted into the ductwork, spaced approximately at a 1.5 metre pitch on vertical ductwork and at a 1 metre pitch for horizontal runs. Jets may be manifolded by a plastic supply pipe and controlled from a valve on the fume cupboard.

It is important to note that a water wash system should only be used for washing the ductwork and removing any condensed acids after a series of experiments. It should not be used during a series of experiments as the spray will contaminate the experimental work. A booster pump may be required if the head of water is not sufficient for the higher jets to operate satisfactorily.

Fume scrubbers

For those extract systems handling perchloric or hydrofluoric acid, fume scrubbers may well be needed. Two types of fume scrubber are generally available — the compact scrubber/demister unit and the tower scrubber.

In the case of perchloric acid, the compact scrubber/demister unit is ideal as it can be fitted adjacent to the fume cupboard. Therefore, all ductwork from the outlet of the scrubber will remain uncontaminated and water washing will not be required. This scrubber can also be used for hydrofluoric acid applications.

The compact scrubber/demister unit is only suitable for connection to single fume cupboards. Where larger volumes of extract air are to be handled, from several cupboards, then a tower scrubber must be used.

Fume scrubbers are normally fabricated from UPVC and, in the case of tower scrubbers, feature GRP reinforcement. All scrubbers comprise three sections:

- the holding or capacity tank for the scrubbing media;
- a packed scrubbing section fitted with wash jets;
- a demist section to remove the washing media before discharge into the extract system.

The installation requires a circulating pump provided with a water supply with a ball valve fitted, together with drain connection.

9.7.2 Air input systems

A factor sometimes overlooked in fitting out a new laboratory is that fume cupboards extract a considerable volume of air from the laboratory area. In non-ventilated laboratories without sealed windows, it may not be necessary to install an air input system if there are only a small number of fume cupboards, as approximately six to eight air changes per hour can be achieved within the laboratory by natural leakage. In modern laboratory blocks with well sealed windows or in those where there are large number of fume cupboards, consideration must be given to the installation of an air input system. In this instance, care must be taken in the siting of the actual input grilles so that turbulence at the fume cupboard face is minimized. As a general principle, no input grille should be within 1.5 m of the face of the fume cupboard.

Ideally, the input grilles or slot diffusers should be on the opposite side of the laboratory to the cupboards in order to 'wash' the laboratory with clean air. The use of grilles or slot diffusers is suitable to achieve room air change rates of up to 20 per hour. If the air change rate is above this, then a perforated ceiling grid should be used.

9.8 Air flow systems

9.8.1 Air-handling for the laboratory

Air management control systems, which when considered at the planning stage of a laboratory, provide economies in both capital investment and operational costs. Variable airflow reduces the entire air requirements which as a result enables the building ventilation system to be designed smaller, thereby reducing investment costs. Operational costs are minimized through continual adjustment of the air flow to meet the current working situation. The rate of all supply and extract air may be computer controlled to optimize plant operation providing lower energy consumption and the opportunity to introduce diversity factors to reduce capital and operational cost.

Construction and components

(a) Airflow controller

The airflow controller is a processor which monitors and regulates the volume of extract air depending on the position of the sash. Upper and lower nominal limits are established for the open and closed sash positions. For all other sash positions the air flow rate is determined as a linear function.

A sensor constantly measures the air volume and adjusts a damper when variations occur until the present value is achieved. The sensor is placed in a bypass system to protect it from aggressive fumes.

Most airflow controllers can be switched to different operational modes: normal operation, night operation (lower amount of air) as well as emergency operation (maximum amount of air with fully opened damper). They can also be provided with volt-free contacts for connection to a building management system (BMS).

(b) Sash controller

The sash controller is a processor responsible for closing the sash when no-one is standing in front of the fume cupboard. Continual controlling of the sash opening ensures an optimum working condition with maximum safety. Typically, a passive infrared detector senses the movement of a person in front of the fume cupboard. When the person moves away, out of range of the detector and following a pre-set time delay, the automatic sash closing function is initiated.

(c) Manual volume control damper

The manual damper maintains a constant pre-set air volume even under varying pressure conditions. Such regulation is found in permanent vented units with constant air volume (cabinets, vented underbench units). The required volume of air for these installations is a burden which has been taken into consideration in balancing the room air.

Temporarily vented units, canopies or local extraction hoods etc. which are either switched on and operated at the full pre-set air volume or off, incorporate a damper which sends a signal to indicate its operational condition enabling this to be taken into account in the process of adjusting the air volume levels.

(*d*) *Group controller*

The controller constantly receives on-line data on the current individual air requirements from all variable extracting units in the laboratory (fume cupboards, temporary running extracting units). It processes this data and sends a control signal in the form of a nominal electrical signal to the supply air damper which adjusts the volume of air. In this way the group controller acts as a link between the extract air dampers of the individual units and the supply air damper of the laboratory.

Where available, the BMS may undertake the function of the group controller.

(*e*) *Supply air dampers*

The damper receives the control signal from the group controller and adjusts the supply for compliance with the applicable specifications for air volume and room pressure.

(*f*) *Supply air grilles*

Sufficient air grilles should be allowed for supply air to the laboratory without draughts.

(*g*) *Supply air and extract air ducting*

These ensure optimal guiding of air in the room.

9.8.2 Air handling efficiency

Within the modern laboratory the emphasis on safety has led to an increase in the number of fume cupboards, local extract hoods and ventilated cabinets. The resultant demand on air flow creates unrealistically high air change rates. The consequences of not addressing the problem could lead to:

- large air handling equipment;
- large ductwork;
- high energy costs;
- complex control systems.

The first three are a product of the air volumes; safety requirements do not permit air recirculation, therefore, all treated air supplied to the laboratory is dumped. Complex controls are necessary to manage the diversity on air volume demand. Depending upon the number of extract units and the operational requirements, systems may incorporate multiple fan and damper arrangements for both supply and extract air, monitored by probes and sensors. There are a number of methods which may be adopted to improve the efficiency of laboratory air flow.

Fume cupboard face velocity control system and laboratory air input controls

Significant savings can be achieved in running costs to heated or cooled air input to the laboratory.

A number of fume cupboards may be served by a single extraction fan. Make up air supply is introduced to the laboratory via a standard Air Handling Unit (AHU) with heating/cooling coils.

The important feature of this type of system is that the extract fan runs at full volume at all times. This ensures that the discharge velocity remains constant and thus the contaminated air is dumped. As the fume cupboard sashes are closed, the total volume extracted through the fume cupboards is reduced to only 15%. Hence, a fresh air bleed damper is built into the system which allows air to be taken from outside the building through the extract fan to make up the 85% reduction and, thus, maintain the discharge velocity. The fresh air bleed damper can be operated by an adjustable weighted arm, or by an actuator controlled by the extract duct pressure.

The air input system is required to provide make up air for the fume cupboards. The air is taken from outside the building and heated or cooled as required. For laboratories where the building fabric is not well sealed, in order to control the air input, a duct probe is used to produce a signal proportional to the extract volume.

For laboratories where the building fabric is well sealed, it is possible to measure the differential pressure between the laboratory and an adjacent area with a stable pressure regime. This measurement can be used to produce an output signal to control the air input to balance the variable extract volume.

A significant feature of this system is the possibility of applying a diversity factor to the air input and extract units. Typically installations may operate where the air input unit and the extract fan are sized to 50% of the maximum design volume of the total fume cupboards.

This means that 50% of the fume cupboards can be open with the other 50% closed or all the fume cupboards can be half open, i.e., any combination of sash openings up to a total for all fume cupboards of 50% opening. Controls and alarms operate by measuring the face velocity on the fume cupboards. An alarm would be activated centrally and/or on individual fume cupboards if the total 50% opening is exceeded. Application of a diversity factor with an integrated control and alarm system can result in very substantial cost savings.

343

Secutromb auxiliary air fume cupboards
Conventional fume cupboards achieve their containment by extracting large volumes of heated laboratory air to provide a sufficiently high face velocity to contain fumes. This process can result in a high rate of room air change and heating or conditioning of this air is often expensive. Furthermore, should additional fume cupboards be required, it may not be possible to supply sufficient air necessary for efficient extraction to these cupboards.

The Secutromb fume cupboard works on a completely new and novel principle which involves auxiliary air from outside the laboratory area being supplied to the cupboard. As a result of the configuration of the cupboard's air input plenum ducts and the positioning of the extract take-off ducts, two contra rotating vortices are formed. As the air in the centre of the vortex is moving faster than the air on the outside, a negative pressure is formed in the centre of the vortex and any fumes generated within the fume cupboard migrate to and into the vortex and are then extracted via the extract take-off ducts.

In practice, this means that there is a vertical extract column at each side of the cupboard over its whole height. This ensures good scavenging of the cupboard and, very importantly, the concentration of fumes within the cupboard is very much lower than that found in conventional cupboards. Up to 70% of the total extract volume can be supplied as auxiliary air with only 30% needing to be extracted from the laboratory itself.

The auxiliary air must be heated and be within 8°C of the laboratory air. In air-conditioned laboratories, no cooling is necessary. Auxiliary air should, however, be filtered to ensure plenum gauzes do not become blocked as a result of atmospheric contamination.

As an additional safety feature, an airflow controller may be incorporated. This microprocessor-controlled system monitors the rate of flow of supply/auxiliary and extract air and controls their flow rates within pre-set limits. Airflow controllers have audible and visual alarms to warn in the event of either auxiliary air or extract system failure.

9.9 Safety and containment

Filters
Fume cupboards used for radio-isotope applications normally require the extract system to be fitted with HEPA filters and, in some instances (for example, isotopes of iodine), carbon filters may also be required.

With minor modification and the addition of a pneumatically operated volume control damper, the unit can be used as a constant face velocity module (i.e. total extract volume is variable dependent on sash position).

Maximum permitted leak concentration of test gas in accordance with DIN 12 924

Front sash closed	0.2 ppm
One-third open	0.5 ppm
Fully open	0.8 ppm

Sash lock/airflow failure alarm module

A combined sash lock/airflow failure alarm module should be designed to satisfy statutory safety standards. Generally the unit would comprise three separate parts:

- alarm airflow sensor;
- Printed Circuit Board (PCB) assembly in an enclosure;
- annunciator front fascia plate.

The alarm airflow sensor is typically a hot wire anemometer device. A sensor uses two signal diodes, one of which is heated. The diode is cooled by ambient air passing over it, its signal then being compared with the second unheated diode, which acts as a comparator for variations in ambient air temperature. Velocity sensors are usually installed in the top panel of the fume cupboard and produce a stable signal, which represents the face velocity.

Volt-free contacts may be included for remote monitoring and fume cupboard status and for fan stop-start relay.

Typically the annunciator face-plate incorporates an analogue meter showing 'Safe-Unsafe' face velocity with green and flashing red indicator lights. It may also feature an audible alarm with a mute button to show low face velocity. The face-plate can incorporate fan 'stop-start' buttons. A sash 'high' release button with red and green indicator lights is used where the sash is raised above its working height for setting up experimental apparatus in the cupboard.

Process development facilities and pilot plants

10

ROY KENNEDY and KEITH PLUMB

10.1 Introduction

Process development facilities and pilot plants are an integral part of research and development operations for all major pharmaceutical companies seeking to provide new products for the future. Their design, construction, commissioning and validation have their own special problems arising out of the individual company's traditional research methods, the class of compounds to be developed and the regulatory requirements.

These facilities are frequently multi-purpose and/or multi-product and the processes used are constantly under development. The design requires a degree of 'crystal ball gazing' because future requirements usually need to be included in the specification.

The full range of pharmaceutical processing needs to be covered by process development facilities and pilot plants from chemical synthesis to production of the active pharmaceutical ingredient, through physical manipulation to formulation, production of the final dosage form, filling and packaging.

The problems for chemical synthesis facilities are often different to those for other facilities. In the case of chemical synthesis facilities, the large number of chemicals used exacerbates the difficulties. In the other facilities it is often the problems of cross-contamination and the variety of machines required that dominate.

Pilot facilities for primary and secondary manufacture require a greater degree of flexibility for the reconfiguration of equipment compared with general production operations. This is easier for secondary operation than for primary as the reactors and other items of chemical apparatus are more difficult to reposition and link into each other.

For secondary operation, although it is necessary to have a dedicated sterile unit, all other operations are usually self-contained.

This chapter summarizes the main design requirements that are necessary in these facilities for development and small scale manufacture. The detailed requirements follow the principles for primary and secondary operation in earlier chapters.

10.2 Primary and secondary processing

The division between primary and secondary processing is to some extent arbitrary and different manufacturers place the dividing line at different places in the total manufacturing process. In general all the chemical stages up to and including the manufacture and purification of the active pharmaceutical ingredient are part of primary processing. In some cases, physical manipulation processes such as milling are also included. All the steps after purification (except in some cases milling) are usually included in secondary processing.

The decision of where to place the dividing line is often based on:

- the type of purification and physical manipulation processes that are required;
- the chemicals used within the purification and physical manipulation processes;
- the need for the primary process to stop at a point where meaningful samples can be taken;
- the type of building and facilities available to the manufacturer.

10.3 Process development

There are a number of stages in the development of pharmaceutical products. These stages are driven by the regulatory process, which is summarized in Chapter 2. The initial research that involves searching for new chemical entities is usually carried out at the laboratory scale and is not discussed further in this chapter.

Once a promising new chemical entity (NCE) has been discovered, tests will begin on the compound to confirm that it has the required activity, stability, and low toxicity. It will also be necessary at this pre-clinical stage to identify that the compound can be synthesized by a practical route and that it can be purified and formulated.

Much of this pre-clinical trials activity will be carried in the laboratory but it may be necessary to carry out some work in a small-scale pilot facility. Once this work has been completed, the clinical trials themselves start. These are carried out in three stages with increasingly large quantities of material. For the first stage, a small-scale pilot facility (in the order of one-hundredth of production scale) will usually be sufficient, whereas normal scale pilot facilities (one-tenth production scale) are usually required for stage three clinical trials.

During process development the whole manufacturing process will need to be both scaled up and optimized. Initially, this will be carried out in the small-scale pilot plant, followed by the normal scale pilot plant. The final scale up work will be carried out using production scale equipment and this may take place some time after the product launch since it is often possible to produce launch quantities of material using the pilot plant.

Process optimization needs to take place early in development because the regulatory authorities require the stage three clinical trials to be carried out using material produced by the same manufacturing process as is used for the full scale.

Stage three clinical trials are usually carried out at one-tenth of production scale because the regulatory authorities expect the scale up from the development scale to the product scale to be no more than ten fold.

10.3.1 Good manufacturing practice

Once the clinical trials start it is necessary to produce all the material required based on the GMP requirements detailed in chapter 3. Although the regulations are directed primarily at the stages after chemical synthesis, the principles should be applied throughout the whole manufacturing process.

Contamination by operating staff

The operating staff in a pharmaceutical facility is likely to be the main source of product contamination. Body particles are continually shed as people move around. Microbiological contamination is always a problem and all stages of production, apart from the early stages of chemical synthesis, will require a hygiene regime. The operating staff can cause cross-contamination between different products and/or different intermediates by material spilt on their clothing. Clean clothing needs to be regularly supplied.

Small-scale facilities involve a large number of manual operations, so contamination by the operating staff may be greater than on a production scale. For particularly sensitive products, this may require a high level of protective clothing for the operators and/or the use of laminar flow booths or glove-box isolators.

Cross-contamination

Cross-contamination between different products and/or different chemical intermediates is a major source of drug adulteration. Since small facilities can be used to make a large variety of products and/or intermediates, the possibility of cross-contamination needs to be addressed at the design stage. Issues to be considered include:

(a) Easy to clean equipment

Small-scale chemical synthesis equipment is cleaned either manually, by using mobile cleaning rigs or by refluxing with solvents. Equipment will need to be accessible and may need to be disassembled to allow access. This is a design requirement.

Mobile cleaning rigs using high-pressure hot water jets with or without the use of detergents can be useful for cleaning. However, these rigs can only be effective if all contaminated surfaces can be accessed. It can be used to clean large bore pipework. Careful consideration will need to be given to the possible safety hazards. Hot water can easily lead to scalding of operating personnel and some strong detergent solutions are particularly corrosive.

Refluxing with solvents can be useful for reaction equipment fitted with a condenser.

Other small-scale equipment will almost certainly need to be stripped down for cleaning.

In all cases, it is important for the equipment to be constructed without crevices that hold chemical, particulate or microbiological contamination and to have a surface finish that is inherently easy to clean. Glass, electropolished stainless steel and PTFE are common easy to clean finishes.

(b) Primary containment

Primary containment is based on the actual equipment used to do the processing, for example, reaction vessels should have a closed top and a seal on the agitator. Full primary containment can be achieved when solids are charged via glove-box isolators and liquid connections are made via hygienic dry break connections.

Primary containment minimizes the need for secondary containment and reduces the building standards required.

(c) Secondary containment

Secondary containment involves placing the manufacturing equipment in some form of ventilated enclosure resulting in a number of conflicting issues:

- segregating flameproof areas from safe areas;
- protecting the product from contamination by the operating staff and the environment;
- protecting the operating staff from highly active materials.

These three requirements can be resolved by the use of air locks, the correct pressurization routines and correct extraction and ventilation regimes. For more details see Chapters 6 and 8.

Some enclosures are large enough for the operating staff to enter via air locks. In other cases the equipment can be enclosed in a down flow booth with an open front that allows the equipment to be operated with sufficient air velocity to protect both the operator and the material being produced.

With the smallest scale equipment, it is possible to place the equipment inside an isolator with equipment being operated via gloves or a half suit. Such a system offers a high level of protection and is frequently used when the material being produced is highly potent or highly active.

(d) Ventilation

Ventilation can be used in laminar down flow or cross flow booths to protect the product from cross-contamination and the operating staff. Also an increase or decrease in pressure in different areas prevents the flow of air into or out of the room.

Ventilation systems require careful design because they can be the cause of cross-contamination themselves, particularly if one ventilation system is used to serve more than one area.

(e) Cross-contamination by the operating staff

This is minimized by making efficient use of the containment system, by ventilation, providing clean clothing at regular intervals and appropriate changing facilities.

Materials of construction

The materials of construction for pharmaceutical equipment are covered by both European and American guidelines and regulations.

Paragraph 3.39 of the European Guide to Good Manufacturing Practice states: 'Production equipment should not present any hazard to the products. The parts of the equipment in contact with the product must not be reactive, additive or absorptive to such an extent that it will affect the quality of the product and thus present any hazard.'

Section 211.65 of the Code of Federal Regulations title 21 states: 'Equipment shall be constructed so that surfaces that contact components, in-process materials, or drug products shall not be reactive, additive, or absorptive so as to alter the safety, identity, strength, quality, or purity of the drug product beyond the official or other established requirements.'

To meet these regulations, it is necessary to specify materials of construction that are corrosion resistant, easy to clean, do not release material into the process by leaching of the material or absorb any of the process materials. The most commonly used materials are glass and stainless steel. These are corrosion resistant and easy to clean if constructed correctly. However, it is possible for glass to absorb ions from some chemicals, which can lead to cross-contamination, and materials can be leached out of stainless steel unless the surface is correctly treated.

Polymeric and elastomeric materials need to be chosen to have the widest range of chemical resistance as well as being able to withstand the range of temperatures likely to be encountered. These materials are particularly problematical with respect to chemicals being leached out of them because they generally include a range of plasticizers to improve their stability or flexibility.

Whatever materials are used, documentation will be required to demonstrate that the specified material has been installed.

Surface finishes

To ensure that equipment is easy to clean, liquids drain easily and solids do not adhere to walls, it is necessary to consider the surface finish of both the inside and outside of equipment.

Metals normally need some form of treatment such as polishing. Mechanical polishing of metals requires the use of grits that are held together by soaps and grease and these can become embedded in the surface and lead to product contamination. At the small scale this problem is best overcome by having the metal surface electro-polished. Mechanical polishing is suitable for the external surface of metal.

Non-metallic materials such as glass and PTFE have an inherent smooth and easy to clean finish both on the outside and the inside. However, other non-metallic materials are much less smooth. Lining these with PTFE may be required. An external finish will need to take into account the likelihood of damage due to manhandling the equipment.

Material storage and handling

Systems to ensure that intermediates and products are not confused are of fundamental importance to good manufacturing practice. At the small scale there is likely to be a large number of materials to be stored, made by different processes. The storage handling system must be able to prevent different materials from being incorrectly identified and must prevent the same material made by differing processes from being mixed.

351

10.4 Small-scale pilot facilities

10.4.1 Chemical synthesis – primary manufacture

Reaction equipment

Small-scale pilot facilities with capacities ranging from 20–100 litres are generally required for the chemical synthesis stage of the manufacturing process. To provide a high level of corrosion resistance such facilities usually use glass equipment that can be configured for the particular processes taking place. In some cases the whole rig is built from scratch and then dissembled when it is not required. This type of rig is often called a 'kilo-lab'.

Solids handling equipment

Simple filters, centrifuges and dryers will be required since most pharmaceutical intermediates are solids. The solids handling equipment will be corrosion resistant and mobile so that it can be connected to the reaction equipment.

Depending on the quantities of material used, stage three clinical trials material may be produced by this equipment, which will replicate the type of equipment that will be used on a production scale and have a similar modus operandi. This solids handling equipment can be hired from equipment vendors.

Small-scale solids handling equipment suffers from the problem that much of the product may be held up in the equipment and consequently the yield is very low.

Multi-purpose equipment

Some specialist vendors provide multi-purpose equipment that can be used for reactions, filtration and drying. These units have the advantage that they reduce handling and, thus, reduce the exposure of the operators to the chemicals. However, as this equipment is complex it is usually a compromise and the result is not cost effective and less than optimal for each unit operation.

Solvents

Most chemical syntheses use flammable solvents which means that the small-scale facility will need to be a flameproof area. Since these facilities are often located within laboratory complexes it is necessary to separate flameproof areas from safe areas. This can be achieved by the use of pressurized air locks and in some cases the pressurization of the safe areas.

Small-scale facilities make use of a large number of solvents usually handled in drums requiring a flameproof drum handling and storage area, outside the building to reduce ventilation needs. A method of safely transferring the solvents from the drums to the manufacturing equipment is required. One method involves moving the drums from the drum store to a dispensary area, where the required quantity is decanted into a safe solvent container that is used to transfer the solvent to the reaction area. In other cases intermediate containers may be used to transfer the liquid from the drum store to the dispensary.

Since a large number of solvents are used in small-scale facilities it is unusual to find solvent recovery facilities included in the area, unless one or more solvents used in larger quantities can be recovered using the equipment used for the chemical processes. The recovery of solvents prevents cross-contamination and enables them to be disposed of safely.

Toxicity

Although the final drug product manufactured may have a low potency, the chemical intermediates that are made during the synthesis of the active pharmaceutical ingredient are often highly potent. The design of the facility must ensure that the operating personnel are protected. This in part may be covered by the building design, but also it will require the use of fume cupboards, local extract ventilation, glove-boxes, rapid transfer ports, contained transfer couplings and air suits.

Environmental considerations

The chemical synthesis route of many pharmaceuticals is highly complex (see Figure 1.1, page 3). In many cases more than 20 intermediates are made before the active pharmaceutical ingredient is prepared. Even if every stage has a high yield the overall yield can be very low. This means that facilities must be provided for all the waste streams to be handled.

The large variety of chemicals produced in low volumes usually precludes the use of an on-site effluent treatment plant for handling all the waste streams. Liquids and solids must be put into groups that can be mixed together for disposal; for example, halogenated solvents will need to be separated from non-halogenated solvents.

Depending on the quantities involved and their toxicity, vapour and gaseous emissions will be treated. Vapours can often be condensed using a low temperature system — the use of a liquid nitrogen cooling system is economical at the small scale. Solvent, acid or alkali scrubbing systems may be required for the gaseous emissions. The choice of equipment will depend on the chemicals used and the flexibility required.

10.4.2 Physical manipulation

Physical manipulation is a process not involving a chemical reaction that changes the purity of the material. It usually involves crystallization, filtration, chromatography, milling, drying or blending for example. This type of process is frequently required to achieve one or more of the following requirements:

- crystal morphology;
- moisture content;
- specific particle size;
- particle surface physico-chemistry.

Depending on the product, the equipment for crystallization, filtration and centrifugation may be the same equipment as is used in the chemical synthesic process, and so most of the comments made in Section 10.4.1 are relevant. However, other equipment is used to carry out a particular operation, such as milling, micronization or granulation.

To achieve maximum flexibility this equipment needs to be mobile. In some cases developers may hire this equipment from the vendor when it is required.

Many organic solids are explosive when finely divided and require explosion protection and it is likely that the most appropriate method will be to use inert gas blanketing.

10.4.3 Manufacturing the final dosage form – secondary manufacture

The first stage of the manufacturing process is formulation. This is the process of adding the drug(s) to one or more excipients (see Chapter 6 for more details of excipients) to provide the correct mixture for the final dosage form. These may be solids or liquids depending on the final dosage form.

Liquids, gels, creams and syrups

If the final dosage form is a liquid, gel, cream or syrup then the equipment used for chemical synthesis may be suitable for the required blending operation. However, some formulations such as those required for aerosols require specialist formulation equipment because the propellants used are pressurized liquids with vapour pressures in the region of 3 to 4 bar g.

Conversely it may be advantageous to use equipment located close to the filling equipment, which may require dedicated formulation equipment, so that it is possible to run the formulated product directly to the filling machine. If this is not then the product would be transferred into one or more intermediate vessels and moved to the filling area. Rapid transfer between formulation and filling is a particular requirement with terminally sterilized products, as these must be formulated, filled and sterilized within 24 hours.

The choice of whether to use equipment directly connected to the filling equipment is determined by the nature of the product and overall facilities available to a company.

Solids

When the final dosage form is a tablet or pellet a solids mixing system is required. Small specialist solids mixing equipment is usually provided for formulation. At the smallest scale this equipment may be hand operated and similar to a modern version of a pestle and mortar.

Solid dosage forms, such as tablets, capsules, suppositories and solid dose inhalers require a second manufacturing stage beyond formulation for their production.

The machinery required is highly specialized and designed to carry out a particular task. Whilst hand operated bench scale equipment exists, this is usually only used to test the formulation and demonstrate that the required final dosage form can be produced. Such equipment is suitable for use at the pre-clinical stage.

Once material is produced for clinical trials, small-scale automatic machines is required. Since these materials are designed to make specific final dosage forms, pharmaceutical companies often specialize in a small range of dosage forms. This reduces the number of machines required.

10.4.4 Filling

Filling is the process of putting the finished pharmaceutical product into its primary container, which may be a bottle, vial, ampoule, tube, aerosol can, or blister pack.

In the early stages of clinical trials, automated filling machines may not be used for tablets and capsules as these can be filled and packed by hand.

Suppositories are filled by machine as they are easily damaged, and solid dose inhalers will almost certainly be filled by machine due to their complexity. However, they will only be simple semi-automatic machines.

For liquid products, the filling operation produces the final dosage form. When only small quantities of these are required hand operated bench scale machines may be used, larger quantities will require automatic machines. These are specialist machines and, as with solid dose machines, companies tend to specialize in a few dosage forms.

With liquid filling it is usual to connect the formulation equipment to the filling machine so that the liquid can be transferred directly. In production facilities it is common to have completely integrated filling lines with filling, check weighing and washing connected together. At the smaller scale flexibility can be increased by keeping the individual machines separate and manually moving the filled packs from one unit to another.

10.4.5 Packing

Most pharmaceutical products are sold in some form of secondary packaging. This gives protection to the primary packaging and allows detailed instructions to be included with the product. Packing is the process of putting the product already in its primary packaging into its secondary packaging.

At the early stages of clinical trials, this can be carried out by hand. However, once the required quantities increase to more than a 1000 containers, a semi-automatic packing machine is usually necessary. If it is expected that the product will be packed by machine during the production process, then the chosen pack(s) will need to be tested on the packing machine during clinical trials to prevent a delay to the product launch.

There are several stages to packing:

- labelling the primary packaging;
- putting the primary packaging and instructions into the secondary packaging;
- printing lot specific information on the secondary packaging;
- fastening a tamper evident label to the secondary packaging;
- over-wrapping the secondary packages into collated parcels;
- packing the over-wrapped parcels into cases.

Maximum flexibility can be achieved by using semi-automatic operations with each machine separated and fed by hand. The placing of the primary package and the instructions in the secondary packaging is usually a manual operation. The machine then folds and closes the secondary packaging, carries out any external printing and attaches the tamper evident label. Case packing is usually carried out by hand at this scale. Hiring the machines from the vendor or using contract packing-companies may be an option.

10.4.6 Building design

To handle the large number of processes reaching the pre-clinical trials stage, the building layout must be flexible and allow the use of mobile equipment. Often the buildings for small-scale facilities consist of a number of processing rooms on the ground floor with a service floor above providing all the required services such as air conditioning. The process rooms may have technical spaces for other general purpose equipment, such as hydraulic power packs, vacuum equipment, or condensers. The rooms can also be used for access to some of the pipework as it enters the process space.

The rooms are fitted out with a minimum amount of furniture and process equipment so that mobile equipment can be moved around and equipment set up.

In some instances, one part of a specialized fixed equipment item is designed to be placed in a clean environment while other parts are designed to be installed in a technical space. Examples of this are horizontal dryers and centrifuges. The materials being handled are fed into the machine in the clean area and discharged in the clean area whereas the mechanical parts of the machine and the solvent handling equipment are located in the technical space. This is achieved by siting the equipment in the wall of the room.

Depending on the level of instrumentation and control, it may also be appropriate to have separate control rooms away from the processing rooms. It is usually advantageous to have the control room adjacent to the processing rooms to be able to observe the operations.

Changing rooms

Changing rooms are an integral part of any pharmaceutical facility. For small-scale facilities these will need to be designed to ensure that the operators can be dressed in suitable clothing, that cross-contamination does not occur and that any highly active materials are not carried out of the building on clothing.

A number of different changing rooms might be required to allow access to different parts of the building.

Equipment store

With small-scale facilities making use of mobile equipment, consideration must be given to the clean equipment store. It must ensure that the equipment is not damaged during storage.

Equipment may need to be stored on GMP pallets so that it can be moved easily and so that multilevel staging can be used to save space.

Each unit should be numbered and have a log book which clearly identifies its status (clean/dirty) and the processes for which it has been used. There should be an appropriate place for signatures of the operators and supervisors.

Access for potable equipment

To be able to move equipment around a building safely, sufficient access for movement should be designed. Consideration should be given to:

- the width of corridors;
- turn areas;
- size of doors;
- size of lifts;
- size of transfer hatches.

Office/write up areas

Experimental work generates large quantities of data and reports. Some writing areas will be required within the development areas adjacent to the equipment. In other cases it is necessary to have an office and write up area out of the main development areas but within the same building. This is because some processes run for a considerable time and only need to be visited for short times but at regular intervals. Often it is necessary to go through several different change areas, one after another, in order to arrive at an area of a higher or lower status within a building and this can take some time. Offices between the changing areas allow this time to be reduced.

Environmental control

Pharmaceutical products need to be handled in controlled environments to prevent contamination. With small-scale, flexible, frequently manually operated equipment, it may be difficult to provide primary containment and, therefore, high quality secondary containment is required. (See Chapter 8 for more details of room environments).

To achieve the required flexibility, it may be necessary to provide the equipment to supply many of the rooms with high quality air. To prevent cross-contamination it may be necessary to provide each room with its own stand-alone system.

Fume extraction and the use of flammable solvents will have an impact on the choice of equipment to be used for environmental control.

Laboratory

Since development requires many experimental tests to be carried out and adjustments are made to the process on the results of these tests, an in-house laboratory is necessary. In some cases this may be close to the process and, to reduce testing time, may be inside the area controlled by the innermost changing area.

Airlocks/pressure regimes

The pressure regime within a building must ensure that air flows in the desired direction. The pressure regime along with the air locks between each area must be designed to prevent the following arising:

- product contamination;
- cross-contamination;
- flammable vapour/dust contacting a non-flameproof and non-explosion proof electrical equipment;
- highly active compounds contacting unprotected operators or the outside environment.

Engineering workshop

Small-scale equipment is often built into test rigs and modified frequently as the process develops. With equipment in controlled environments and operators having passed through a number of change areas, it is often appropriate to have a small engineering workshop close to the process rooms. This area must be carefully designed to ensure that tools are not lost and that the area does not become a source of contamination.

Movable walls

Processing areas can be made more flexible if movable walls are used. To achieve this, the services need to come through the ceiling where possible. With the correct choice of materials it is possible to have movable walls even when a very high quality environment is required.

Communication between areas

With the need for operators to be dressed in appropriate clothing for different areas and with need to protect the product, it is not possible to walk around a pharmaceutical facility with ease. This means that communication between areas can be difficult.

Consideration should be given to speech panels, intercom systems, transfer hatches, visual panels and CCTV system to improve communication. Consideration should also be given to the safety of personnel working in areas that may be 'remote' from other areas within the building. This is particularly relevant where hazards exist.

Equipment cleaning

Dedicated equipment cleaning areas will be required. In some cases solvents are used for cleaning and this will require explosion proof electrical equipment.

Automated washing machines can be used and these have the advantage of reducing the labour requirements, producing reproducible results and keeping all the liquids handling equipment in the technical spaces.

Building services

For a flexible small-scale facility it will be necessary to provide a wide range of services to some or all of the process areas. The services will depend on the processes being carried out, but are likely to include:

- water for injection (not usually required for the early stages of chemical syntheses);
- purified water;
- potable water;
- compressed air;
- breathing air;
- nitrogen;
- vacuum;
- air conditioning with temperature and humidity control;
- fume extraction (usually only required for chemical syntheses or where solvents are used);
- steam;
- cooling water;
- single fluid heat transfer fluid;
- services for solvents used in high volumes (e.g. recovery for safe disposal).

10.4.7 Controls and instrumentation

The control and instrumentation requirements for a small-scale facility will depend on the range of products being made and the equipment being used. The following considerations will need to be taken into account:

- the equipment selected will have to be compatible with the environment in which it will be used;
- the instrumentation should be suitable for in-house calibration so that it is not affected by the many processes used;
- control systems loops should be short, simple and flexible.

10.5 Chemical synthesis pilot plants

10.5.1 Introduction

According to a senior executive from one of the pharmaceutical industry's major multinationals, the future of the pharmaceutical industry will be 'moulded by science, shaped by technology and powered by knowledge'. His views would no doubt be shared by the bosses of the other top nine pharmaceutical companies who, in the previous 12 months spent between them over £10 billion on research and development.

Pilot plants are an essential component of the R&D operations of all major pharmaceutical companies seeking to provide new products for the future. The particular requirements for the design of each individual pilot plant will depend very much on the company's traditional research methods, the class of compounds likely to be developed and the regulatory requirements. However, there are some features that must be considered in every case.

A typical pilot plant for primary chemical manufacture will normally be used to transform chemical processes from the original laboratory bench procedure towards practical industrial scale manufacturing facilities. Alternative process routes will be compared and evaluated until the optimum mix of process safety and operability, product quality and manufacturing cost are achieved.

The pilot plant will also be used for the synthesis of samples and supplies to be used for formulation development, clinical trials, safety assessment and stability testing. It will normally comprise facilities and equipment for dispensing, reaction, separation, filtration and drying and finishing and will, therefore, normally include downflow booths, reactors, filtration equipment, a range of different types of dryers, and sieving, milling or micronizing equipment.

When the engineer is asked to produce a design for a new chemical pilot plant, the main challenges will include:

- scope definition;
- multi-product and multi-process capability;
- flexibility;
- GMP operation;
- layout;
- regulatory requirements;
- political aspects.

The following sections look at each of these areas in more detail.

10.5.2 Scope definition

Each pharmaceutical manufacturer has their own ideas on the best pilot plant to suit their needs. For example, when asked their opinion following a tour of a competitor's highly complex fully automated plant, the pilot plant manager from a major pharmaceutical company replied: 'I would be much happier with a glass bucket and a thermometer!'

The point is that it is extremely important to adopt a team approach when working on scope definition. The team must include the ultimate user(s), bearing in mind that these people are normally chemists or pharmacists and are not always aware of the impact of seemingly small changes on the overall engineering design.

When plant facilities to handle novel processes are being designed, it is unrealistic to expect that the user's needs would be fully specified from the start. The process parameters are generally unknown, so the only way to proceed is to develop a capacity model by considering sample processes.

The capacity model can then be reviewed against previous pilot plant activity and the perceived business needs.

The useful life of a pilot plant should be at least ten years, so it pays to spend time at the front-end of the project speaking to the business managers and considering how the company's future products may evolve.

10.5.3 Multi-product capability

The plant must have the capability to permit the handling of future unknown compounds. This may be obtained by:

- using simple (manual) material handling systems;
- using materials of construction for the equipment and pipework that have a high resistance to corrosion;
- providing a high degree of product segregation to prevent cross-contamination;
- providing a high level of containment to protect the operators and the environment;
- providing cleaning systems that allow rigorous decontamination between different product runs.
- using materials of construction that do not react with the product contact parts.

10.5.4 Multi-process capability

In order to provide this capability, the pilot plant will need:

- a speculative range of vessel sizes (typically 50 to 2000 litres) in a suitable mix of materials of construction, based on the capacity model developed earlier;
- vessels with variable volume capability, e.g. double jacket reactors;
- variable temperature capabilities for the reactors, possibly via the use of a single heat transfer fluid system. A typical plant provides heating/cooling in the range of 150°C to −30°C;
- portable/mobile equipment, which allows equipment to be brought closer together avoiding complex piping runs and provides better utilization of available space;
- services such as water, air, steam, nitrogen, heat transfer fluid and perhaps solvents, should be piped to all areas where it is remotely possible that processing will take place, including areas set aside for future expansion. This will allow maximum flexibility and provide a hedge against changes of function due to market forces;
- a high quality de-mineralized water system providing a supply to purified water requirements;
- equipment that is suitable for Cleaning In Place (CIP), in order to reduce downtime between processes.

10.5.5 Uncharacterized products/processes

The very purpose of a chemical pilot plant, i.e. to synthesize New Chemical Entities (NCEs), means that the potential hazards of the processes and compounds involved are not normally known at the time the facility is being designed. It is, therefore, necessary to provide high levels of primary and secondary containment.

The dispensary design will have to allow for raw materials with widely differing hazard potential, which are received in a wide variety of packaging sizes and shapes.

Most pilot plants have down-flow booths for operator protection during dispensing and a local extract ventilation system provided across all other areas. Other containment options, depending on the severity of the hazard, include glove boxes and full air suits.

It is a key part of the design function to classify the types of compound that will be entering the facility and adjust containment levels accordingly.

Another aspect of containment is the need to restrict atmospheric or other emissions of harmful substances to levels that are acceptable to the Environment Agency. For example, releases of Volatile Organic Compounds (VOCs) such as solvents must be prevented and will require the installation of a scrubbing or recovery system.

Good operating procedures in compliance with the legislation require that the volume of all waste materials is kept to a minimum and that all hazardous waste is disposed of in a safe, legal and traceable way.

The multi-function basis and the lack of a defined process, may mean that novel methodology will be required to allow meaningful Safety, Health and Environmental (SHE) reviews to be carried out. Typically, this would involve the development of system envelopes (including control systems), which would be reviewed against guidewords to ensure that the design is sound. Such reviews would be expected to highlight those issues that are chemistry specific. These areas would have to be noted, and then developed in more detail prior to the introduction of each new process into the plant.

10.5.6 Operation

The way in which a chemical pilot plant is operated depends very largely on its designated purpose, but also on the traditions of the client/owner. However, because of the unknown and potentially hazardous nature of the compounds and processes to be employed, many major companies prefer to have the reaction areas of their pilot plant normally unmanned.

This is of course contrary to the chemist's preference for reaction visibility. Typically they like to observe changes of colour or state as the reaction proceeds.

On a smaller capacity plant, which is operated at medium temperatures and pressures, this requirement may be satisfied by using borosilicate glass equipment allowing the operators to observe the reaction areas through windows.

On larger plants where the processes involve more onerous conditions, the use of glass is not tenable. In this situation, some companies have provided the chemists with the possibility to make real time observations of the reactor contents by using closed circuit television cameras.

10.5.7 Layout

As with most of the other topics discussed in this section, the type of layout adopted by the design team will very much depend on the owner's past experience and culture.

Free access is highly desirable to allow easy maintenance and enable the inevitable plant modifications.

Many modern pilot plants have adopted a vertical modular arrangement (see Figure 10.1) which allows gravity feed to be used in processing and is well suited to moving products between the modules via flow stations. However, this type of arrangement is by no means universal. A large number of manufacturers still prefer the traditional 'reactor hall' arrangement with separate areas for the finishing steps including filtration, drying and particle size reduction.

One point worth mentioning is the high level of HVAC that chemistry pilot plants will require in order to provide the required level of air filtration, pressure differentials and clean environments. This means that the routing of process

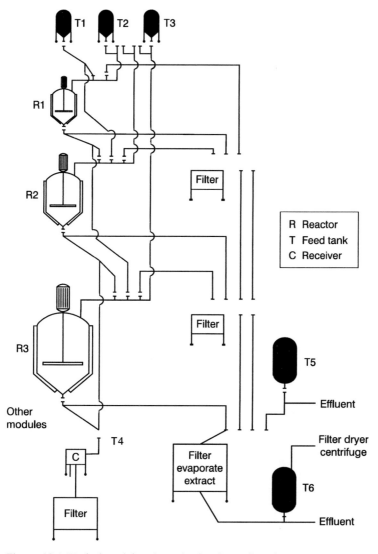

Figure 10.1 Typical module schematic chemistry pilot plant

pipework and building services ductwork will be a critical task. It is wise to decide at an early stage in the project to separate these two major services to avoid possible clashes.

10.5.8 Controls

If you ask a typical pilot plant user what type of instrumentation and control system they prefer, they will invariable reply 'simple!' This is fine when you are working with passive substances and reactions, but totally unsuited to the needs of the modern pharmaceutical research establishment.

The main factors affecting the choice of control system are:

- data acquisition and storage;
- operational safety;
- multi-functional requirements;
- environmental aspects;
- regulatory compliance.

The raison d'être of the pilot plant is to research and develop alternative process routes for the preparation and scale-up of NCE's for pharmaceutical products. In order to achieve this mission, it must have a system for the recording, storage, retrieval and collation of the critical parameters observed during each process run.

Many pharmaceutical products are themselves highly active, or are manufactured from highly active materials. This requires high levels of containment to protect the pilot plant operators. If containment fails, the control system must stop the process and activate a fail-safe alarm procedure to direct uninvolved personnel away from the area of risk.

In addition to highly active substances, the controls will be required to alert the operators to runaway exothermic reactions and possibly detect leakage of flammable compounds.

Some pharmaceutical products have a hydrogenation step in their manufacture. Hydrogen has very wide explosive limits and very low minimum ignition energy. A suitable control package in this case would, include at least hydrogen detectors and a trip system.

The multi-purpose capabilities required of most modern pilot plants can also have a major impact on the choice of control system. If the plant is reconfigurable the control system must allow for these changes. A pilot plant recently completed for a major pharmaceutical manufacturer has around 250 valid equipment configurations. In order to ensure that the configuration set-up is correct, a system of electronic tagging is scanned and checked by the control system for the required 'recipe'. If the arrangement is correct, the system reveals a password that must be manually entered into the process control computer before process operations can begin.

During the design of the above plant, it was found that one of the most economic ways of providing flexibility whilst still meeting processing and containment requirements was to use mobile equipment that could be installed at various locations throughout the plant. Each item of equipment has its own instrumentation and control requirements, which are identified, powered, controlled and recorded by the control system. When the equipment is correctly located, an umbilical cable is connected using a plug and socket, which provides the necessary signal and control for that item of equipment. At the same time the control system re-assigns the internal address of that equipment item to suit the new location.

It is not always easy to find instrumentation that will operate across the full temperature range of the pilot plant whilst still meeting GMP requirements. Often detailed studies must be undertaken to identify and select the most appropriate type of sensors to be installed.

The control system must monitor and control equipment that is installed to ensure that the emission limits laid down by the Environment Agency are not exceeded. It must take GMP into account and be suitable for validation to meet the requirements of the regulatory authorities.

10.5.9 Legal and regulatory requirements

Chemical synthesis pilot plants for the pharmaceutical industry must be designed to be safe and not pollute the environment. The multi-function basis and the lack of defined process will probably mean that novel methodology will have to be developed and agreed with the legislative authorities prior to Safety, Health and Environmental (SHE) reviews being carried out.

Due to the multi-process nature of the plant, safety reviews need to take place throughout the life of the facility. The initial reviews take place during the engineering design phase, then during commissioning and following that, whenever a new process configuration is required during operation.

As the plant will normally be used to manufacture small quantities of product for clinical trials and potentially subsequent marketing purposes, it must be designed to meet current Good Manufacturing Practice (cGMP) and be suitable for validation by the appropriate regulatory authority.

The pilot plant may also be used to demonstrate the suitability of the selected manufacturing process for industrialization. The normal scale-up factor permitted/accepted by the regulators is 10:1.

10.5.10 Cost

There is no precise guidance on the relative costs of pilot plants when compared to typical manufacturing facilities, other than that the unit cost of the pilot plant will always be higher.

The reasons for the higher costs are simply put down to the wide range of features previously described, which are employed to obtain maximum flexibility and benefit from what normally represents a major investment without guaranteed returns.

It should be expected that the ratio of engineering costs to overall costs would also be higher than for conventional manufacturing units.

The complexity of the pilot plant design to increase as engineering progresses should be expected and allowed for. This will be brought about as solutions are evolved to problems, and by new technology coming available which improve the general usefulness of the plant.

Validation cost is very significant and must be considered from the outset.

10.5.11 Political aspects

A new chemical pilot plant will often be of major strategic importance to the owner, not only because it provides the vital link in developing promising, newly discovered products to market, but also because it demonstrates to investors that this is very much a research-led organization, planning for future growth.

The new facility may often be the only facility of its kind within the company, so the design, layout and its worldwide location may be subject to thorough vulnerability analysis to ensure its security and availability.

10.6 Physical manipulation pilot plants

The equipment used for physical manipulation includes:

- crystallizers;
- filters;
- filter/dryers;
- centrifuges;
- dryers;
- mills;
- micronizers.

This equipment makes use of solvents and gravity flow and is used within the same facility as a chemical synthesis pilot plant. For equipment that falls into this category, most of the detail given in Section 10.5 will be appropriate. However, it should be remembered that physical manipulation is being applied to an active pharmaceutical ingredient (API) and that the equipment will need to be compatible with the GMP requirements.

In a few cases the physical manipulation equipment does not make use of solvents and gravity flow is of no particular advantage. In these cases this equipment may be included in a final formulation facility and the information contained in Section 10.7 will be appropriate.

10.7 Final formulation, filling and packing pilot plants

The equipment used in this type of pilot plant is a smaller version of the production scale equipment. Facilities are usually built to cope only with certain types of products. For example, a facility to manufacture tablets is likely to be able to cope with a large variety of different products because the processes involved in making tablets are similar even if the active ingredient is completely different. However, this facility would be completely different to one making inhalation products even if the tablet and the inhalation product contained the same active ingredient.

The design of these types of pilot plants is discussed in detail.

10.7.1 Cross-contamination

With the potential to use a large number of products within a pilot plant, cross-contamination is a problem, which means that containment is important. With automatic equipment dedicated to specific purposes it is possible to make use of primary containment to some extent, but with the need to make frequent changes and modifications it is probably wise to provide secondary containment. The secondary containment may be in the form of isolators around the equipment but it may be appropriate to have each piece of equipment in its own room.

To maintain flexibility it will be necessary to have easily cleaned equipment. Some use may be made of Clean In Place techniques, but it is inevitable that equipment will have to be disassembled. This can be one of the greatest sources of airborne particulates, which can lead to cross-contamination, so this need must be considered at the design stage. Rooms and isolation cabinets will need to be designed with easy cleaning in mind.

10.7.2 Material flow and storage

Due to the potential to use many different products and with processes being under development, it is easily possible to mix up materials. Materials flows need to be simple and prevent incompatible materials coming into contact with one another.

Good housekeeping is a major priority. Storage facilities must have sufficient space for easy access and materials must be readily identifiable. Separate areas for raw materials, quarantine materials and passed finished products are required.

10.7.3 Flexibility

At the production scale, equipment for the final formulation, filling and packaging is often connected directly together. This is good for the high production levels required at the full scale, and it allows a high level of automation and minimizes labour requirements. However, such systems are not flexible.

Flexibility can be increased by having stand-alone machines and moving the output from one machine to the next by hand. This requires a number of suitable mobile containers to be included in the design.

It is also possible that some of the smaller machines can be made mobile, which allows the facility to have a reduced number of processing areas with equipment not in use stored in an appropriate place.

10.7.4 Automation

With filling and packing it is necessary to automate the machines at one-tenth the production scale. However, to enable easy change between different products, the automation should be kept as simple as possible. Changes to the system must be possible without reconfiguring the computer software which would require a high level of documentation to validate software changes.

10.7.5 Building requirements

The building requirements for final formulation, filling and packing pilot plants is similar to that required for small-scale facilities with the following differences:

- fewer rooms are required but the rooms will be larger;
- if mobile equipment is used it will be larger. It may only be possible to move the equipment by having very large doorways or by having removable walls;
- fewer building services will be required and it is likely that each room will only be supplied with the services appropriate to the equipment used in that room.

10.8 Safety, health and environmental reviews

The requirement to carry out a number of different processes makes a SHE review difficult at the design stage. It is necessary to carry out some form of generic review and to examine those processes that are currently known.

The introduction of each new process will require further SHE audits to ensure no new problems have been introduced.

10.9 Dispensaries

Dispensaries are an important part of pharmaceutical processing and are described in Chapter 6. Since small-scale facilities and pilot plants use a large number of products, dispensaries are a major area of risk from cross-contamination.

Dispensaries need to be considered at the design stage and integrated with the operation of the facility. Sufficient space must be allowed to ensure operations are safe and efficient.

10.10 Optimization

Processes carried out within small-scale facilities and pilot plants are not usually optimized, because the facility is multi-functional. It is usually necessary to sacrifice speed of processing and product recovery in order to achieve flexibility.

Equipment should be chosen to ensure that it is:

- quick and easy to change between products;
- easy to clean;
- retains the minimum amount of product;
- simple to operate;
- conforms to GMP requirement.

10.11 Commissioning and validation management

The User Requirement Specification is always difficult to define for these facilities, however once the Design Qualification has been agreed and signed off Installation Qualification is similar to that for production scale commissioning except that it is an ongoing operation as new processes are being continually introduced.

Performance Qualification is more of a problem because data will need to be added to the validation files each time a new process comes on-line.

Pilot manufacturing facilities for the development and manufacture of bio-pharmaceutical products

11

TINA NARENDRA-NATHAN

11.1 Introduction

Biotechnology, 'the application of biological systems and organisms, to technical and industrial processes and products' is not a new discipline. The fermentation of grain using yeast to produce alcohol has been taking place for centuries in most cultures throughout the world. However, advances over the past 20 or so years in the field of molecular biology and hybridoma technology have provided us with many new opportunities for improved processes and products. Human healthcare in particular is now beginning to benefit from these rapid advances in modern biotechnology, proving that it offers much more than just the promise of new drugs to solve many of the serious health issues facing mankind. The first bio-pharmaceuticals reached the market nearly a decade ago and are making a significant contribution not only to health care around the world, but also to the finances of the companies manufacturing them.

Bio-pharmaceuticals, which generally include vaccines, blood and blood products, allergenic extracts, and biological therapeutics, are regulated under a whole range of guidelines from a variety of regulatory authorities. These authorities require that bio-pharmaceuticals be manufactured and prepared at a facility holding an unsuspended and unrevoked licence. Lack of clarity about licensing requirements can lead one to make major investments in large-scale manufacturing facilities before initiating the clinical trial(s) necessary to demonstrate the safety and effectiveness of the products. Such investments can result in significant financial loss if the product is not ultimately brought to market. This chapter will attempt to clarify the regulatory requirements for the use of small-scale and pilot facilities. For details of regulatory aspects see Chapters 2, 3 and 4.

The principals that apply to small-scale and pilot plant facilities equally apply to manufacturing facilities.

11.2 Regulatory, design and operating considerations

11.2.1 Regulatory considerations

The development of important new biological products is expensive and time-consuming and companies must be able to forecast and evaluate their expenditures for this process. Constructing a new large-scale facility to manufacture a product that has not been fully tested in clinical trials could result in a major financial loss, with the company being unable to recover a major capital expenditure if the product is not ultimately brought to market. For some companies the best financial option may be the use of a pilot facility where the product may be manufactured at a smaller scale than would be for an approved product. While regulatory authorities do not object to the use of pilot production facilities for the manufacture of clinical material, provided such manufacture is in compliance with the requirements applicable to investigational drugs, many companies are concerned that these facilities and the products manufactured in them would not be eligible for establishment licensure.

Although the advances in the technology have been staggering, it must be recognized that the same basic regulations and requirements are still applicable to the manufacture and control of bio-pharmaceuticals as for 'conventional' pharmaceuticals. The regulatory requirements for taking a conventional pharmaceutical through clinical trials to the market, however, emphasize the physico-chemical analysis of the 'final dosage form', which is then correlated with a suitable bio-assay to provide assurances of product uniformity. On the other hand, with a bio-pharmaceutical which cannot be totally defined by simple analyses of its physico-chemical characteristics and biological activity, most of the complexities occur during the bulk manufacturing process, while the preparation of the final dosage form for most part is rather 'uncomplicated'.

For this reason, the bio-pharmaceutical industry, together with the regulatory authorities, decided to focus upon the entire manufacturing process and not simply on the monitoring and analysis of the final dosage form. This is important as the quality, safety, and efficacy attributes of a bio-pharmaceutical for which end-product controls alone are inadequate, can only be assured by having comprehensive controls over the entire manufacturing process. Therefore, as well as validating the consistency of manufacture and characterizing the final product, constant monitoring throughout processing is also stressed. This results in much work needing to be completed even before the clinical trials could commence. For example, over 750 different separate 'in-process tests' are carried out in the manufacture of a recombinant human growth hormone, whereas only about 60 tests are required in the chemical synthesis of a conventional peptide hormone such as the thyroid hormone.

In order to further streamline the approval process, the regulatory authorities have recently changed their procedures to eliminate the requirement for a separate establishment licence for certain 'well-defined' classes of biological products. Recent scientific advances, both in methods of manufacture and analysis, means that some products developed through biotechnology can be characterized in ways not historically considered possible, thereby enabling the authorities to allow well-characterized biological products to be regulated under a single application.

The guiding principle is that an application for establishment licensure can be made for any facility (regardless of the scale of manufacture) which has been fully qualified, validated, operates in accordance with current good manufacturing practices (cGMPs) and which also complies with applicable local laws and regulations. These facilities should be distinguished from facilities used in research and development that may not operate under appropriate current good manufacturing practices (cGMPs). When manufacture of a product is transferred from a pilot to a different facility, a demonstration of product consistency, as well as data comparing the two products, together with the relevant process validation data should be submitted to the regulatory authorities. This should include a description of the manufacturing changes that have occurred, a protocol for comparing the products made in each facility, and the data generated using this protocol, as well as documentation on process validation and all stability data for the product manufactured in the new facility. It would be expected that the methods of cell expansion, harvest, and product purification would be identical except for the scale of production. For each manufacturing location, a floor diagram should be included that indicates the general production facility layout, as well as information on product, personnel, equipment, waste and air flow for production areas; an illustration or indication of which areas are served by each air handling unit; and air pressure differentials between adjacent areas.

It is, therefore, quite obvious how important it is that the manufacturer discusses with the regulatory authorities what data are necessary to compare products, as such data may range from simple analytical testing to full clinical trials, and could well be required even before the product made using the new facility or process is allowed to be included in any further clinical trials.

11.2.2 Design considerations

The cost of building facilities that are fully validated and in compliance with cGMP can be overwhelming to biotechnology companies with limited finances. The basic design and construction costs are driven higher by the various regulatory, containment, process utilities and waste treatment requirements. In addition, companies also demand increased value from their clinical production facilities. The facility design must, therefore, allow for flexibility of operations, diverse process utility requirements, as well as for campaigning different products in the same facilities.

It is possible to build such facilities in a cost effective, flexible manner, while satisfying the regulatory requirements as well as ensuring that the completed facility will provide all the functions intended. The most effective techniques used to manage such a project would be the use of the concept of 'total project management'. 'Total project management' means integrating regulatory requirements, design and engineering, validation, as well as construction requirements on one single schedule, to determine the critical path (least time to completion). This leads to more effective management, permitting 'what if' scenarios that can result in substantial savings in time and cost, especially if cost estimation is implemented early in the design phase.

The key element is to begin with the careful analysis of manufacturing process needs and to define the facility requirements specifically. Careful site selection is important to eliminate any costly surprises. It is also important to avoid over-specifying very expensive process utilities. This would be followed by the implementation of modular facilities design and construction. A well thought-out facility design using pre-engineered, self-contained elements can in many cases be the most cost effective, flexible solution to clinical production. The application of modular clean rooms to create the cGMP facilities for different products can therefore be achieved.

Buildings and facilities used in the manufacture, processing, packing, or holding of bio-pharmaceuticals should be of suitable design, size, construction and location to facilitate cleaning, maintenance and proper operations. Adequate space should be provided for the orderly placement of equipment and materials, to prevent mix-ups and contamination among different raw materials, intermediates, or the final product. The flow of raw materials, intermediates and the product through the building or buildings, should be designed to prevent mix-ups and contamination. To prevent mix-ups and contamination, there should be defined areas and/or other control systems for all the important activities. Also, facility design must be integrated in support of the process in order to comply with cGMP and other regulatory requirements such as:

- flow of personnel, materials, product, equipment or glassware, and waste flows;
- product separation and/or segregation;
- aseptic and/or sterile processing;
- sanitary design — cleaning and decontamination and spill containment;
- bio-hazard containment and/or isolation;
- special clean utilities;
- solvent recovery, handling, and storage;
- HVAC zoning, pressurization, and filtration;
- drain and exhaust systems.

11.2.3 Operating considerations

Implementing cGMP

The current Good Manufacturing Practices (cGMPs) mentioned above are those practices designed to demonstrate that the control over the process, the facility, and the procedures used in the manufacture, maintains the desired quality of the product, be it a conventional drug or a bio-pharmaceutical, and consequently protects the product's integrity and purity. The implementation of cGMP is now a legal requirement and certainly makes for better quality products and sound economic sense.

As technology and scientific knowledge evolve, so does understanding of critical material, equipment and process variables that must be defined and controlled to ensure end product homogeneity and conformity with appropriate specifications. The cGMP regulations would not achieve their statutory mandated purposes if they were not periodically reassessed to identify and eliminate obsolete provisions or to modify provisions that no longer reflect the level of quality control that current technology dictates and that the majority of manufacturers have adopted. cGMP regulations are based on the fundamental concepts of quality assurance:

- quality, safety, and effectiveness must be designed and built into a product;
- quality cannot be inspected or tested into a finished product;
- each step of the manufacturing process must be controlled to maximize the likelihood that the finished product will be acceptable.

Even though cGMPs have been known and have been evolving for over 20 years, many pharmaceutical and biotechnology companies (both established companies and those just starting operations) still need to achieve a sound basic understanding and implementation of the fundamental rationale and requirements of cGMP. There is still a persistent lack of understanding among a limited number of manufacturers with respect to certain of the cGMP regulations. Some pharmaceutical firms have not subjected their procedures to sufficient scrutiny, while others have failed to update such procedures to accommodate changes or advances in the manufacturing process. In some cases, manufacturers may be relying on methods and procedures that were acceptable at some time in the past, but that are not acceptable in light of current standards. The regulatory authorities have also encountered serious deficiencies particularly with validation procedures designed to ensure the quality of the manufacturing process.

Those implementing cGMPs in the design of bio-pharmaceutical facilities must recognize the inherent variability in the manufacturing processes. A distinction can be drawn between the application of cGMPs to well-characterized operations, such as filling and finishing, and the nature of the early stages of biotech product manufacturing typified by the attributes below:

- raw material variances;
- product yields;
- non-linear process flow, reprocessing;
- process complexity.

The role of process validation

cGMP regulations specify the nature and extent of validation that is necessary to ensure that the resulting products have the identity, strength, quality and purity characteristics that they purport to possess. The term validation is used for those elements of the manufacturing process under the control of the manufacturer, while the term qualification is used for those items produced by a person other than the manufacturer, or otherwise not under the control of the manufacturer. Process validation is the establishment of documentary evidence to provide a high degree of assurance that a specifically defined process, using specified equipment and systems, which when in control, will consistently and reliably yield a product meeting its pre-determined specifications and quality attributes or characteristics.

So what does validation actually mean to the ordinary scientist responsible for putting together a process for the manufacture of a bio-pharmaceutical. Validation is simply the formal process of establishing with a high degree of assurance, and demonstrating to the relevant authorities, through a programme of documented tests, challenges, and results, that an item of equipment, system, or process actually and consistently does what it claims to do. Because it guarantees the ability to achieve and routinely maintain a product of a quality which meets all its pre-determined specification, it provides for a better understanding of how the equipment, system, or the process works, as it highlights potential weaknesses and enables corrective action to be taken. Also, by demonstrating reliable and consistent performance, validation also ensures profitability, because a validated process should be under control to such an extent that any deviation could be detected and enable corrective action to be taken.

So how and when do the regulatory authorities recommend that process validation be carried out? The validation programme should begin with the raw material in the warehouse or stores, and finish when the final product is fully packaged and ready for use. When any new manufacturing formula or method of preparation is adopted, steps should be taken to demonstrate its suitability for routine processing. However, validation is required not just when a totally new and untried item of equipment or system is adopted, but on every occasion that any of the above is substantially amended, as product quality and/or the reproducibility of the process may be affected. Also, processes and procedures should undergo periodic critical re-validation to ensure that they remain capable of achieving the intended results.

Experience has shown that a simple, logical, well-planned approach is the key to achieving success with process validation. Not only will this minimize the mountain of documentation required, but will also provide the training for process, plant and maintenance personnel, as well as providing the basis for any calibration and preventative or routine engineering maintenance programmes required. Also, if validation is planned, interfaced and integrated with the design and construction phase of the operation, then user requirements can be addressed, enabling the overall timelines to completion to be shortened. This in turn will minimize expensive duplication of effort, by identifying and enabling correction of potential design mistakes or omissions.

Validation strategy

The validation programme should embrace steps in the process that are critical to the quality and purity of the final product and should include all associated facilities, operating utilities and equipment. All critical process operations and facilities are required to be systematically investigated to ensure that the product can be manufactured reliably and reproducibly using all the pre-defined production and control methods. It is important to remember, however, that the level of validation should be appropriate to the end use of the product. The requirements become less stringent, but no less important, further away from the final process step. A final dosage filling facility for a parenteral will require a much higher degree of validation than an intermediate bulk production facility.

Validation begins with the development of the Master Validation Plan. It is important to combine the MVP with the construction schedule to ensure that validation is a focus of the total effort and that validation documentation is available as necessary and prepared concurrently with construction, and to ensure that the overall time to complete validation is minimized. The VMP should include and cover the following:

- a summary of the validation philosophy, its approach and rationale;
- a definition of the product in terms of its critical quality attributes, including purity, qualitative and quantitative impurity profiles, physical characteristics such as particle size, density, polymorphic forms, moisture and solvent content, if appropriate, homogeneity, and whether the product is susceptible to microbial contamination;
- a summary of the methodologies and techniques to be used;
- identification of process steps and parameters that could affect the critical quality attributes of the product, and the range for each critical process parameter expected to be used during routine manufacturing and process control. These should be determined by scientific judgment, and typically be based on knowledge derived from research and scale-up batches, unless a specific parameter can only be determined from manufacturing experiences gained from a production-scale batch;
- validation planning worksheet identifying individual tasks;
- list of available resources — both internal and external; and resource levelling to establish the time required for the project based on the available resources.

The documentation related to the validation programme is as important as the execution of the programme itself, if not more so. The design and implementation of the documentation system involves the preparation, review (audit), and authorization of all required validation protocols for the standard operating procedures (SOPs), and manufacturing instructions, including calibration methods (metrology programmes), acceptance and certification criteria, as well as the assignment of responsibility. The validation protocol is the blueprint of the validation process for a particular drug product. It is the written plan describing the process to be validated, including the equipment used, and how validation will be conducted. The protocol should specify a sufficient number of replicate process runs to demonstrate reproducibility, and provide an accurate measure of variability among successive runs.

Execution of validation field activities

This begins with installation qualification (IQ), followed by operational qualification (OQ) and finishes with performance qualification (PQ), which covers both equipment (or system) validation, and process qualification, including establishing critical circumstances for re-validation.

Installation qualification (IQ) is the formal process of verifying and establishing confidence that an item of equipment or system was received and installed, meets the specification as ordered and intended, that the proper utilities are available and supplied, that it is installed as recommended by the manufacturer, any local or state codes, standards and cGMP, and is capable of consistently operating within established limits and tolerances.

It is clear, therefore, that the 'as-built' drawings and other documents supplied by the manufacturer are essential to successfully carry out installation qualification (IQ).

Operational qualification (OQ) is the formal process of verifying and establishing that such an item of equipment or system, once installed, is capable of satisfactory operation as specified and intended, over the entire range of operational parameters such as pressures, temperatures, etc. It involves water commissioning to check the various ancillaries such as motors and valves, and usually follows installation qualification (IQ), but can also be carried out concurrently.

Performance qualification (PQ) is the formal process of verifying and demonstrating confidence by rigorous challenges and testing, that this item of equipment or system, once installed and operationally qualified, is capable of operating effectively and reproducibly in the process step for which it is intended. This is normally carried out in two parts:

- equipment (or system) validation;
- process qualification.

Equipment (or system) validation involves the following as appropriate:

- sterilization validation by using temperature mapping techniques, followed by the verification of asepsis, or sterility testing;
- containment validation, using the host organism or another 'safe' organism;
- calibration of instruments and certification;
- validation of computer hardware and software used in the process;
- cleaning validation, particularly important in multi-product facilities.

380

Next comes process qualification. Process qualification is the major component of the whole validation effort, as it relates directly to the changes the raw material undergoes during its transformation to the final product. Process qualification is where each critical process step in the manufacture is defined with sufficient specificity and each such step is suitably challenged and tested to determine its adequacy and capability. It is essential that the validation runs are as representative as possible to routine manufacturing steps in terms of activities, conditions and characteristics, to ensure that the results obtained are relevant to routine production. The performance of the various challenges and the compilation of the results must confirm conclusively that the equipment or system involved in the process step is capable of providing the pre-described confidence levels. Manufacturers are also expected to have validation reports for the various key process steps. For example, if an ion-exchange column is used to remove endotoxins, there should be data documenting that this process is consistently effective. By determining endotoxin levels before and after processing, a manufacturer should be able to demonstrate the validity of this process. It is important to monitor the process before, during and after to determine the efficiency of each key purification step. Spiking the preparation with a known amount of a contaminant to demonstrate its removal is a useful method to validate such a procedure.

Prospective, concurrent and retrospective validation

Prospective validation covers activities that should be conducted prior to the commercial distribution of the product manufactured by either a new or substantially modified process. When carrying out prospective validation, data from laboratory and/or pilot-scale batches should identify critical quality attributes and specifications, critical process steps, control ranges, and in-process tests. Scale-up batches can be used to generate data to confirm or refine earlier work, however production-scale batches are needed to provide data showing consistency of the process, using validated analytical methods. The number of consistent process runs would depend on the complexity of the process or the magnitude of the process change being considered. Although three consecutive, successful production batches should be used as a guide, there may be situations where additional process runs are warranted to prove consistency of the process, for example, for products with complex processes such as a recombinant cell fermentation, or for processes with prolonged completion times, such as with an animal cell culture.

Regulatory authorities consider concurrent validation to be a sub-set of prospective validation. They recognize that in a limited number of cases it may not be possible to complete validation of a process in a timely manner before distribution of the product, when data from replicate production runs are unavailable, possibly because only a limited number of batches intended for clinical or orphan drug products have been produced. In such cases, the manufacturer should do all the following:

- perform all the elements of prospective validation, exclusive of replicate production run testing, before releasing any batch for distribution;
- document the reasons for not completing process validation;
- batch production records, in-process controls, and analytical data from each process run should be evaluated thoroughly to determine whether or not each batch should be released.

This approach should not be viewed as a viable alternative if the number and frequency of production batches permit timely completion of process validation prior to product distribution. Also, if analysis of the data shows that the process used to manufacture the distributed batches was not, in fact, validated, no additional batches should be distributed until corrections have been implemented and the process is deemed to be validated.

Retrospective validation may be conducted for a well-established process that has been used without significant changes, such as changes in raw materials, equipment, systems, facilities, or in the production process, that affect the critical quality attributes of the product. This validation approach should only be used when there is sufficient history on past production batches to demonstrate that the process consistently produces acceptable products, and where:

- critical quality attributes and critical process parameters have been identified and documented;
- appropriate in-process specifications and controls have been established and documented;
- there have not been excessive process or product failures attributable to causes other than operator error or equipment failure unrelated to equipment suitability;
- impurity profiles have been established for the existing product.

The number of batches to review will depend on the process, but, in general, data from 10 to 30 consecutive batches should be examined to assess process consistency. All batches within the selected review period should have been manufactured by the same process and have the same documented history of controls and tests as the current products.

Cost of validation

So why does validation cost so much, take so long, and what can be done about it? Validation of a bio-pharmaceutical facility is based on the time-consuming accumulation of details and sometimes the cost of validation can exceed the total cost of a project's architecture and engineering fees. Precious validation time could be spent trying to obtain information from designers, engineers, contractors and manufacturers, which could have been specified and provided if it were considered an integral part of the project. Additionally, most project managers are more concerned with completing the facility than with completing validation. The key is, therefore, to make validation an integral part of the project and include the validation master plan, preparation of protocols and SOPs, and their execution, as a series of tasks on the critical path in the total project schedule.

In conclusion

It is clear that process validation represents a sizeable investment in time and resources, usually taking place during a time period when the scientist and plant personnel are already heavily involved in start-up related activities. The resulting time constraints can often affect the quality of the work needed, so it is important to identify the pitfalls normally encountered during the process of validation so that they can be avoided.

Under-estimating or under-resourcing the amount of work required is the most common problem; a simple, well planned, and logical approach to validation is the key to overcoming this problem.

Surprisingly, too much validation can also be a problem; however, by identifying the critical conditions for each step in the process, it should be possible to avoid this pitfall and save valuable resource and effort.

Re-validation and change control

Once the validation and certification procedure is completed, the equipment, system or process is considered acceptable for use, but only under those conditions and functions specified in the validation protocol. To preserve the validated status of a process, measures must, therefore, be taken that will allow any significant process changes to be recognized and addressed promptly. For example, a slight change in the physical characteristics of an ingredient, or in the order of adding ingredients, may alter the specification of a product. Because of such effects, re-validation is necessary after any change in process or product characteristics or control procedures. Such a change control programme should provide for a classification procedure to evaluate changes in raw materials, manufacturing sites, scale of manufacturing, manufacturing equipment and production processes. Regulatory authorities categorize changes to an approved application as major, moderate, or minor, depending on the nature and extent of the changes, and on their potential to have an adverse effect on the identity, strength or concentration, quality, purity, or the potency of the product, and on the process, as they may relate to the safety or effectiveness of the product.

A major change is defined as one that could significantly affect the critical quality attributes of the product. Such changes that have a substantial potential to have an adverse effect on the product and require submission of a supplement for approval by the regulatory authorities prior to the distribution of the product made using the change, should be justified by additional testing and if appropriate, re-validation. Some examples include:

- process-related changes, such as the extension of culture growth time leading to a significant increase in the number of cell doublings beyond validated parameters; new or revised recovery procedures; new or revised purification process, including a change in a column; a change in the chemistry or formulation of solutions used in processing; a change in the sequence of processing steps, or addition, deletion, or substitution of a process step; reprocessing of a product without a previously approved reprocessing protocol;
- changes relating to the manufacturing processes or analytical methods that results in changes of specification limits or modifications in potency, sensitivity, specificity, or purity; establishes a new analytical method; deletes a specification or an analytical method; eliminates tests from the stability protocol; or alters the acceptance criteria of the stability protocol;
- scale-up requiring a larger fermenter, bioreactor or purification equipment (applies to production stages up to the final purified bulk);
- changes in the composition or the final dosage form of the biological product or even of ancillary components, such as new or different excipients, carriers, or buffers;
- new or different lot of, or source for, in-house reference standard or reference panel, resulting in the modification of reference specifications and/or an alternative test method;
- extension of the expiration dating period and/or a change in storage temperature, container/closure composition, or other conditions, other than changes based on real time data in accordance with a stability protocol in the approved licence application;
- installation of a new Water for Injection (WFI) system, or modifications to an existing WFI system that would have a significant potential to stress or

challenge the system, such as lengthy or complicated distribution system extensions to service new or remote production areas, use of components of lesser quality or function, expansions of ambient temperature water distribution loops, or conversion from hot loop to ambient loop;

- change of the sites at which manufacturing, other than testing, is performed; addition of a new location; contracting of a manufacturing step in the approved licence to be performed at a separate facility;
- conversion of production and related areas from single into multiple product manufacturing areas, especially as there may be changes to the approved and validated cleaning procedures as well as additional containment requirements;
- changes in the location (room, building, etc.) of steps in the production process, which could affect contamination or cross-contamination precautions;
- major construction, or changes in location, involving or affecting environmentally controlled manufacturing or related support areas such as new buildings; new production areas or rooms in existing build-in-support systems with significant potential to affect air, water, or steam quality; installation of a new HVAC system involving or affecting environmentally controlled manufacturing or related support areas; modifications to an existing HVAC system that supplies aseptic processing areas.

Moderate changes have a moderate potential to adversely affect the product and require a supplementary submission to the regulatory authorities at least 30 days prior to distribution of the product made using the change. Some examples include:

- automation of one or more process steps without a change in process methodology;
- addition of duplicated process chain or unit process, such as a fermentation process or duplicated purification columns, with no changes to the in-process parameters;
- addition or reduction in number of pieces of equipment (e.g., centrifuges, filtration devices, blending vessels, columns) to achieve a change in purification scale not associated with a process change;
- change in the fill volume (per vial or syringe) from an approved production batch size and/or scale, excluding those that involve going from a single dose to a multi-dose vial, or changes in product concentration, both of which should be submitted as a supplement requiring prior approval;
- changes in responsible individuals specified in the approved application, including manufacturers' representatives, responsible experts and other individuals designated to communicate with the authorities;

- modification of an approved manufacturing facility or room that is not likely to have an adverse effect on safety, sterility assurance, purity or potency of product, such as adding new interior partitions or walls to increase control over the environment;
- manufacture of an additional product in a previously approved multiple-product manufacturing area using the same equipment and/or personnel, if there have been no changes to the approved and validated cleaning procedures and there are no additional containment requirements;
- change in the site of testing from one facility to another, such as from a contract laboratory to the licence holder, from an existing contract laboratory to a new contract laboratory, or from the licence holder to a new contract laboratory;
- change in the structure of a legal entity that would require issuance of new licences, or a change in name of the legal entity or location;
- addition of release tests and/or specifications, or tightening of specifications for intermediates;
- minor changes in fermentation batch size using the specifications of the bulk or final product;
- modifications to an existing HVAC system involving or affecting environmentally controlled manufacturing or related support areas, but not aseptic processing areas, with no change in air quality.

Minor changes are those that are unlikely to have a detectable impact on the critical attributes of the product. Such changes would not shift the process in any discernible manner and might be implemented with minimal testing and revalidation. For example, like-for-like equipment replacements where identical or similar equipment is introduced into the process, is unlikely to affect the process if adequately installed and qualified. Such changes should be described and reported by the manufacturer on an annual basis. Examples would include:

- addition of equipment for manufacturing processes which is identical to the primary system and serves as an alternate resource within an approved production room or area;
- upgrade or minor corrective change to production air handling, water, or steam supply systems using equipment of the same or similar materials of construction, design and operating parameters, and not affecting established specifications; such as the removal of dead legs in the WFI system. This, however, does not include replacement of parts or routine repair and maintenance, which would not be changes to an approved application and would not need to be reported;
- relocation of analytical testing laboratories between areas specified in the licence;

- room upgrades, such as the installation of improved finishes on floors/walls;
- installation of non-process-related equipment or rooms to improve the facility, such as warehousing refrigerators or freezers;
- modifications in analytical procedures with no change in the basic test methodology or existing release specifications provided the change is supported by validation data;
- change in harvesting and/or pooling procedures, which does not affect the method of manufacture, recovery, storage conditions, sensitivity of detection of adventitious agents or production scale;
- replacement of an in-house reference standard or reference panel (or panel member) according to SOPs and specifications in an approved licence application;
- tightening of specifications for existing reference standards to provide greater assurance of product purity, identity and potency;
- establishment of an alternative test method for reference standards, release panels or product intermediates, except for release testing of intermediates licensed for further manufacture;
- establishment of a new Working Cell Bank (WCB) derived from a previously approved Master Cell Bank (MCB) according to a SOP on file in the approved licence application;
- change in the storage conditions of in-process intermediates based on data from a stability protocol in an approved licence application, which does not affect labelling, except for changes in storage conditions, which are specified by regulation;
- change in shipping conditions, such as temperature, packaging or custody, based on data derived from studies following a protocol in the approved licence application;
- a change in the stability test protocol to include more stringent parameters, such as additional assays or tightened specifications;
- addition of time points to the stability protocol;
- replacement of equipment with that of identical design and operating principle involving no change in process parameters;
- upgrade in air quality, material, or personnel flow where product specifications remain unchanged. Involves no change in equipment or physical structure of production rooms;
- relocation of equipment within an approved operating room, rearrangement of the operating area or rooms where production is performed or relocation of equipment to another approved area to improve product/personnel/raw material flow and improve segregation of materials with no change in room air classification;

- modifications to the pre-treatment stages of a WFI system, including purified water systems used solely for pre-treatment in WFI production;
- change in the simple floor plan that does not affect production process or contamination precautions;
- trend analyses of release specification testing results for bulk drug substances and drug products obtained since the last annual report.

Change control procedures

No change that could affect performance in any way should be allowed without the written approval of at least the production, QA and engineering departments. Such changes should only be handled through a change control procedure with protocols for initiating and proving the change, together with procedures for re-validation. Such change control measures may apply to equipment, SOPs, manufacturing instructions, environmental conditions, or any other aspect of the process or system that has an effect on its state of control and, therefore, on the state of validation and should include procedures to:

- prevent unauthorized modifications to a validated system;
- evaluate proposed changes against development and technology transfer documents;
- identify and evaluate all proposed changes to assess their potential effects on the process and determine if, and to what extent, re-validation is needed;
- ensure that all documents affected by changes are promptly revised;
- determine the impact of changes on the critical chemical and physical attributes of the product, such as its impurity profiles, stability, etc.

Changes implemented to improve process yields should be evaluated carefully to determine if they result in new or higher levels of impurities; impurity profiles of resulting batches should be comparable to the batches used in drug safety and clinical testing, and evaluated to ensure that these do not have an adverse effect on analytical methods, due to increased interference caused by new or higher levels of impurities and by-products; and analytical methods should be modified as necessary to ensure that they are capable of detecting and quantifying impurities.

11.3 Primary production

The manufacture of bio-pharmaceuticals involves certain specific considerations arising from the nature of the products and the processes. Unlike conventional pharmaceuticals, which can be manufactured, analysed and characterized using chemical and physical techniques capable of a high degree of consistency, the production of bio-pharmaceuticals involves processes and materials that display an inherent variability, resulting in variability in the range and nature of the by-products. Moreover, the control and characterization of bio-pharmaceuticals usually involves bio-analytical techniques, which have a greater variability than the usual physico-chemical determinations. In addition, genetically modified cells, although providing special opportunities for producing novel protein sequences that exhibit improved activity compared to that of the natural molecule, necessitate special considerations of process design and operation.

Therefore, the methods used in the manufacture, control, and use of these bio-pharmaceuticals make certain precautions necessary, and are a critical factor in shaping the appropriate regulatory control. Bio-pharmaceuticals manufactured by such methods include vaccines, immune-sera, antigens, hormones, cytokines, enzymes and other products of fermentation, including monoclonal antibodies and products derived from r-DNA, and can be largely defined by reference to their method of manufacture:

- microbial cultures, excluding those resulting from r-DNA techniques;
- microbial and cell cultures, including those resulting from recombinant DNA or hybridoma techniques;
- extraction from biological tissues;
- the propagation of live agents in embryos or animals.

This chapter applies to the production, extraction, purification and control of such bio-pharmaceuticals manufactured for use in clinical trials or for marketing, as human or veterinary medicines, and applies to the point where the product is rendered sterile — i.e. the bulk active substance.

11.3.1 Starting materials

The source, origin and suitability of starting materials should be clearly defined. In instances, where the necessary tests take a long time, it may be permissible to process starting materials before the results of the tests are available. In such cases, release of a finished product is conditional on satisfactory results of these tests. Where sterilization of starting materials is required, it should be carried out where possible by heat, although other appropriate methods may also be used for inactivation of biological materials, such as gamma-irradiation for serum supplements used in the culture of animal cells.

Control of raw materials

Many of the raw materials used in fermentation processes can have significant impact on the subsequent recovery. As they are usually derived from animal sources (such as serum, transferrin, etc.), they represent potentially variable sources of contaminants such as viruses, mycoplasma, or even hydrolytic enzymes. Pre-treatment of these raw materials by heating, acidification or sterile filtration is often necessary to avoid contaminating the production cells as well as the product. For example, contamination of the seed train by serum borne mycoplasma or virus may irreversibly repress cell growth and product titre; and once the cells are contaminated, they will produce poorly, and the harvest fluid will most likely contain degradative enzymes that decrease the quality of the purified product.

A monoclonal antibody (mAb) may also be a raw material when used for purification of the product. In such cases, the standards for their production should be at least as rigorous as those for the product it is used to purify. The manufacturer must fully characterize the mAb-producing cell line, establish that it is free from adventitious agents, assess the purity of the mAb and validate its purification process for the removal of nucleic acids and viruses, as well as minimize residual levels of the mAb in the product of interest.

Raw materials should be handled and stored in a manner to prevent contamination and cross-contamination. Identifying labels should remain legible, and containers should be appropriately cleaned before opening to prevent contamination. Written procedures should be established describing the purchase, receipt, identification, quarantine, storage, handling, sampling, testing and approval or rejection of raw materials, and such procedures should be followed. Bagged and boxed raw materials should be stored off the floor and suitably spaced to allow cleaning and inspection, and those stored outdoors should be in suitable containers. For solvents or reagents delivered in bulk vessels, such as in tanker trucks, a procedural or physical system, such as valve locking or unique couplings, should be used to prevent accidental discharge of the solvent into the wrong storage tank. Each container or grouping of containers of raw materials should be assigned and identified with a distinctive code, lot or receipt number with a system in place to identify each lot's status. Large containers, such as tanks or silos, which are used for storing raw materials, including their attendant manifolds, filling and discharge lines, should also be appropriately identified.

Receipt, sampling, testing, and approval of raw materials

Upon receipt and before acceptance, each container or grouping of containers of raw materials should be examined visually for appropriate labelling, container damage, seal integrity (where appropriate) and contamination. Raw materials should be held under quarantine until they have been sampled, tested or examined as appropriate and released for use. Representative samples of each shipment of each lot should be collected for testing or examination in accordance with an established procedure. The number of containers to sample and the sample size should be based upon appropriate criteria, such as the quantity needed for analysis, sample variability, degree of precision desired, and past quality history of the supplier, and the sample containers properly identified.

At least one test should be conducted to verify the identity of each raw material. A supplier's certificate of analysis may be used in lieu of performing other testing, provided the manufacturer has a system in place to evaluate vendors (vendor audits) and establishes the reliability of the supplier's test results at appropriately regular intervals. For hazardous or highly toxic raw materials, where on-site testing may be impractical, suppliers' certificates of analysis should be obtained showing that the raw materials conform to specifications. However, the identity of these raw materials must be confirmed by examination of containers and labels, and the lack of on-site testing for these hazardous raw materials should be documented.

Use and re-evaluation of approved raw materials

Approved raw materials should be stored under suitable conditions and, where appropriate, rotated so that the oldest stock is used first. Raw materials should be re-evaluated, as necessary, to determine their suitability for use, for example, after prolonged storage, or after exposure to heat or high humidity.

Rejected raw materials

Rejected raw materials should be identified and controlled under a quarantine system designed to prevent their use in manufacturing or processing operations for which they are unsuitable, and if necessary discarded by appropriate methods.

11.3.2 Cell culture, fermentation and process control

Cell bank system and cell culture

The starting material for manufacturing a bio-pharmaceutical includes bacterial, yeast, insect or mammalian cell culture which expresses the protein product or monoclonal antibody (mAb) of interest. In order to prevent the unwanted drift of characteristics which might ensue from the repeated subcultures or multiple generations, the production of biological medicinal products obtained by microbial or animal cell culture should be based on a system of master and working cell banks (MCB, WCB) consisting of aliquots of a single culture. Also known as seed lots, such cell bank systems are used by manufacturers to assure the identity and purity of the starting raw material.

The MCB is derived from a single colony of prokaryotic (bacteria, yeast), or a single eukaryotic (mammalian, insect) cell stored cryogenically, and is composed of sufficient ampoules of culture to provide source material for the WCB. The WCB is defined as a quantity of cells derived from one or more ampoules of the MCB, stored cryogenically, and used to initiate a single production batch. Both the MCB and the WCB must be stored in conditions that assure genetic stability. Generally, cells stored in liquid nitrogen or its vapour phase are stable longer than cells stored at $-70°C$.

Establishment of cell banks should be performed in a suitably controlled environment to protect the cells and, where applicable, the personnel handling them. During the establishment of the cell banks, no other living or infectious material such as viruses, other cell lines or cell strains, should be handled simultaneously in the same area or by the same persons. Only authorized personnel should be allowed to handle the material, and this handling should be done under the supervision of a responsible person. It is desirable to split the cell banks and to store the parts in more than one location so as to minimize the risks of total loss. All ampoules containing the cell banks should be treated identically during storage.

Cell banks should be established, stored and used in such a way as to minimize the risks of contamination or alteration. They should be adequately characterized and tested for contaminants and shown to be free of adventitious agents such as fungi, bacteria, mycoplasma, and exogenous viruses; tested for tumourigenicity; and probed for the expression of any endogenous retroviral sequences by using conditions known to cause their induction; and their suitability for use demonstrated by the consistency of the characteristics, and quality of the successive batches of product. The number of generations (or doublings or passages) between the cell bank and the finished product should be as low as is practicable.

Inoculation and aseptic transfer

Inoculation of the seed culture into the fermenter or bioreactor, as well as all transfer and harvesting operations must be done using validated aseptic techniques. Additions or withdrawals from fermenter or bioreactors are generally done through steam sterilized lines and steam-lock assemblies. Steam may be left on in situations where the heating of the line or the vessel wall would not be harmful to the culture. If possible, the media used should be sterilized in-situ, using a Sterilization in Place (SIP) or a continuous sterilization system (CSS), and any nutrients or chemical added beyond this point must be sterile. Additions of materials or cultures, and the taking of samples, should be carried out under carefully controlled conditions to ensure that the absence of contamination is maintained. Care should be taken to ensure that vessels are correctly connected when additions or samplings take place. In-line sterilizing filters should be used where possible for the routine addition of air and other gases, media, acids or alkalis, and defoaming agents, to the fermenter or bioreactor.

Process monitoring and control

It is important for a fermenter or bioreactor to be closely monitored and tightly controlled to achieve the proper and efficient expression of the desired product. The parameters for the fermentation process, including information on growth rate, pH, waste by-product levels, addition of chemicals, viscosity, density, mixing, aeration, and foaming, must, therefore, be specified and monitored. Other factors that may affect the finished product, such as shear forces, process-generated heat, should also be considered. Many growth parameters can influence protein production. Although nutrient-deficient media are used as a selection mechanism in certain cases, media deficient in certain amino acids may cause substitutions. The presence of such closely related products may cause difficulties later on during the separation and purification stages, and may have implications both for the application of release specifications and the effectiveness of the product purification process.

Containment considerations

Bioreactor systems designed for recombinant microorganisms require not only that a pure culture is maintained, but also that the culture be contained within that system. Such containment can be achieved by the proper choice of a host-vector system that is less capable of surviving outside a laboratory environment, as well as by physical means, when this is considered necessary. For the cultivation of recombinant cell lines, there are defined and established physical containment levels. Good Large-Scale Practice (GLSP) level of physical containment is recommended for large-scale production involving viable, non-pathogenic and non-potent recombinant strains derived from host organisms that have an extended history of safe large-scale use, and for organisms that have built-in environmental limitations that, although allowing optimum growth in the fermenter, have limited survival outside in the environment. Biosafety level 1 (BL1) level of physical containment is recommended for large-scale production of viable recombinant organisms that require BL1 containment at the laboratory scale. Similar recommendations exist for BL2 and BL3. No provisions are made for the large-scale research or production of viable recombinant organisms that require BL4 containment at the laboratory scale.

Personnel considerations
The immunological status of personnel should be taken into consideration for product safety. All personnel engaged in the production, maintenance and testing should be vaccinated where necessary with appropriate specific vaccines and have regular health checks. Apart from the obvious problem of staff exposure to infectious agents, potent toxins, or allergens, it is necessary to avoid the risk of contaminating a production batch with infectious agents. Therefore, visitors are generally excluded from production areas. Furthermore, in the course of a working day, personnel should not pass from areas where exposure to live organisms or animals is possible to areas where other products or different organisms are handled. If such passage is unavoidable, clearly defined decontamination measures including change of clothing and shoes and, where necessary, showering should be followed by staff involved in any such production.

11.3.3 Product recovery and purification
Once the fermentation process is completed, the desired product is extracted, isolated, separated and, if necessary, refolded to restore configurational integrity, and then purified. Whether the product is intra-cellular or extra-cellular, soluble, insoluble or membrane bound or located in a subcellular organelle will influence the choice of extraction method and buffer components used. Typically, manufacturers develop downstream processes on a small scale and determine the effectiveness and limitations of each particular processing step. Allowances must, therefore, be made for several differences when the process is scaled-up. Longer processing times can adversely affect product quality since the product is exposed to various reaction conditions, such as pH and temperature, for longer periods. Product stability under such varying purification conditions must, therefore, be carefully defined.

Product recovery
Determining the optimal time of harvest is an important area of interaction between fermentation and recovery. Often, allowing a culture to run longer results in an increase in titre, but with a concomitant increase in cellular debris and degraded forms of the product. Although it may be simple to overcome the effect of increased cell debris by increasing the capacity of the downstream equipment, it is much more difficult to purify away the slightly altered or degraded forms of the product.

With extra-cellular products, it is possible to achieve a high degree of purification by simply removing the cells. For the recovery of extra-cellular proteins, the primary separation of product from producing organisms is accomplished by centrifugation or membrane filtration. Ultra filtration is commonly used to remove the desired product from the cell debris. The porosity of the membrane filter is calibrated to a specific molecular weight, allowing molecules below that weight to pass through while retaining molecules above that weight. Centrifugation can be open or closed, although the adequacy of the environment must be evaluated for open centrifugation. Following centrifugation, other separation methods, such as ammonium sulphate precipitation and aqueous two-phase separation, can also be employed to concentrate the product.

With extra-cellular products, cell breakage is unnecessary and undesirable. Cell disintegration not only releases membrane fragments that can foul process equipment, but also undesirable impurities derived from the cell cytoplasm, particularly host cell proteins and DNA. The harvest/cell separation operation is more difficult with mammalian and other animal cells, as they are much more fragile than bacterial or yeast cells. Consequently, high-speed centrifuges may not be appropriate and these cells must be harvested with special low shear, low centrifugal field centrifuges. Harvesting can also be carried out effectively and efficiently using depth or tangential flow filtration. The advantage of filtration is its ability to achieve quantitative increases in product yield by washing (diafiltration) the cells.

Intra-cellular or membrane-bound products will require detergents or organic solvents to solubilize them. For the recovery of completely intra-cellular products, the cells must be disrupted after fermentation, which can be achieved by chemical, enzymatic or physical methods. Following disruption, the cellular debris is removed either by centrifugation or filtration.

Purification

Further purification steps primarily involve a variety of chromatographic methods to remove impurities and to bring the product closer to final specifications. One or more of the following column chromatography techniques usually achieves this:

- affinity chromatography;
- ion-exchange chromatography (IEC);
- gel filtration or size-exclusion chromatography (SEC);
- hydrophobic interaction chromatography (HIC);
- reverse-phase HPLC (RP-HPLC).

A prior knowledge of the protein stability and its sensitivity to temperature, extremes of pH, proteases, air and metal ions will also aid the design of a purification procedure. If the product to be purified is an enzyme or receptor it may be possible to exploit its activity by affinity purification on a substrate or ligand, or an analogue. Knowledge of the size and pH of the protein will indicate suitable matrices and conditions for gel filtration and ion-exchange chromatography. The final use of the product will define how much of the purified protein is required, whether loss of activity can be tolerated, how pure it should be, and the time and cost of purifying it. If it is for research use, the quantities required are reasonably small, whilst in terms of purity the removal of interfering activities becomes essential. In contrast, for therapeutic applications, purity is of the utmost importance and quantities required are relatively small.

Selection and sequence of the downstream processing steps

Each protein has a unique combination of properties that can be exploited for purification. Thus by combining a series of steps that exploit several of these properties, the protein can be purified from a mixture. Each technique should be evaluated for its capacity, resolving power, probable product yield and cost, and would use a different property of the product, such as charge or hydrophobicity, to effect adsorption and separation. These factors must be balanced against one another and the requirement for each stage of the purification. Moreover, the number of steps in a purification process should be limited by ensuring that the product from one technique can be applied directly onto the next step without further manipulations.

The capacity of the technique is defined as the amount of sample (in terms of volume and protein concentration) that can be handled. A key requirement early in the purification is often to reduce the volume when high capacity techniques such as precipitation methods, which can handle the large initial volumes and protein concentrations, are often used first. Of the chromatography steps, those involving adsorption have the highest capacity. Gel filtration or size exclusion chromatography has a low capacity and is, therefore, usually inappropriate for early stages and is mostly used as a final clean up.

The resolution of a technique determines how efficiently it separates proteins from one another. Precipitation steps have low resolution, whilst chromatography steps are more highly resolving. Affinity chromatography often shows extremely high resolution and it is possible to frequently achieve purification factors of greater than 1000 fold.

Due to the nature of the various interactions and the conditions used, each technique will show a range of average yields. Precipitation with ammonium sulphate and aqueous two-phase extraction usually gives yield of more than 80%, whilst affinity methods often result in lower yield (~60%) due to the harsh conditions required for the elution of the product.

With respect to cost, affinity techniques are usually expensive and so not often used as an initial purification step. A cheaper technique such as ion exchange chromatography is usually used first to remove the bulk of the contaminants such as particulate matter, lipids and DNA.

Integration with upstream operations

In the narrowest definition, downstream processing is the purification of proteins from conditioned media or broths. However, many controllable factors that influence purification occur early in the production process. The integration of downstream processing with upstream operations such as molecular biology and fermentation can, therefore, provide significant downstream opportunities.

The interaction between molecular biology and recovery can take several forms. With recombinant DNA products, purification can be influenced before the starting material is even available. Given the gene sequence, it is possible to predict how the product will behave on size separation media and ion exchange resins, although the actual ionic properties of the protein may be influenced by its tertiary structure. Leading or tail sequences can be added to impart properties that will make the protein easier to purify. It is common practice in bacterial systems to employ fusion proteins to enhance expression, secretion or the subsequent recovery. In mammalian and animal cell systems, the tools of molecular biology are used to enhance expression levels and to alter the biological properties of the final product. Higher titres provide a direct benefit to the recovery process by increasing the ratio of product to contaminant, thereby reducing the fold purification that is ultimately required and also by enabling reductions in the operation volumes of early steps.

Perhaps the most important examples of process integration occur in the interaction between recovery and fermentation. One of the primary areas of interaction between these disciplines is the development of suitable media for cell growth. In cases where the expression system uses an amplified selectable marker, it may be necessary to maintain selective pressure during some or all stages of cell culture. The use of media supplements such as serum may release this selection pressure, resulting in a decrease in expression level, as well as adversely affecting the overall recoverability by leading to complex formation and product degradation. This problem can be overcome by the use of low serum, fractionated serum, or even serum-free medium.

11.3.4 Primary production facilities

The risk of cross-contamination between biological medicinal products, especially during those stages of the manufacturing process in which live organisms
are used, may require additional precautions with respect to facilities and
equipment, such as the use of dedicated facilities and equipment, production on
a campaign basis and the use of closed systems, until the inactivation process is
accomplished. The degree of environmental control of particulate and microbial contamination of production premises should, therefore, be adapted to the
product and the production step, bearing in mind the level of contamination of
the starting materials and the risk to the finished product.

Production on a campaign basis may be acceptable for spore-forming
organisms provided that the facilities are dedicated to this group of products,
and not more than one product is processed at any one time. Simultaneous
production in the same area using closed systems such as fermenters may be
acceptable for products such as monoclonal antibodies and products prepared
by recombinant DNA techniques. Processing steps after harvesting may be
carried out simultaneously in the same production area provided that adequate
precautions are taken to prevent cross-contamination. For killed vaccines and
toxoids, such parallel processing should only be performed after inactivation of
the culture or after detoxification. Equipment used during the handling of live
organisms should be designed to maintain cultures in a pure state and
uncontaminated by external sources during processing.

Positive pressure areas should always be used to process sterile products, but
negative pressure in specific areas at point of exposure of pathogens is
acceptable for containment reasons. Where negative pressure areas or safety
cabinets are used for aseptic processing of pathogens, they should be
surrounded by a positive pressure sterile zone. Air filtration HVAC units
should be specific to the processing area concerned and recirculation of air
should not occur from areas handling live pathogenic organisms. The layout
and design of production areas and equipment should allow effective cleaning
and decontamination. The adequacy of cleaning and decontamination procedures should be validated. Pipework systems, valves and vent filters should be
properly designed to facilitate cleaning and sterilization. The use of CIP and
SIP systems should be encouraged. Primary containment should be designed
and tested to demonstrate freedom from leakage risk. Effluents that may
contain pathogenic microorganisms should be effectively decontaminated.

Genetically engineered organisms

When handling genetically engineered materials, the biosafety controls
required should include testing facilities that adequately provide a controlled

environment and separation of test systems, as well as adequate and appropriate areas for receipt and storage of both the host organism and test substance, as well as for any other materials, such as any stocks of plants, feed and soils used in the study, as well as facilities for waste disposal. Both the laboratory facilities and any separate outdoor testing facilities, such as greenhouses and field sites, that are used for testing the genetically engineered substance should be of sufficient design (layout, size and location) to provide the necessary containment of appropriate biosafety level to protect personnel and the environment. They should be designed to provide a barrier to the unintended release of any organisms if a spill or application accident were to occur, and the decontamination facilities should be separated from the other areas of the facility. The laboratory should have decontamination procedures for containing or killing genetically engineered organisms and host organisms. Moreover, the facility should have proper ventilation, so that air flows from areas of low contamination to areas of higher contamination, and complete air containment and decontamination should be provided. Environmental conditions such as temperature, humidity and ventilation should be monitored using appropriate instruments, and recorded and specified in the protocol for the ongoing study.

Animal quarters and care

Animals are used for the manufacture of a number of biological products, for example polio vaccine (monkeys), snake anti-venoms (horses and goats), rabies vaccine (rabbits, mice and hamsters) and serum gonadotropin (horses). Animals may also be used in the quality control of most sera and vaccines, such as for pertussis vaccine (mice), pyrogenicity (rabbits), BCG vaccine (guinea-pigs). Quarters for animals used in the production and control of biological products should be separated from the production and control areas. The health status of animals from which some starting materials are derived, and of those used for quality control and safety testing, should be routinely monitored and recorded. Staff employed in such areas must be provided with special clothing and changing facilities.

11.3.5 Safety issues

The presence of process-related contaminants in a bio-pharmaceutical is chiefly a safety issue. The sources of contaminants are primarily the cell substrate (DNA, host cell proteins and other cellular constituents, viruses), the media (proteins, sera and additives) and the purification process (process-related chemicals and product-related impurities).

Residual host cells

In the early days, there were concerns about the safety of immortal transformed cell lines since, by definition, they were thought to contain oncogenic DNA or proteins. In addition to the issues arising from the transformed nature of these cells, there were also concerns regarding the contamination of these cell lines by adventitious agents such as viruses, fungi and mycoplasma. Furthermore, there were also concerns about the immunogenicity resulting from residual host cell proteins, in patients who received drugs that were purified from recombinant sources. There are various regulatory guidelines for the characterization of the cells used in the manufacture of bio-pharmaceuticals. The exhaustive characterization of the cell banks by diverse methods provides at least an initial degree of confidence that the resultant products can be safely injected into humans. Concerns over the presence of residual host cell proteins have largely been put to rest by relying on well-established techniques for sterile filtration, as well as in advances in analytical method development.

Residual contaminating proteins

Due to the concern about the safety of proteins from non-human sources with respect to the generation of immune responses, recombinant proteins are generally being brought to unprecedented levels of purity. It can be as difficult to quantitate and prove the levels of purity as it is to achieve them. For example, whereas the purity of albumin preparations is commonly about 95–99%, the purity of recombinant products such as human growth hormone, human insulin or even hepatitis B vaccine is greater than 99.99% with respect to host proteins. In order to measure impurities at this level, two major analytical strategies have been developed. The first method, which is uniquely applicable to all recombinant products, is the use of a blank run. This involves fermentation and recovery using a host cell containing the selectable marker but lacking the gene for the product, thereby enabling the manufacturer to specifically prepare and quantitate the host cell derived impurities. The second approach is the direct measurement of the impurities. The most general method uses an immuno-assay based on antibodies to the host cell proteins. Although this type of assay is complex both in its development and composition, it provides an extremely sensitive way to quantitate protein impurities in each batch of product.

400

Residual nucleic acids

When immortalized mammalian cells were first considered as host systems for recombinant protein, there was substantial theoretical concern about the possibility of DNA from recombinant immortal cell lines causing oncogenic events in patients receiving products from these cell substrates. However, various scientists have shown that DNA does not induce any oncogenic events when injected into immuno-suppressed rodents, even at levels at least eight orders of magnitude greater than that expected in a dose of human therapeutic protein such as t-PA. It is most likely the naked DNA is degraded very quickly to inactive fragments and nucleotides by circulating nucleases. With current technology it is possible to directly measure the DNA content of clarified cell culture fluid and the early processing steps with a DNA dot blot assay using 32P labelled DNA derived from the host cell line. For some products, especially those that are administered in multi-milligram quantities, it is necessary to demonstrate a reduction to assure a level of DNA of less than 10 pg per human dose. This can be further validated by spiking 32P labelled DNA into aliquots of process fluid and then purifying the samples on representative scaled-down versions of the recovery process operations.

Viruses

The presence of retro-viruses in continuous mammalian and other animal cell lines has received a great deal of attention because of concern that these particles can potentially cause oncogenic events in man. However, the approach of demonstration of freedom from functional retro-viruses in the culture is not usually sufficient to answer regulatory concerns, because it is always possible that there might be levels of retro-virus just below the sensitivity limit of these assays or that the specificity of the retro-virus assays might not be broad enough to pick up some unusual potential contaminant. To address this issue, the authorities require the testing of the harvested culture fluid directly for the presence of retro-viruses, or following concentration by ultracentrifugation before analysis, to increase the sensitivity of electron microscopy past the estimated detection limit of 106 particles per ml. These direct measurements can be supplemented by validated process procedures for removal and/or inactivation of putative retro-viruses. Only with steps that are truly independent is it legitimate to determine the total clearance as a result of the clearances from the individual steps. Moreover, the use of more than one model virus and the assay of the virus by more than one technique would also serve to strengthen the believability and validity of this approach.

Pyrogen and endotoxins

In contrast to bacterial fermentations, especially of gram-negative bacteria such as *E.coli*, mammalian and other animal cell fermentations should contain little or no pyrogen, and the recovery process should not need to incorporate steps to remove pyrogens. The process strategy thus becomes oriented more towards keeping pyrogens out rather reducing their levels, and it is much more important to keep raw materials and equipment pyrogen-free.

11.4 Secondary production

One of the more difficult processes to regulate, and one which has presented considerable problems over the years, is that of the manufacture of sterile bio-pharmaceuticals. During the past few years, a number of sterile batches from different manufacturers have been reported to have exhibited microbiological contamination. One manufacturer had approximately 100 batches contaminated in a six month time period, whilst another had approximately 25 batches contaminated in a similar period; other manufacturers have had recalls due to the lack of assurance of sterility. Not surprisingly, the manufacture of sterile bio-pharmaceuticals is subjected to special requirements relating to the minimizing of risks of microbiological, as well as of particulate and pyrogen contamination.

The manufacture of a sterile pharmaceutical must be performed in closed systems with minimal operator handling, although much of this depends on the skills, training and attitudes of the personnel involved. Quality assurance is particularly important and this type of manufacture must strictly follow carefully established and validated methods of preparation and procedure. Most bio-pharmaceuticals cannot be terminally sterilized and must, therefore, be manufactured by aseptic processing. Thus, it is important to recognize that as there is no further processing to remove contaminants or impurities such as particulates, endotoxins and degradants, sole reliance for sterility or other quality aspects, must not be placed on any terminal process or finished product test.

11.4.1 Starting materials

The manufacture of a sterile bio-pharmaceutical should be performed and supervised by competent people. The purchase of starting materials is an important operation, which should involve staff who have a thorough knowledge of the suppliers and who should only purchase from approved suppliers named in the relevant specification. The source, origin and suitability of starting materials should be clearly defined; the various components, containers and closures that are received, identified, stored, handled, sampled, tested and approved or rejected should be regularly inspected, and the system should be challenged to test if it is functioning correctly. There must be written procedures describing how these operations are done and if the handling and storage of components are computer controlled, the programme must be validated.

Control of raw materials

Written procedures should be established describing the purchase, receipt, identification, quarantine, storage, handling, sampling, testing and approval or rejection of raw materials, and such procedures should be followed. In fact, it is beneficial for all aspects of the manufacture and control of the starting material in question, including handling, labelling and packaging requirements, as well as complaints and rejection procedures to be discussed with the supplier. All materials and products should be handled and stored under the appropriate conditions established by the manufacturer, in an orderly fashion to permit batch segregation and stock rotation, as well prevent contamination or cross-contamination. The manufacturer must be able to show that the containers and closures are compatible with the product, will provide adequate protection for the product against deterioration or contamination, are not additive or absorptive, and are suitable for use.

Receipt, sampling, testing and approval of raw materials

Incoming materials should be physically or administratively quarantined immediately on receipt, until they have been sampled, tested or examined as appropriate and released for use or distribution. They should be checked to ensure that the consignment corresponds to the order, and examined visually for integrity of package and seal, for correspondence between the delivery note and the supplier's labels, for damage to containers and any other problem that might adversely affect the quality of a material. The receiving records must be traceable to the component manufacturer and supplier and should contain the name of the component, manufacturer, manufacturer's lot number, supplier if different from the manufacturer, and carrier. All handling of starting materials, such as receipt and quarantine, sampling, storage, labelling, dispensing, processing, packaging and distribution should be done in accordance with written procedures or instructions and, where necessary, recorded.

The number of containers to sample and the sample size should be based upon appropriate criteria, such as the quantity needed for analysis, sample variability, degree of precision desired and past quality history of the supplier, and the sample containers properly identified. At least one test should be conducted to verify the identity of each raw material. A supplier's certificate of analysis may be used instead of performing other testing, provided the manufacturer has a system in place to evaluate vendors (vendor audits) and establishes the reliability of the supplier's test results at appropriately regular intervals. For hazardous or highly toxic raw materials, where on-site testing may be impractical, suppliers' certificates of analysis should be obtained showing that the raw materials conform to specifications. However, the identity of these raw materials must be confirmed by examination of containers and labels, and the lack of on-site testing for these hazardous raw materials should be documented. Intermediate and bulk products purchased as such should also be handled as though they were starting materials.

Starting materials in the storage area should be appropriately labelled and should only be dispensed by designated persons, following a written procedure, to ensure that the correct materials are accurately weighed or measured into clean and properly labelled containers. Materials dispensed for each batch should be kept together and conspicuously labelled as such. Information on the labels should provide traceability from the component manufacturer to its use in the finished product, and should bear at least the following information:

- the designated name of the product and the internal code reference where applicable;
- a batch number given at receipt;
- the status of the contents (e.g. in quarantine, on test, released, rejected) where applicable;
- an expiry date or a date beyond which re-testing is necessary, if appropriate.

When fully computerized storage systems are used, all the above information need not necessarily be in a legible form on the label.

Use and re-evaluation of approved raw materials

Approved raw materials should be stored under suitable conditions and, where appropriate, rotated so that the oldest stock is used first. Raw materials should be re-evaluated as necessary to determine their suitability for use, for example, after prolonged storage or after exposure to heat or high humidity. Sanitary conditions in the storage area, stock rotation practices, re-test dates and special storage conditions, such as protection from light, moisture, temperature and air, should be checked regularly.

Rejected raw materials

Rejected raw materials should be identified and controlled under a quarantine system designed to prevent their use in manufacturing or processing operations for which they are unsuitable.

404

11.4.2 Final processing operations

Sterile products are usually produced by dissolving the non-sterile bulk active substance in a solvent and then filtering the solution through a sterilizing filter. After filtration, the sterile bulk material is separated from the solvent by crystallization, precipitation and spray-drying or lyophilization. During these final processing operations, all necessary in-process controls and environmental controls should be carried out and recorded, and any significant deviation from the expected yield should be recorded and investigated.

Critical manufacturing steps

Each critical step in the manufacturing process should be done by a responsible individual and checked by a second responsible individual. If such steps in the processing are controlled by automatic mechanical or electronic equipment, its performance should be verified. Critical manufacturing steps not only include the selection, weighing, measuring and identifying of components, and addition of components during processing, but also the recording of deviations in the manufacturing record, testing of in-process material and the determination of actual yield and percent of theoretical yield. These critical manufacturing steps should be fully validated and documented when done. At all times during processing, all materials, bulk containers, major items of equipment and, where appropriate, the rooms used, should be labelled or otherwise identified with an indication of the product or material being processed, its strength (where applicable), batch number and the stage of production. Labels applied to containers, equipment or premises should be clear, unambiguous and in the company's agreed format. It is often helpful in addition to the wording on the labels to use colours to indicate status, such as quarantined, accepted, rejected and clean.

Preparation

Before any processing operation is started, steps should be taken to ensure that the work area and equipment are clean and free from any starting materials, products, product residues or documents that are not required for the operation being planned. Intermediate and bulk products, and all starting materials should be kept under appropriate conditions. Checks should be carried out to ensure that pipelines and other pieces of equipment used for the transportation of products from one area to another are connected in a correct manner. Non-combustible gases, and all solutions, in particular large volume infusion fluids, should be passed through a microorganism retaining filter if possible, immediately prior to filling. Any components, containers, equipment and any other article required in the clean area where aseptic work takes place should be sterilized and passed into the area through double-ended sterilizers sealed into the wall, or by a procedure which achieves the same objective of not introducing contamination. Bioburden and contamination levels should be monitored before sterilization and where appropriate, the absence of pyrogens should also be monitored. The interval between the washing, drying and the sterilization of components, containers and equipment, as well as between their sterilization and use should be minimized and subject to a time-limit appropriate to the storage conditions.

Batching

Many of these bio-pharmaceutical products lack preservatives, inherent bacteriostatic or fungistatic activity. Obviously, the batching or compounding of bulk solutions should, therefore, be controlled in order to prevent any potential increase in microbiological levels that may occur up to the time that the bulk solutions are filter sterilized. One concern with any microbiological level is the possible increase in endotoxins that may develop. Good practice would, therefore, include working in a controlled environment, and in sealed tanks to control accessibility, particularly if the non-sterile product solutions are to be stored for any period prior to sterilization.

Filling

The filling of bio-pharmaceuticals into ampoules or vials presents many of the same problems as the processing of conventional pharmaceuticals. The batch size of a bio-pharmaceutical is likely to be small and the validation of aseptic processes presents special problems when the batch size is small. In these cases, the number of units filled may be the maximum number filled in production and because of the small batch size, filling lines may not be as automated as for other products typically filled in larger quantities. Moreover, filling and sealing will often be a hand operation, presenting great challenges to sterility; and with more involvement of people filling these products, more attention should be given to environmental monitoring. Typically, vials to be lyophilized are partially stoppered by machine. However, some filling lines have even been observed using an operator to place each stopper on top of the vial by hand. The immediate concern in this case is the avenue of contamination offered by the operator. Due to the active involvement of people in filling and aseptic manipulations, the number of persons involved in these operations should be kept to a minimum, and the environmental programme should include an evaluation of microbiological samples taken from people working in such aseptic processing areas. Some of the problems that are routinely identified during filling include inadequate attire, deficient environmental monitoring programmes and failure to validate some of the basic sterilization processes.

One major concern is the use of inert gas to displace oxygen during both the processing and filling of the solution, and therefore, limits for dissolved oxygen levels for the solution must be established for products that may be sensitive to oxidation, and parameters such as line speed and the location of the filling syringes with respect to their closures should be defined. In the absence of inert gas displacement, the manufacturer should be able to demonstrate that the product is not affected by oxygen. Another major concern with the filling operation of a lyophilized product is the assurance of fill volumes. Obviously, a low-fill would represent a sub-potency in the vial. Unlike a powder or large volume liquid fill, a low-fill would not be readily apparent after lyophilization, particularly for a product where the active ingredient may be only a milligram. Due to its clinical significance, sub-potency in a vial can potentially be very serious.

Lyophilization (freeze drying) or spray drying

Many bio-pharmaceuticals are lyophilized because of stability concerns. Unfortunately, the cGMP aspects of the design of lyophilizers have lagged behind the sterilization and control technology employed for other processing equipment. It is not surprising that many problems with the lyophilization process have been identified. These problems are not limited to bio-pharmaceuticals, but generally pertain to lyophilization of all products including bio-pharmaceuticals. With regard to bulk lyophilization, concerns include air classification, aseptic barriers for loading and unloading the unit, partial meltback, uneven freezing and heat transfer throughout the powder bed, and the additional aseptic manipulations required to break up the large cake. For bulk lyophilization, unlike other sterile bulk operations, media challenges can be performed, and hence suitable validation studies must be carried out.

There are also concerns over the spray drying of sterile bio-pharmaceuticals, including the sterilization of the spray dryer, the source of air and its quality, the chamber temperatures, and the particle residence or contact time. In some cases, charring and product degradation have been found for small portions of a batch. These should all be assessed during process validation.

Sterile filtration of products which cannot be sterilized in their final container

If the product cannot be sterilized in the final container, then solutions or liquids must be filtered through a sterile filter of nominal pore size of 0.22 micron (or less), or with at least equivalent microorganism retaining properties, into a previously sterilized container. Such filters can remove most bacteria and moulds, but not all viruses or mycoplasmas, so consideration should be given to complementing the filtration process with some degree of heat treatment. Moreover, if other means of sterilization in the final container were possible, then final sterile filtration alone is not considered sufficient. The specification for the filters should include information such as its fibre shedding character-istics, the criteria used for the selection of the filter, as well as the procedures used for integrity testing of the filters. The integrity of the sterilized filter should be verified before use, and should be confirmed immediately after use by an appropriate method such as a bubble point, diffusive flow, or the pressure hold test. The time taken to filter a known volume of bulk solution, the maximum filtration pressures and the pressure differential across the filter should also be determined during validation, and any significant differences from this should be noted and investigated. The same filter should never be used for more than one working day unless such use has been validated. If filters were not changed after each batch is sterilized, there should be data to justify the integrity of the filters for the time periods utilized and prove that grow-through has not occurred.

Terminally sterilized products

Steam sterilization is the preferred method of those currently available. However, before any sterilization process is adopted, its suitability for the product and its efficacy in achieving the desired sterilizing conditions in all parts of each type of load to be processed should be demonstrated by physical measurements and by the use of biological indicators where appropriate. There should also be a clear means of differentiating products which have not been sterilized from those which have, with each basket, tray or other carrier of products or components clearly labelled with the name of the product, its batch number, and an indication of whether or not it has been sterilized. Typically, a sterile pharmaceutical contains no viable micro-organisms and is non-pyrogenic. Parenteral drugs in particular must be non-pyrogenic because the presence of pyrogens can cause a febrile reaction in human beings. Pyrogens are the products of the growth of microorganisms, so any condition that allows microbial growth should be avoided in the manufacturing process. Pyrogens may develop in water located in storage tanks, dead legs and pipework, or from surface contamination of containers, closures or other equipment, and may also contain chemical contaminants that could produce a pyretic response in humans or animals even though there may be no pyrogens present.

Therefore, the procedures used to minimize the hazard of contamination with microorganisms and particulates of sterile bio-pharmaceuticals become extremely important. Preparation of components and other materials should be done in at least a grade D environment in order to give low risk of microbial and particulate contamination, suitable for filtration and sterilization. Where the bio-pharmaceutical is at a higher than usual or an unusual risk of microbial contamination; for example, because the product actively supports microbial growth, or must be held for a long period before sterilization, or needs to be processed in other than closed vessels, then all the preparation should be carried out in a grade C environment. Filling of a bio-pharmaceutical for terminal sterilization should be carried out in at least a grade C environment. Where the product is at an unusual risk of contamination from the environment because, for example, the filling operation is slow or the containers are wide-necked or is necessarily exposed for more than a few seconds before sealing, the filling should be done in a grade A zone, with at least a grade C background.

Finishing of sterile products

Filled containers of bio-pharmaceuticals should be closed by appropriately validated methods. Containers closed by fusion, for example, glass or plastic ampoules, should be subject to 100% integrity testing, while those closed by other means should be checked for integrity according to appropriate procedures. Containers sealed under vacuum should be tested for maintenance of that vacuum after an appropriate, pre-determined period. Filled containers should be inspected individually for extraneous contamination or other defects, and if inspection is done visually, it should be done under suitable and controlled conditions of illumination and background. Where other methods of inspection are used, the process should be validated and the performance of the equipment checked at intervals with the results recorded.

Some sterile bio-pharmaceuticals may be filled into different types of containers, such as sterile plastic bags. For sterile bags, sterilization by irradiation is the method of choice because it leaves no residues, although some manufacturers use formaldehyde. A major disadvantage is that formaldehyde residues may, and frequently do appear in the sterile product. If multiple sterile bags are used, operations should be performed in an aseptic processing area. Since all the inner bags have to be sterile, outer bags should also be applied over the primary bag containing the sterile product in the aseptic processing area. One manufacturer was found to apply only the primary bag in the aseptic processing area, resulting in the outer portion of this primary bag being contaminated when the other bags were applied over this bag in non-sterile processing areas! Important validation aspects of the sterile bag system include measurement of residues, testing for pinholes, foreign matter (particulates), as well as for sterility and endotoxins.

11.4.3 Secondary (sterile) production facility

Manufacturing operations are divided into two categories — those where the product is terminally sterilized and those which are conducted aseptically at some or all stages. The design, validation and effective operation of clean rooms for the manufacture and testing of pharmaceuticals, biotechnology and medical device products is among the most exacting and challenging activities. Patient's lives, product integrity, company profitability and regulatory compliance all factor into the risks inherent if the clean room is not built right and does not function right. The manufacture of sterile products should be carried out in clean areas, entry to which should be through airlocks for personnel and/or for equipment and materials, and maintained to an appropriate standard of cleanliness, and supplied with air that has passed through filters of an appropriate efficiency. Adequate space must be provided for the placement of equipment and materials to prevent mix-ups for operations such as the receiving, sampling, and storage of raw materials; manufacturing, processing, packaging and labelling; storage for containers, packaging materials, labelling and finished products; as well as for production and control laboratories. Facility design features for the aseptic processing of sterile bulk active products should include temperature, humidity and pressure control, and there must be adequate lighting, ventilation, screening and proper physical barriers for all operations including dust, temperature, humidity and microbiological controls, with the various operations of component preparation, product preparation and filling carried out in separate areas within the clean area.

Area classification and monitoring of controlled environments

Clean areas for the manufacture of sterile products are classified according to the required characteristics of the environment. Each manufacturing operation requires an appropriate level of cleanliness in the operational state, in order to minimize the risks of particulate or microbial contamination of the product or materials being handled. In order to meet in-operation conditions, these areas should be designed to reach certain specified air-cleanliness levels in the at-rest occupancy state. The at-rest state is the condition where the installation is installed and operating, and is complete with production equipment, but has no operating personnel present. The in-operation state is the condition where the installation is functioning in the defined operating mode with the specified number of personnel.

For the manufacture of sterile medicinal products there are normally four grades of clean areas. The requirement and limit for these areas depend on the nature of the operations carried out. Grade A is for the aseptic preparation and filling of products, and the local zone for high risk operations such as the filling

410

zone, stopper bowls, open ampoules and vials, making aseptic connections. Normally such conditions are provided by a laminar airflow workstation, which should provide a homogeneous air speed of $0.45\,\mathrm{m\,s}^{-1} \pm 20\%$ (guidance value) at the working position. Grade B is for aseptic preparation and filling, and the background environment for grade A zone. Grade C is for the preparation of solutions to be filtered and the filling of products that are at high risk. Grade D is a clean area for carrying out less critical stages in the manufacture of sterile products, for the handling of components after washing, and for the preparation of solutions and components for subsequent filling.

11.4.4 Safety issues

Contamination control

Manufacturing on a campaign basis is typical in the bio-pharmaceuticals industry. Whilst this may be efficient with regard to system usage, it can present problems when it is discovered in the middle of a campaign that a batch is contaminated. Frequently, all the batches processed in a campaign in which a contaminated batch is identified are suspect. Such failures should be investigated and reported, and the release of any other batches in the campaign should be justified. Some of the more significant product recalls have occurred because of the failure of a manufacturer to conclusively identify and isolate the source of a contaminant.

When working with dry materials and products, special precautions should be taken to prevent the generation and dissemination of dust. This could result in the risk of accidental cross-contamination arising from such uncontrolled release of dust, gases, vapours, sprays or organisms from materials and products in process, from residues on equipment and from operators' clothing. The significance of this risk varies with the type of contaminant, and the product being contaminated. Amongst the most hazardous contaminants are highly sensitizing materials, biological preparations containing living organisms, certain hormones, cytotoxics, and other highly active materials. Products in which contamination is likely to be most significant are those administered by injection and those given in large doses and/or over a long time.

Environmental control

Containers and materials liable to generate fibres should be minimized in clean areas. All components, containers and equipment should be handled after the final cleaning process in such a way that they are not re-contaminated. After washing, all components should be handled in at least a grade D environment.

411

The handling of sterile starting materials and components, unless subjected to sterilization or filtration through a micro-organism-retaining filter later in the process, should be done in a grade A environment with grade B background. However, the handling and filling of aseptically prepared products should be done in a grade A environment with a grade B background. The preparation of solutions that are to be sterile filtered during the process should be done in a grade C environment; however, if not filtered, the preparation of materials and products should be done in a grade A environment with a grade B background. The preparation and filling of sterile suspensions should be done in a grade A environment with a grade B background if the product is exposed and is not subsequently filtered. Prior to the completion of stoppering, the transfer of partially closed containers, as used in lyophilization (freeze drying) should be carried out either in a grade A environment with grade B background, or in sealed transfer trays in a grade B environment.

Prevention of cross-contamination
In clean areas, and especially when aseptic operations are in progress, all activities should be kept to a minimum, and the movement of personnel should be controlled and methodical to avoid excessive shedding of particles and organisms due to over-vigorous activity. The production of non-medicinal products should not be carried out in areas or with equipment destined for the final processing of bio-pharmaceuticals. Certainly, operations on different products should not be carried out simultaneously, or consecutively in the same room, unless there is no risk of mix-up or cross-contamination, and preparations of microbiological origin should not be made or filled in areas used for the processing of other sterile medicinal products; however, vaccines of dead organisms or of bacterial extracts may be filled, after inactivation, in the same premises as other sterile medicinal products. Manufacture in segregated areas is required for products such as penicillins, live vaccines, live bacterial preparations and certain other specified biologicals, or manufacture by campaign (separation in time) followed by appropriate cleaning. Precautions to minimize contamination should be taken during all processing stages including the stages before sterilization. These include using closed systems of manufacture, as well as appropriate air-locks and air extraction; using cleaning and decontamination procedures of known effectiveness, as ineffective cleaning of equipment is a common source of cross-contamination; as well as keeping protective clothing inside areas where products with special risk of cross-contamination are processed.

412

Control of sterility

Manufacturers are expected to validate all critical aseptic processing steps in the manufacture of bio-pharmaceuticals with at least three consecutive validation runs. Such validation must encompass all parts, phases, steps and activities of any process where components, fluid pathways or in-process fluids are expected to remain sterile. Furthermore, such validation must include all probable potentials for loss of sterility as a result of processing and account for all potential avenues of microbial ingress associated with the routine use of the process.

Sterility testing

The sterility test applied to the finished product should only be regarded as the last in a series of control measures by which sterility is assured. The test should be fully validated for the product(s) concerned with any examples of initial sterility test failures thoroughly investigated. In those cases where parametric release has been authorized, special attention should be paid to the validation and the monitoring of the entire manufacturing process. Samples taken for sterility testing should be representative of the whole of the batch, but should in particular include samples taken from parts of the batch considered to be most at risk of contamination. For example, for products that have been filled aseptically, samples should include containers filled at the beginning and at the end of the batch, and after any significant intervention. For products that have been heat sterilized in their final containers, consideration should be given to taking samples from the potentially coolest part of the load.

Media fill validation

Validation of aseptic processing should include simulating the process using a nutrient medium, the form of which is equivalent to the dosage form of the product, although suitable microbiologically-inert non-media alternatives would also be acceptable. This process simulation test should imitate as closely as possible the routine aseptic manufacturing process and include all the critical subsequent manufacturing steps, and should be repeated at defined intervals and after any significant modification to the equipment and process. The number of containers used for a medium fill should be sufficient to enable a valid evaluation. For small batches, the number of containers for the medium fill should at least equal the size of the product batch. The contamination rate should be less than 0.1% with 95% confidence level, and care should be taken that any validation does not compromise the processes, although the limitations of 0.1% media fill contamination rate should be recognized for the validation of aseptic processing of a non-preserved single dose bio-pharmaceutical, stored at room temperature as a solution. Any alternative proposals for the validation of the aseptic processing of bio-pharmaceuticals may be considered by the regulatory authorities, but only on a case-by-case basis. For example, it may be acceptable to exclude from the aseptic processing validation procedure certain stages of the post-sterilization bulk process that take place in a totally closed system. Such closed systems should, however, be Sterilized in Place by a validated procedure, integrity tested for each lot, and should not be subject to any intrusions whereby there may be the likelihood of microbial ingress. Suitable continuous system pressurization would be considered an appropriate means for ensuring system integrity.

Control of pyrogens and endotoxins

Typically, a sterile pharmaceutical contains no viable microorganisms and is non-pyrogenic. Parenteral drugs must be non-pyrogenic because the presence of pyrogens can cause a febrile reaction in human beings. As pyrogens are the products of the growth of microorganisms, any condition that allows microbial growth should be avoided. Parenterals may also contain chemical contaminants that could produce a pyretic response in humans or animals, even if there are no pyrogens present. Moreover, in addition to pyrogens, microorganisms could contaminate the process stream with by-products such as glycosidases and proteases, which irreversibly alter or inactivate the product and as a result could adversely affect product stability.

The manufacturing process strategy, therefore, should be oriented more towards keeping endotoxins and pyrogens out as well as trying to reduce their levels. In some instances, where pipework systems for aqueous solutions have been shown to be the source of endotoxin contamination in sterile products, the manufacturer should be able to give assurance that there are no 'dead legs' in the system. In addition, water sources, water treatment equipment and treated water should be monitored regularly for such chemical and biological contamination and, as appropriate, for endotoxins.

Some manufacturers have argued that if an organic solvent is used in the manufacture of a sterile product, then the endotoxins levels are reduced at this stage. As with any operation, this may or may not be correct, and should be proven. For example, one manufacturer who conducted extensive studies using organic solvents for the crystallization of a non-sterile pharmaceutical to the sterile product observed no change from the initial endotoxin levels. In the validating the reduction or removal of endotoxins, challenge studies can be carried out on a laboratory or pilot scale to determine the efficiency of the step. However, since endotoxins may not be uniformly distributed, it is also important to monitor the bioburden of the non-sterile product(s) being sterilized. For

414

example, gram negative contaminates in a non-sterile bulk drug product prior to sterilization are of concern, particularly if the sterilization (filtration) and crystallization steps do not reduce the endotoxins to acceptable levels.

11.4.5 Out of specification

Regulatory authorities require that suitable process controls be established using scientifically sound and appropriate specifications, standards, sampling and re-sampling, testing and re-testing. These should be designed to ensure that all materials relating to the bio-pharmaceutical manufacture, such as components, containers, closures, in-process materials, labelling, including the product conform to appropriate standards of identity, strength, quality and purity. These controls should be used for the determination of conformity to applicable specification, for the acceptance of each batch (or lot) of material relating to manufacture, processing, packing, or the holding of the pharmaceutical. 'Out of specification' is defined as an examination, measurement, or test result that does not comply with such pre-established criteria. cGMP guidelines require written procedures to be in place to determine the cause of any apparent failure, discrepancy, or out of specification result. Out of specification results can be caused by laboratory error, non-process or operator error, or by process-related error, such as personnel or equipment failures. If, however, the result could not be clearly attributed to sampling or laboratory error, then there should be scientifically sound procedures and criteria for the exclusion of any test data found to be invalid and, if necessary, for any additional sampling and testing.

Re-testing

Although re-testing may be an appropriate part of the investigation, an investigation consisting solely of repeated re-testing is clearly inadequate. If quality is not built into a product, re-testing cannot make it conform to specifications. The number of re-tests performed before it can be concluded that an unexplained out of specification laboratory result is invalid, or that a product is unacceptable, is a matter of scientific judgment. There are no regulations on specific re-testing procedures, although manufacturers are expected to have written investigation and re-testing procedures, applying scientifically sound criteria. A variety of written and unwritten practices and procedures have been observed, under which manufacturers have disregarded out of specification laboratory results after minimal re-testing, re-sampling, inappropriate averaging of results or inappropriate testing. Some manufacturers then proceeded to release a product without a thorough investigation or an adequate justification for disregarding an out of specification result. Regulatory authorities recognize the distinction between the limited investigation that may be necessary to identify a laboratory error and the more extensive investigation and testing necessary when out of specification results may be attributed to another cause. The manufacturer may impose additional criteria beyond those required to ensure identity, strength, quality and purity under cGMP regulations or as required for licensure. Although such internal controls are encouraged, under some circumstances it is possible to have test results that violate the internal standards, without being out of specification, as defined by regulations. The investigation should extend to other batches of the same product, and other products that may have been associated with the specific failure or discrepancy.

Re-testing for pyrogens and endotoxins

As with sterility, re-testing for pyrogens or endotoxins can be performed and is only acceptable if it is known that the test system was compromized and the cause of the initial failure is known, thereby invalidating the original results. It cannot be assumed that the initial failure is a false positive without sufficient documented justification. Again, any pyrogen or endotoxin test failures, the incidence, procedure for handling, and final disposition of the batches involved, should be investigated thoroughly, and the reasons for re-testing fully justified.

Sterility re-testing

The release of a batch, particularly of a sterile bio-pharmaceutical, which fails an initial sterility test and passes a re-test is very difficult to justify. Sterility re-testing is only acceptable if the cause of the initial non-sterility is known, and thereby invalidates the original results. It cannot be assumed that the initial sterility test failure is a false positive. This conclusion must be justified by sufficient documented investigation, and repeated sampling and testing may not identify any low level contamination. Sterility test failures, the incidence, procedures for handling, and final disposition of the batches involved should be routinely reviewed.

Reprocessing

The term reprocessing describes steps taken to ensure that the reprocessed batches will conform to all established standards, specifications and characteristics, and relates to steps in the manufacturing process that are out of the normal manufacturing processing sequence or that are not specifically provided for in the manufacturing process. As with the principal manufacturing process, reprocessing procedures should be validated. All the data pertaining to the reprocessed batches, as well as the data used to validate the process, should be reviewed and detailed investigation reports, including the description, cause and corrective action taken, should be available for the batch. The number and frequency of process changes made to a specific process or step can be an indicator of a problem experienced in a number of batches. For example, a number of changes in a short period of time can be an indicator that that particular process step is experiencing problems.

Rejection

The demonstration of the adequacy of the process to control other physico-chemical aspects is an important aspect of validation. Depending upon the particular bio-pharmaceutical, these include potency, impurities, particulate matter, particle size, solvent residues, moisture content and blend uniformity. For example, if the product is a blend of two active products or an active product and an excipient, then there should be some discussion and evaluation of the process for assuring uniformity. The process validation report for such a blend should include documentation for the evaluation and assurance of uniformity. Manufacturers occasionally reject the product following the purification process or after final processing. As with all pharmaceutical products, it is expected that any batch failing specifications is investigated thoroughly, and reports of these investigations are complete. For example, during one production campaign it was noted that approximately six batches of a bio-pharmaceutical product were rejected because of low potency and high levels of impurities. The problem was finally attributed to a defective column and, as a result, all the batches processed on that particular column were rejected.

11.5 Design of facilities and equipment

11.5.1 Facility design

When designing facilities for bio-pharmaceutical manufacture, the following activities should be considered as areas to control contamination:

- the receipt, identification, storage and withholding from use of raw materials or process intermediates, pending release for use in manufacturing; as well as the quarantine storage of intermediates and final products pending release for distribution;
- the holding of rejected raw materials, intermediates and final products before final disposition;
- the storage of released raw materials, intermediates and final products;
- manufacturing and processing operations;

- packaging and labelling operations;
- all laboratory operations.

Control of microbiological, physical, and chemical contamination

The regulatory authorities require the establishment of, and adherence to, written procedures designed to prevent microbiological contamination of pharmaceuticals purporting to be sterile. These requirements also cover such procedures as the validation of any sterilization process, and are intended to reflect the fact that whether aseptic processing techniques or terminal sterilization methods are used, either technique must be validated. Where microbiological specifications have been established for the product, then facilities should also be designed to limit objectionable microbiological contamination, especially if different bio-pharmaceuticals are handled in the same premises and at the same time. For the production of the same products, campaign working may be acceptable in place of dedicated and self-contained facilities.

Products can become contaminated with physical or chemical contaminants in a variety of ways. For example, ineffective cleaning procedures may leave residues of the product or cleaning agents in the equipment; production workers may fail to take proper precautions while transporting a substance from one area to another thereby introducing a contaminant to the second production area; or particles may become airborne and travel to production areas throughout the facility. A number of substances such as dust, dirt, debris, toxic products, infectious agents, or residue of other drugs or drug components can also contaminate products.

Experience indicates that the potential dangers of contamination are more extensive and varied than once believed. For example, adulteration of the sterile product with sensitizing substances (such as penicillin, cephalosporins), substances having high pharmacological activity or potency (such as steroids, cytotoxic anti-cancer agents), infectious agents (such as spore-bearing organisms), and products that require viral inactivation or reduction (such as live viruses, products from animal cells), may pose health risks to humans or animals, even at minimal levels of exposure. Preventing cross-contamination of such potentially active substances is the goal and manufacturers are expected to identify any such substances posing a serious threat of contamination and to control it through dedicated production processes. Moreover, because the identity or even the presence of some of these contaminants may not be known, health care professionals providing care to a patient suffering from such an adverse effect may be unable to provide appropriate medical intervention.

Most contamination, however, can be controlled to an acceptable level through measures such as proper planning and implementation of cleaning and sanitation processes, employee training, gowning, and air filtration. cGMP guidelines require that manufacturers set contamination limits on a substance-by-substance basis, according to both the potency of the substance and the overall level of sensitivity to that substance, and prohibit the release of the product for distribution if these limits were exceeded. Depending on the product, a variety of measures may be acceptable to eliminate cross-contamination; there may, however, be situations where nothing short of dedicated facilities, air handling and process equipment would be sufficient, especially if there are no reasonable methods for the cleaning and removal of a substance or compound residues from buildings, facilities and equipment. For example, a manufacturer might develop a hypothetical product of high therapeutic potential that also poses a high risk of contamination and if it posed a special danger to human health, dedicated facilities would be required. If, however, experience demonstrated that the product did not pose such a risk, or if changes in manufacturing technology greatly reduced the risk, then dedicated facilities might no longer be required.

Sanitation

The sanitation of clean areas is particularly important. Any building used in the manufacture, processing, packing or holding of bio-pharmaceuticals and their intermediates should be maintained in a clean and sanitary condition. Sanitation procedures should apply to work performed by contractors or temporary employees as well as work performed by full-time employees during the ordinary course of operations. Written procedures should, therefore, be established, assigning responsibility for sanitation, and describing the cleaning schedules, methods, equipment and materials to be used in cleaning buildings and facilities, and for the use of suitable rodenticides, insecticides, fungicides, fumigating agents, or other cleaning and sanitizing agents to prevent the contamination of equipment, raw materials, packaging and labelling materials, as well as the final product. Where disinfectants are used, more than one type should be employed, and monitoring should be undertaken regularly to detect the development of resistant strains. Disinfectants and detergents should be monitored for microbial contamination, and those used in grades A and B areas especially should be sterile prior to use.

Monitoring programmes in controlled environments

It is the responsibility of the manufacturer to develop, initiate and implement an environmental monitoring programme tailored to specific facilities and condi-

419

tions and capable of detecting any adverse drift in microbiological conditions in a timely manner, allowing meaningful and effective corrective action. Such microbiological monitoring programmes should be utilized to assess the effectiveness of cleaning and sanitization practices and of personnel that could have an impact on the bioburden of the controlled environment. Routine microbial monitoring, regardless of how sophisticated the system may be, will not and need not identify and quantify all microbial contaminants present in the controlled environment. It can only provide information to demonstrate that the environmental control systems are operating as intended. The objective of microbial monitoring is, therefore, to obtain representative estimates of bioburden in the environment.

The environmental monitoring programme for the manufacture of sterile bio-pharmaceuticals should include the daily use of surface plates and the monitoring of personnel, with alert or action limits established, and appropriate follow-up corrective action taken when they are reached. Where critical aseptic operations are performed, monitoring should be frequent using methods such as settle plates, volumetric air and surface sampling (such as swabs and contact plates). Additional microbiological monitoring is also required outside production operations, for example, after validation of systems, cleaning and sanitization. The particulate conditions for the at-rest state should be achieved in the unmanned state after a short clean-up period of about 15–20 minutes (guidance value) after completion of operations. The particulate conditions for grade A in operation should be maintained in the zone immediately surrounding the product, whenever the product or open container is exposed to the environment. It may not always be possible to demonstrate conformity with particulate standards at the point of fill when filling is in progress due to the generation of particles or droplets from the product itself.

Some manufacturers utilize UV lights in operating areas. Such lights are of limited value as they may mask a contaminant on a settle or aerobic plate or may even contribute to the generation of a resistant (flora) organism. Therefore, the use of surface contact plates is preferred, as they will provide more information on levels of contamination. There are some manufacturers that set alert/action levels on averages of plates. For the sampling of critical surfaces, such as operators' gloves, the average of results on plates is unacceptable. The primary concern is any incidence of objectionable levels of contamination that may result in a non-sterile product. Since processing is commonly carried out around the clock, monitoring of surfaces and personnel during the second and third shifts should also be routine.

In the management of a sterile operation, periodic (weekly/monthly/ quarterly) summary reports of environmental monitoring should be generated. Trained personnel should evaluate any trends when data are compiled and analysed. While it is important to review environmental results on a daily basis, it is also critical to review results over extended periods to determine whether trends are present, as they may be related to decontamination procedures, housekeeping practices, personnel training, cross-contamination and the potential for microbial build up during production. A full investigation should, therefore, include a review of area maintenance documentation, sanitization documentation, the inherent physical or operational parameters, and the training status of personnel involved, while a limited investigation triggered by an isolated, small excursion might include only some of these areas. Based on the review of the investigation and testing results, the significance of the event and the acceptability of the operations or products processed under that condition can be ascertained. Any investigation and the rationale for the course of action should be documented and included as part of the overall quality management system.

11.5.2 Laboratory design

The design of a laboratory that handles any bio-pharmaceutical, which may include infectious agents, should provide secondary containment to protect the people as well as the environment outside the laboratory from exposure to any infectious materials. Laboratory design should take into account the nature of the material being handled, the process step or study being planned for investigation, and the degree of biosafety necessary. They must be sufficient to enable the proper conduct of the study and must provide appropriate space, environmental conditions, containment, decontamination areas and support systems, such as air and water, for the study being conducted.

There are three types of laboratory designs that provide four different levels of containment. They all consist of three elements: laboratory practices and techniques, safety equipment, and laboratory facilities. The first two elements are considered primary containment, since they provide protection within the laboratory to personnel and the immediate environment. The third element, the design of the laboratory itself, is considered secondary containment since it protects persons and the environment outside of the facility. Changes in vendor and/or the specifications of major equipment and reagents would require re-validation. Each laboratory should have documentation and schedules for the maintenance, calibration and monitoring of all laboratory equipment involved in the measurement, testing and storage of raw materials, product, samples, and reference reagents, and more importantly the laboratory personnel should be adequately trained for the jobs they are performing. Important characteristics of each of the biosafety levels are summarized below.

Basic laboratory

These are appropriate for Biosafety levels 1 and 2. They are used for studies where there is a minimum level of hazard, the personnel are able to achieve sufficient protection from the implementation of standard laboratory practices, and the organisms used in the study are not associated with any diseases in healthy adults.

Biosafety level 1

The organisms involved are defined and characterized strains, which are of minimal hazard and are not known to cause disease in healthy human adults. Although access to the laboratory may be restricted, the facility is generally not closed off from the rest of the building. The laboratory is designed to facilitate cleaning, with space between equipment and cabinets, and bench tops that are impervious to water and resistant to solutions. Personnel should be knowledgeable in all laboratory procedures and supervised by a scientist trained in microbiology or a related science. Most work is conducted on open bench tops, with procedures performed in a manner that limits the creation of aerosols, and special containment equipment is not usually needed. Decontamination of work surfaces should be done daily and after spills, and all contaminated wastes should be decontaminated before disposal. Each laboratory has a hand-washing sink. Personal safety equipment, such as laboratory coats or uniforms, should be worn and hands washed before and after handling viable materials. Any contaminated materials that will be decontaminated at another location should be transported in a durable leak proof container that is sealed before removal from the area.

Biosafety level 2

Work done under Biosafety level 2 involves organisms of moderate potential hazard. Many of the characteristics of this level are the same as those for Biosafety level 1. However, for Biosafety level 2, laboratory access is limited while work is being conducted, and only persons informed of the potential hazards of the environment and who meet any other entry restrictions developed by the organization should be allowed entry. Biological safety cabinets (Class I or II) should be used for containment when procedures with a high potential for creating infectious aerosols such as centrifugation or blending are conducted or when high concentrations or large volumes of infectious agents are used. An autoclave should be available for use in decontaminating infectious wastes. Personnel should be trained in handling pathogenic agents and be under the direction of skilled scientists. Before leaving the area, personnel should either remove any protective clothing and leave it in the laboratory, or cover it with a clean coat. Skin contamination with infectious materials should be avoided and gloves worn when such contact is unavoidable. Spills and accidents causing overt exposure to infectious materials should be reported promptly with appropriate treatment provided and records of the incident maintained. If warranted by the organisms at use in the laboratory, baseline serum samples for all at-risk personnel should be collected and stored.

Containment laboratory

Containment laboratories qualify as Biosafety level 3 facilities and are designed with protective features to allow for the handling of hazardous materials in a way that prevents harm to the laboratory personnel, as well as the surrounding persons and environment. These may be freestanding buildings or segregated portions of larger buildings, as long as they are separated from public areas by a controlled access zone. Containment laboratories also have a specialized ventilation system to regulate airflow.

Biosafety level 3

Work done under Biosafety level 3 conditions can occur in clinical, diagnostic, teaching, research or production facilities, and involves organisms that may cause serious or potentially lethal disease following exposure through inhalation. The laboratory is, therefore, segregated from general access areas of the building, and two sets of self-closing doors must be passed through to enter the laboratory from access hallways. Access should be limited to persons who must be present for programme or support functions, and the doors remain closed during experiments. Protective clothing should be worn in the laboratory and removed before exiting the facility, and all such clothing should be decontaminated before laundering. All work with infectious materials should be conducted in a biosafety cabinet (Class I, II or III) or other physical containment device, or by personnel wearing the necessary personal protection clothing. Upon completing work with infectious materials, all work surfaces should be decontaminated. Walls, ceilings and floors should be water-resistant to facilitate cleaning, and windows should be closed and sealed. The laboratory sinks should be operable by foot, elbow or automation, and be located near the exit of each laboratory area. Vacuum lines should be protected with high efficiency particulate air (HEPA) filters and liquid disinfectant traps. The HEPA-filtered exhaust air from Class I or II biosafety cabinets may be discharged directly to the outside, or through the building exhaust system, or

423

be recirculated within the laboratory if the cabinet is appropriately certified and tested.

Maximum containment laboratory

These laboratories are Biosafety level 4 facilities. Maximum containment laboratories are designed to provide a safe environment for carrying out studies involving infectious agents that pose an extreme hazard to laboratory personnel, or may cause serious epidemic disease. These facilities have secondary barriers, including sealed openings into the laboratory, air locks, a double door autoclave, a separate ventilation system, a biowaste treatment system, and a room for clothing change and showers that adjoins the laboratory.

Biosafety level 4

This safety level is necessary for work with organisms that present a high individual risk of life-threatening disease. These facilities are usually located in an independent building, or in a separate, isolated, completely segregated, controlled area of a larger building. Access to the facility should be controlled by the use of locked doors. All personnel entering should sign a logbook, must enter and leave the facility through the clothing change and shower rooms, and must shower before exiting. Any supplies or materials that do not enter through the shower and change rooms must enter through a double door autoclave, fumigation chamber, or airlock that is decontaminated between each use. All organisms classified as Biosafety level 4 should be handled in Class III biosafety cabinets, or in Class I or II biosafety cabinets used in conjunction with one-piece positive pressure personnel suits ventilated by a life support system. All biological materials removed from a Class III cabinet, or the maximum containment laboratory in a viable condition, should be placed in a non-breakable, sealed primary container and enclosed in a secondary container that is removed through a disinfectant dunk tank, fumigation chamber, or airlock. All other materials must be autoclaved or decontaminated before removal from the facility. Walls, floors and ceilings of the facility together should form a sealed internal shell, with any windows resistant to breakage. Most importantly, the facility should be available for the quarantine isolation and treatment of personnel with potential or known laboratory-related illnesses.

11.5.3 Equipment design

The types of equipment commonly used in a bio-pharmaceutical facility will vary based not only on the types of processes and organisms used, but also on whether the equipment is used during development, during testing, or during manufacture of material for clinical trials and marketing. Types of equipment commonly used include bioreactors, air compressors, sterilization equipment, product recovery systems such as centrifuges and cell disrupters, waste recovery and decontamination equipment, sampling and analysis instruments, safety equipment such as biosafety cabinets and protective clothing, equipment for transporting biological materials such as sealed containers, and environmental control equipment.

Equipment capacity and location

As always, the equipment used in the manufacture, processing, packing or holding of the bio-pharmaceutical product or any of the process intermediates should be of appropriate design, adequate size and construction, and suitably located to facilitate operations for its intended use and for its cleaning and maintenance. Closed equipment should be used when feasible to provide adequate protection of the bulk-active and any intermediates, and always in the case of sterile products. When equipment is opened or open equipment is used, appropriate precautions should be taken to prevent contamination or cross-contamination of bulk active substance and intermediates. New equipment must be properly installed and operate as designed, and must be cleaned before use according to written procedures, with the cleaning procedures documented and validated.

Equipment construction and installation

Equipment should be constructed and installed, to enable easy cleaning, adjustments and maintenance. Equipment should be constructed so that surfaces that come into contact with raw materials, intermediates, bulk active substances or sterile products, are not reactive, additive, or absorptive, so as to alter the quality, purity, identity, or strength of the product beyond the established specifications. Similarly, any substances required for the operation of equipment, such as lubricants, heating fluids or coolants, should not contact raw materials, packaging materials, intermediates, or the bulk active, so as to alter its quality and purity beyond established specifications. If the equipment requires calibration, there must written procedures for calibrating the equipment and documenting the calibration. With filters, the type of filter, its purpose, how it is assembled, cleaned, and inspected for damage, and if a microbial retentive filter, methods used for integrity testing, should be specified. Qualification of equipment should ensure that it is installed according to approved design specifications, regulatory codes, and the equipment manufacturers' recommendations, and that it operates within the limits and tolerances established for the process.

Biosafety cabinets

These are common primary containment devices for work involving infectious organisms. Their primary function is to protect the laboratory worker and the immediate environment by containing any infectious aerosols produced during the manipulation of organisms within the cabinet. Biosafety cabinets are classified into three types (I, II and III) based on their performance character-istics. Class I and II cabinets are appropriate for use with moderate and high-risk micro-organisms. They have an inward face velocity of 75 linear feet per minute and their exhaust air is filtered by HEPA filters. They can be used with a full width open front, an installed front closure panel, or an installed front closure panel equipped with arm-length rubber gloves. The Class II cabinet is a vertical laminar-flow cabinet with an open front. In addition to the protection provided by the Class I cabinet, these cabinets also protect materials inside the cabinet from extraneous airborne contaminants since the HEPA filtered air is recirculated within the workspace. The Class III cabinet is a totally enclosed, ventilated, gas tight cabinet used for work with infectious organisms. Work in a Class III cabinet is conducted through connected rubber gloves. The cabinet is maintained under negative pressure with supply air drawn in through HEPA filters, and exhaust air filtered by two HEPA filters and discharged to outside the facility using an exhaust fan that is generally separate from the facility's overall exhaust fan. However, it is important to remember that each of the cabinet types is only protective if it is operated and maintained properly by trained personnel.

Organism preparation

Other commonly used laboratory equipment in a biotechnology laboratory or facility includes culture plates, roller bottles, shake flasks, and a seed fermenter. These are used to bring the organism or the cell line from its origination in the master cell bank through its preparation for growth and/or propagation.

Bioreactors or fermenters

Fermenters or bioreactors play a central role in biotechnological processes, with their main purpose being to grow and/or propagate a microorganism or a cell line in a controlled, aseptic environment. The most popular type is the mechanical fermenter, which uses mechanical stirrers to agitate the culture, and one of the most commonly used mechanical fermenters is the stirred tank reactor. In order to satisfy the metabolic requirements of the microorganism or the cell line, aeration must be adequate to provide sufficient oxygen, and those using agitation need to be designed to maintain a uniform environment within the bioreactor. Major attributes of a good bioreactor are that it should be economical, robust, of simple mechanical design, easy to operate under aseptic

426

conditions, of reasonably flexible design with respect to the various process requirements, with no dead zones giving good control to bulk flow, and have good heat and mass transfer.

The level of sophistication involved in the design of a fermenter is largely a function of the requirements of the process. Stainless steel is commonly chosen as the material of construction for the fermenter, as it can withstand repeated cycles of sterilization (121°C for at least 30 min) without breakage and has better heat transfer than glass. Other sterility considerations include smooth and crevice free welded joints; short, straight pipework with appropriate slopes to avoid accumulation of pockets of liquid during operation; all wetted internals polished to 180–200 grit finish, and all other materials used amenable to steam sterilization.

There should be adequate monitoring and control equipment to control the metabolic processes, by monitoring parameters such as pH, temperature, agitation, and aeration rates within the bioreactor. For off-line systems, a sample is taken from the bioreactor at specified intervals and chemically analysed using automated laboratory instruments — these can have a lengthy turnaround time for analytical results and do not provide a high level of containment. For on-line systems, sampling and analysis are done continuously, often requiring additional secondary containment. In-line or at-line systems, however, provide a continuous, non-invasive indication of bioreactor conditions, through the use of probes, sensors, and sampling devices that directly contact the material.

Temperature within the fermenter is maintained by circulating water at a controlled temperature through the jacket of the fermenter, which envelops the complete level of liquid in the shell. Baffle plates are provided inside the jacket for effective circulation of the cooling or heating medium in the jacket, with a drain port provided at the bottom for efficient removal of condensate at the end of sterilization, and a vent at the top of the jacket. Bioreactor aeration system is designed for supplying sterile moisture-free air rate at 0–3 vvm (volume of air per volume of liquid per minute), although an aeration rate of 0.2–0.3 vvm is commonly used. Medical air (compressed air) at 1.5 bar g, from which moisture and oil vapours are stripped, is supplied from an air compressor, passed through a pressure regulator, flowmeter and a steam sterilizable air filter to remove undesirable organisms and particles from the air. This sterile filtered air is sparged into the fermenter through the sparger, which usually consists of an open-ended stainless steel pipe discharging directly under the agitator. The fermenter requires a versatile agitation system to ensure optimal mixing at low shear. The agitator port is sealed, either with a double mechanical seal with a sterile condensate lubrication system, or a magnetically coupled seal system.

The seal assembly is selected primarily with consideration of the cell line used, the heavy wear and tear and the repeated sterilization cycle the system undergoes. The main elements of the agitation system consist of the baffles on the shell wall for breaking vortex during peak agitation and impellers with adjustable height on the vertical shaft.

Product recovery

A product recovery or purification system is required to separate and concentrate the desired product from the contents of the bioreactor. Such systems include centrifugation, cell disruption, broth conditioning, filtration, extraction, chromatography, and drying and freezing techniques — the type of equipment depending on the type(s) of product handled.

Centrifuges are used to separate viable cells from liquid culture broth and include batch-operated solid bowl machines, semi-continuous solids-discharging disc separators, or continuous decanter centrifuges. Batch centrifuges include the solid-bowl disc centrifuge, one-chamber centrifuges (used for protein fractionation from blood plasma), zonal centrifuges (used to separate intracellular and extra-cellular products such as in virus purification or cell constituent isolation), and tubular centrifuges (used to separate liquid phases). Biosafety cabinets must be used during solids removal from batch centrifuges. Semi-continuous solids-discharging machines generally provide the best containment and are the most widely used type for biotechnology applications. Filtration units are also used to separate cellular, intra-cellular or extra-cellular, solids from broth. Types of filtration units include continuous rotary drums, continuous rotary vacuum filters or tangential flow filtration systems using either microporous or ultrafiltration membrane filters. The type of filtration unit used depends on the type of product being recovered.

Cell disruption is used to recover intra-cellular products and can be performed using mechanical or non-mechanical methods. Mechanical methods include ball mills and high-speed homogenizers, whilst non-mechanical methods include chemical or enzymatic lysis, heat treatment, freeze-thaw or osmotic shock. Non-mechanical methods are easily contained and are most often used in biotechnology laboratories. Chromatography processes such as affinity or gel filtration are used to purify intra-cellular or extra-cellular products, using an eluting solvent in a packed column and collected in a fraction collector. If adequate containment is provided, such as a biological safety cabinet, product recovery using chromatography can be used to purify hazardous organisms. Other purification equipment includes centrifugal extractors (used for liquid–liquid extraction), spray packed, mechanically agitated, or pulsed columns. Either freezing or drying may be used to facilitate the handling and storage of products. Organisms to be frozen are placed in vials and frozen. The most common types of dryers used are freeze dryers and vacuum tray dryers, and since freezing provides primary containment and produces less aerosols than dryers, it is more appropriate for product storage. If drying is performed, proper filtration and ventilation systems must be provided.

Isolator technology

The use of isolator technology to minimize human interventions in processing areas usually results in a significant decrease in the risk of microbiological contamination of aseptically manufactured products from the environment. There are many possible designs of isolators and transfer devices. The isolator and the background environment should be designed so that the required air quality for the respective zones can be realized. The air classification required for the background environment depends on the design of the isolator and its application and for aseptic processing it should be at least grade D. In general, the area inside the isolator is the local zone for high-risk manipulations, although it is recognized that laminar airflow may not exist in the working zone of all such devices. The transfer of materials into and out of the unit is one of the greatest potential sources of contamination. Such transfer devices may vary from a single door to double door designs to fully sealed systems incorporating sterilization mechanisms. Isolators should be introduced only after appropriate validation. Validation should take into account all critical factors of isolator technology, such as the quality of the air inside and outside (background) the isolator, sanitization of the isolator, the transfer process and isolator integrity. Isolators are constructed of various materials more or less prone to puncture and leakage. Monitoring should be carried out routinely and should include frequent leak testing of the isolator and glove/sleeve system.

Computer and related automatic and electronic systems

These are used in the control of critical manufacturing steps in bio-pharmaceutical manufacture. They should be appropriately qualified and validated to demonstrate the suitability of the hardware and software, to perform assigned tasks in a consistent and reproducible manner. The depth and scope of the validation programme would depend on the diversity, complexity and criticality of the system. All changes should be approved in advance and performed by authorized and competent personnel, and records kept of all changes, including modifications and enhancements to the hardware, software and any other critical components of the system, to demonstrate that the modified system is maintained in a validated state.

Appropriate controls over computer or related automatic and electronic systems should be exercised to ensure that only authorized personnel make changes in master production and control records. Procedures should be established to prevent unauthorized entries or changes to existing data. Systems should identify and document the persons entering or verifying critical data. Input to and output from the computer or related system should be checked for accuracy at appropriate intervals and where critical data are entered manually, there should be an additional check on the accuracy of the entry. This may be performed by a second operator, or by the system itself.

A back-up system should be available to respond to system breakdowns or failures that result in permanent loss of critical records. Back-ups may consist of hard copies or other forms, such as tapes or microfilm, that ensure back-up data are exact, complete and secure from alteration, inadvertent erasure or loss. The current regulations also require that a 'back-up file of data entered into the computer or related system shall be maintained except where certain data, such as calculations performed in connection with laboratory analysis, are eliminated by computerization or other automated processes'. If computerization or another automated process has eliminated such calculations 'then a written record of the programme shall be maintained along with data establishing proper performance' emphasizing that the manufacturer must actually establish proper performance.

Regulatory authorities require additional information to be available for pre-approval inspection. The information provided should include a brief description of procedures for changes to the computer system. For each of the systems, a list of the manufacturing steps that are computer-controlled should be provided, together with the identity of the system's developer (i.e. developed in-house or by an external contractor). The validation summary should include:

- a narrative description of the validation process (or protocol), including acceptance criteria;
- certification that IQ and OQ have been completed;
- an explanation of the parameters monitored and tests performed;
- a validation data summary;
- an explanation of all excursions or failures;
- deviation reports and results of investigations for all excursions or failures.

11.5.4 Sterilization methods

All the equipment used in the processing of bio-pharmaceuticals should be capable of being sterilized and maintaining sterility. Sanitization rather than sterilization of critical equipment such as crystallizers, centrifuges, filters, spray and freeze dryers is totally unacceptable. All sterilization processes should be validated, with particular attention given when the adopted sterilization method is not described in the current edition of the Pharmacopoeia, or when it is used for a product that is not a simple aqueous or oily solution. Where possible, heat sterilization is the method of choice.

Biological indicators

If biological indicators are used, strict precautions should be taken to avoid transferring microbial contamination from them. In some cases, testing of biological indicators may become all or part of the sterility testing. Various types of indicators are used as an additional method for monitoring the sterilization and assuring sterility, including lag thermometers, peak controls, Steam Klox, test cultures and biological indicators. Biological indicators are of two forms, each of which incorporates a viable culture of a single species of microorganism. In one form, the culture is added to representative units of the lot to be sterilized, or to a simulated product that offers no less resistance to sterilization than the product to be sterilized. In the second form, the culture is added to disks or strips of filter paper, metal, glass or plastic beads, and used when the first form is not practical, as is the case with solids. If using indicators, there should be assurances that the organisms are handled so they do not contaminate the manufacturing area or the product, and they should be stored and used according to the manufacturer's instructions, and their quality checked by positive controls.

Sterilization by moist heat

The method of choice for the sterilization of equipment and transfer lines is saturated clean steam under pressure. In the validation of the sterilization of equipment and transfer systems, temperature sensors and biological indicators should be strategically located in cold spots where condensate may accumulate, such as the point of steam injection and steam discharge, and in low spots such as the exhaust line. Steam must expel all the air from the sterilizer chamber to eliminate cold spots, and from the drain lines connected to the sewer by means of an air break to prevent back siphoning. After the high temperature phase of a heat sterilization cycle, precautions should be taken against contamination of a sterilized load during cooling. There should be frequent leak tests on the chamber when a vacuum phase is part of the cycle. One manufacturer utilized a steam-in-place system, but only monitored the temperature at the point of discharge and not in low spots in the system where condensate accumulated and caused problems. Care should be taken to ensure that steam used for sterilization is of suitable quality and does not contain additives at a level that could cause contamination of product or equipment. Any cooling fluid or gas in contact with the product should be sterilized unless it can be shown that any leaking container would not be approved for use.

Both temperature and pressure should be used to monitor the process. Control instrumentation should normally be independent of monitoring instrumentation and recording charts. Where automated control and monitoring systems are used, they should be validated to ensure that critical process requirements are met. Each heat sterilization cycle should be recorded on a time/temperature chart with a sufficiently large scale, or by other appropriate equipment with suitable accuracy and precision. The position of the temperature probes used for controlling and recording should be determined during the validation, and where applicable checked against a second independent temperature probe located at the same position. Chemical or biological indicators may also be used, but should not take the place of physical measurements. The time required to heat the centre of the largest container to the desired temperature must be known, and sufficient time must be allowed for the whole of the load to reach the required temperature before measurement of the sterilizing time-period is commenced. Charts of time, temperature and pressure should be filed for each sterilizer load. The items to be sterilized, other than products in sealed containers, should be wrapped in a material which allows removal of air and penetration of steam but which prevents recontamination after sterilization.

Sterilization by dry heat
There are some manufacturers who sterilize processed bulk bio-pharmaceutical powders by the use of dry heat. As a primary means of sterilization, its usefulness is questionable because of the lack of assurance of penetration into the crystal core of a sterile powder, although some sterile bulk powders can withstand the lengthy times and high temperatures necessary for dry heat sterilization. Process validation should cover aspects of heat penetration and heat distribution, times, temperatures, stability (in relation to the amount of heat received) and particulates. Any air admitted to maintain a positive pressure within the chamber should be passed through a HEPA filter. Where this process is also intended to remove pyrogens, challenge tests using endotoxins should be used as part of the validation.

Sterilization by radiation
Radiation sterilization is used mainly for the sterilization of heat sensitive materials and products, although ultra-violet irradiation is not normally an

acceptable method of sterilization. Many medicinal products and some packaging materials are radiation-sensitive, so this method is permissible only when the absence of deleterious effects on the product has been confirmed experimentally. Validation procedures should ensure that the effects of variations in density of the packages are considered, and biological indicators may be used as an additional control. Materials handling procedures such as the use of radiation sensitive colour disks should also be used on each package to differentiate between irradiated and non-irradiated materials and prevent mix-ups. During the sterilization procedure the radiation dose should be measured, and the total radiation dose should be administered within a predetermined time span. For this purpose, dosimetry indicators that are independent of dose rate should be used, giving a quantitative measurement of the dose received by the product itself. These should be inserted in the load in sufficient numbers and close enough together to ensure that there is always a dosimeter in the irradiator. Where plastic dosimeters are used they should be used within the time limit of their calibration, and dosimeter absorbances should be read within a short period after exposure to radiation.

Sterilization with ethylene oxide
There are some manufacturers who still use ethylene oxide for the surface sterilization of powders as a precaution against potential microbiological contamination during aseptic handling, even though a substantial part of the sterile pharmaceutical industry has discontinued its use as a sterilizing agent. Its use is now in decline because of residual ethylene oxide in the product and the inability to validate ethylene oxide sterilization, as well as employee safety considerations. As a primary means of sterilization, its use is questionable because of the lack of assurance of penetration into the crystal core of a sterile powder, and therefore, this method should only be used when no other method is practicable. Process validation should show that there is no damaging effect on the product and that the conditions and time allowed for degassing are such as to reduce any residual gas and reaction products to acceptable limits for the type of product or material. The nature and quantity of packaging materials can significantly affect the process, so materials should be pre-conditioned by being brought into equilibrium with the humidity and temperature required by the process before exposure to the gas. The time required for this should be balanced against the opposing need to minimize the time before sterilization. For each sterilization cycle, records should be made of the time taken to complete the cycle, of the pressure, temperature and humidity within the chamber during the process, the gas concentration, and the total amount of gas used. After sterilization, the load should be stored in a controlled manner under

ventilated conditions to allow residual gas and reaction products to reduce to the defined level.

Sterilization with formaldehyde
The use of formaldehyde is a much less desirable method of equipment sterilization. A major problem with formaldehyde is its removal from pipework and surfaces and it is rarely used primarily because of residue levels in both the environment and the product. Since formaldehyde contamination in a system or in a product is not going to be uniform, merely testing the product as a means of demonstrating and validating the absence of formaldehyde levels is not acceptable; there should be some direct measure, or determination of the absence of formaldehyde. Key surfaces should be sampled directly for residual formaldehyde. One large pharmaceutical manufacturer had to reject the initial batches coming through the system because of formaldehyde contamination. Unfortunately, they relied on end product testing of the product, and not on direct sampling to determine the absence of formaldehyde residues on equipment.

Sterilization In Place (SIP)
SIP systems require considerable maintenance, and their malfunction has directly led to considerable product contamination and recall. One potential problem with SIP systems is condensate removal from the environment. Condensate and excessive moisture can result in increased humidity, and increases in levels of microorganisms on surfaces of equipment. Therefore, environmental monitoring after sterilization of the system is particularly important. Another potential problem is the corrosive nature of the sterilant, whether it is clean steam, formaldehyde, peroxide or ethylene oxide. In two recent cases, inadequate operating procedures have led to weld failures. Therefore, particular attention should be given to equipment maintenance logs, especially to non-scheduled equipment maintenance, and the possible impact on product quality. Suspect batches manufactured and released prior to the repair of the equipment should be identified.

11.5.5 Cleaning procedures and validation
Regulatory authorities requiring that all equipment and facilities be clean prior to use and be maintained in a clean and orderly manner, are nothing new. Of course, the main rationale for requiring clean equipment and facility is to prevent contamination or adulteration of medicinal products. Historically, authorities have looked for gross insanitation due to inadequate cleaning and maintenance of equipment and/or poor dust control systems, and were more

concerned about the contamination of non-penicillin drug products with penicillins, or the cross-contamination of drug products with potent steroids or hormones. Certainly, a number of products have been recalled over the past decade due to actual or potential penicillin cross-contamination.

Rationale and procedures

Cleaning, and its validation, including facility disinfection, personnel control and equipment cleaning, has recently come under increasing scrutiny. Numerous regulatory actions and comments have been issued, resulting in many questions regarding the selection, use, testing, documentation and validation of cGMP sanitation programmes. Regulatory authorities now expect manufacturers to have written procedures detailing the cleaning processes used for various pieces of equipment. If manufacturers have only one cleaning process for cleaning between different batches of the same product, and use a different process for cleaning between product changes, then the written procedures should address these different scenarios. Similarly, if manufacturers have one process for removing water-soluble residues and another process for non-water soluble residues, the written procedure should address both scenarios and make it clear when a given procedure would be followed. Some manufacturers may decide to dedicate certain equipment for certain process steps that produce residues that are difficult to remove from the equipment. Any residues from the cleaning process itself, such as detergents and solvents, also have to be removed from the equipment.

Equipment should be cleaned, held and, where necessary, sanitized at appropriate intervals to prevent contamination or cross-contamination that would alter the quality or purity of the product beyond the established specifications. Even dedicated equipment should be cleaned at appropriate intervals to prevent the build-up of objectionable material or microbial growth. As processing approaches the purified bulk active substance, it becomes important to ensure that incidental carry-over of contaminants or degradants between batches does not adversely impact the established impurity profile. However, this does not always apply to a bio-pharmaceutical, where many of the processing steps are accomplished aseptically, and where it is often necessary to clean and sterilize equipment between batches. Non-dedicated equipment should be thoroughly cleaned between different products and, if necessary, after each use. If cleaning a specific type of equipment is difficult, the equipment may need to be dedicated to a particular bulk active substance or intermediate. Moreover, because the potency of some of these materials may not be fully known, cleaning becomes particularly important.

The microbiological aspects of equipment cleaning consist largely of preventive measures rather than removal of contamination once it has occurred. There should be some evidence that routine cleaning and storage of equipment does not allow microbial proliferation. For example, equipment should be dried before storage, and under no circumstances should stagnant water be allowed to remain in equipment. Subsequent to the cleaning process, equipment should be sterilized or sanitized where such equipment is used for sterile processing, or for non-sterile processing where the products may support microbial growth. Thus, the control of the bioburden through adequate cleaning and storage of equipment is important to ensure that subsequent sterilization or sanitization procedures achieve the necessary assurance of sterility. This is also particularly important from the standpoint of the control of pyrogens in sterile processing, since equipment sterilization processes may not be adequate to achieve significant inactivation or removal of pyrogens.

In sterile secondary production areas, all the equipment, fittings and services, as far as is practicable, should be designed and installed so that operations, maintenance and repairs can be carried out outside the clean area. If sterilization is required, it should be carried out after complete re-assembly wherever possible. The practice of re-sterilizing equipment if sterility has been compromised is important. When equipment maintenance has been carried out within the clean area, the area should be cleaned, disinfected and/or sterilized where appropriate before processing recommences if the required standards of cleanliness and/or asepsis have not been maintained during the work. A conveyor belt should not pass through a partition between a grade A or B area and a processing area of lower air cleanliness unless the belt itself is continually sterilized (for example, in a sterilizing tunnel).

Equipment must be clearly identified as to its cleaning status and content. The cleaning and maintenance of the equipment should be documented in a logbook maintained in the immediate area. Establishing and controlling the maximum length of time between the completion of processing and each cleaning step is often critical in a cleaning process. This is especially important for operations where the drying of residues will directly affect the efficiency of a cleaning process. In all cases, the choice of cleaning methods, cleaning agents and levels of cleaning should be established and justified. When selecting cleaning agents, the following should be considered:

- the cleaning agent's ability to remove residues of raw materials, precursors, by-products, intermediates, or even the bulk active substance;
- whether the cleaning agent leaves a residue itself;
- compatibility with equipment construction materials.

Validation of cleaning methods

Validation of cleaning procedures has generated considerable discussion since the regulatory authorities started to address this issue. The first step is to focus on the objective of the validation process, and some manufacturers fail to develop such objectives. It is not unusual to see manufacturers use extensive sampling and testing programmes following the cleaning process without really evaluating the effectiveness of the steps used to clean the equipment. Several questions need to be addressed when evaluating the cleaning process. For example, at what point does a piece of equipment or system become clean? Does it have to be scrubbed by hand? What is accomplished by hand scrubbing rather than just a solvent wash? How variable are manual cleaning processes from batch to batch and product to product? What other methods for cleaning can be utilized — wipe clean, spray, fog, immersion, ultrasonic, re-circulating spray? Is the contamination viable or non-viable? Are there identifiable baseline bioburden and residue levels? The answers to these questions are obviously important to the inspection and evaluation of the cleaning process, and to determine the overall effectiveness of the process. They may also identify steps that can be eliminated for more effective measures and result in resource savings for the manufacturer.

In general, cleaning validation efforts should be directed to situations or process step where contamination or incidental carry-over of degradants poses the greatest risk to the product's quality and safety. The manufacturer should have determined the degree of effectiveness of the cleaning procedure for each bio-pharmaceutical or intermediate used in that particular piece of equipment. In the early stages of the operation, it may be unnecessary to validate cleaning methods if it could be shown that subsequent purification steps can remove any remaining residues. It must be recognized that for cleaning, as with any other processes, there may be more than one way to validate the process. In the end, the test of any validation process is whether the scientific data shows that the system consistently does as expected and produces a result that consistently meets pre-determined specifications. Moreover, cleaning should also be shown to remove endotoxins, bacteria, active elements and contaminating proteins, while not adversely affecting the performance of the equipment. In cases where cleaning reagents are required for decontamination or inactivation, validation should also demonstrate the effectiveness of the decontamination/inactivating agent(s).

Validation of cleaning methods should, therefore, reflect the actual equipment use patterns. For example, if various bulk actives or intermediates are manufactured using the same equipment, and if the same process is used to clean the equipment, a worst-case bulk active or intermediate can be selected for the purposes of cleaning validation. The worst-case selection should be based on a combination of potency, activity, solubility, stability and difficulty of cleaning. In addition, such cleaning and sanitization studies should address microbiological and endotoxin contamination for those processes intended or purported to reduce bioburden or endotoxins in the bulk active substance or other processes where such contamination may be of concern, for example with non-sterile substances used to manufacture parenteral products.

Documentation

Depending upon the complexity of the system and the cleaning process, and the ability and training of operators, the amount of detail and specificity in the documentation necessary for executing various cleaning steps or procedures will vary. Some manufacturers use general SOPs, while others use a batch record or log sheet system that requires some type of specific documentation for performing each step. When more complex cleaning procedures are required, it is important to document the critical cleaning steps, including specific documentation on the equipment itself and information about who cleaned it and when. However, for relatively simple cleaning operations, the mere documentation that the overall cleaning process was performed might be sufficient. Other factors such as history of cleaning, residue levels found after cleaning and variability of test results may also dictate the amount of documentation required. For example, when variable residue levels are detected following cleaning, particularly for a process that is believed to be acceptable, the manufacturer must establish the effectiveness of the process and operator performance.

Protocols

Cleaning validation protocols should have general procedures on how cleaning processes will be validated. It must describe the equipment to be cleaned; methods, materials and extent of cleaning; parameters to be monitored and controlled; and validated analytical methods to be used. The protocol should also indicate the type of samples (rinse, swabs) to be obtained, and how they are collected, labelled and transported to the analysing laboratory. Validation procedures should address who is responsible for performing and approving the validation study, the acceptance criteria and when re-validation will be required. Validation studies should be conducted in accordance with the protocols, and the results of the studies documented. There should be a detailed written equipment cleaning procedure that provides details of what should be done and the materials to be utilized. Some manufacturers list the specific solvent for each bio-pharmaceutical and intermediate. For stationary vessels,

Clean in Place (CIP) apparatus is often encountered. Diagrams, along with identification of specific valves, will be necessary for evaluating these systems.

Sampling

After cleaning, there should be some routine testing to assure that the surface has been cleaned to the validated level, and to ensure these procedures remain effective when used during routine production. Where feasible, equipment should be examined visually for cleanliness. This may allow detection of gross contamination concentrated in small areas that could go undetected by analytical verification methods. Sampling should include swabbing, rinsing, or alternative methods such as direct extraction, as appropriate, to detect both insoluble and soluble residues. The sampling methods used should be capable of quantitatively measuring levels of residues remaining on the equipment surfaces after cleaning. There are two general types of sampling that have been found acceptable — the most desirable is the direct method of sampling the equipment surface, and the other is the use of rinse solutions.

Direct surface sampling

The advantages of direct sampling are that areas hardest to clean, but which are reasonably accessible, can be evaluated, leading to the establishment of a level of contamination or residue per given surface area. Additionally, residues that are dried out, or are insoluble, can be sampled by physical removal. Swab sampling may be impractical when product contact surfaces are not easily accessible due to equipment design and/or process limitations, such as the inner surfaces of hoses, transfer pipes, reactor tanks with small ports or handling active materials, and small intricate equipment such as micronizers and micro-fluidizers. One major concern is the type of sampling material used and its impact on the test data, since the sampling material may interfere with the test. For example, the adhesive used in swabs has been found to interfere with the analysis of samples. Therefore, it is important to assure early in the validation programme that the sampling medium and the solvent used for extraction from the medium are satisfactory and can be readily used.

Rinse samples

This is the analysis of the final rinse water or solvent for the presence of the cleaning agents last used in that piece of equipment. Two advantages of using rinse samples are that a larger surface area may be sampled, and inaccessible systems or ones that cannot be routinely disassembled can be sampled and evaluated. However, the disadvantage of rinse samples is that the residue or contaminant may not be soluble or may be physically occluded in the equipment. An analogy that can be used is the dirty pot — in the evaluation of cleaning of a dirty pot, particularly with dried out residue, one does not look at the rinse water to see that it is clean; one looks at the pot. A direct measurement of the residue or contaminant should be made for the rinse water when it is used to validate the cleaning process. For example, it is not acceptable to simply test rinse water for water quality (does it meet the compendia tests?), rather than test it for potential contaminates. In addition, indirect monitoring such as conductivity testing may be of some value for routine monitoring once the cleaning process has been validated. This would be particularly true, where reactors and centrifuges and pipework between such large equipment can only be sampled using rinse solution samples.

Analytical methods and establishment of limits

How do you evaluate and select analytical methods to measure cleaning and disinfection effectiveness in order to implement basic cleaning validation and to establish routine in-use controls. Regulatory authorities do not set acceptance specifications or methods for determining whether a cleaning process is validated because it is impractical for them to do so due to the wide variation in equipment and products used throughout the industry.

With advances in analytical technology, residues from the manufacturing and cleaning processes can be detected at very low levels. The sensitivity of some modern analytical apparatus has lowered some detection thresholds to below parts per million (ppm) down to parts per billion (ppb). Some limits that have been mentioned by industry representatives in literature or presentations, include analytical detection levels such as 10 ppm, biological activity levels such as 1/1000 of the normal therapeutic dose, and organoleptic levels such as no visible residue. The residue limits established for each piece of apparatus should, therefore, be practical, achievable and verifiable. If levels of contamination or residual are not detected, it does not mean that there is no residual contaminant present after cleaning; it only means that levels of contaminant greater than the sensitivity or detection limit of the analytical method are not present in the sample. The manufacturer's rationale for establishing specific residue limits should be logical, based on their knowledge of the materials involved, be practical, achievable and verifiable, have a scientifically sound basis, and be based on the most deleterious residue. Limits may, therefore, be established, based on the minimum known pharmacological or physiological activity of the product or its most deleterious component.

Another factor to consider is the possible non-uniform distribution of the residue on a piece of equipment. The actual average residue concentration may be more than the level detected. It may not be possible to remove absolutely every trace of material, even with a reasonable number of cleaning cycles. The permissible residue level, generally expressed in parts per million (ppm), should be justified by the manufacturer. The manufacturer should also challenge the analytical method in combination with the sampling method(s) used, to show that the contaminants can be recovered from the equipment surface, and at what levels, i.e. 50% or 90% recovery. This is necessary before any conclusions can be made based on the sample results. A negative test may also be the result of poor sampling technique.

Clean In Place methods

Where feasible, Clean In Place (CIP) methods should be used to clean process equipment and storage vessels. CIP methods might include fill and soak/agitate systems, solvent refluxing, high-impact spray cleaning, spray cleaning by sheeting action, or turbulent flow systems. CIP systems should be subjected to cleaning validation studies to ensure that they provide consistent and reproducible results, and once they are validated, appropriate documentation should be maintained to show that critical parameters, such as time, temperature, turbulence, cleaning agent concentration, rinse cycles, are achieved with each cleaning cycle. However, the design of the equipment, particularly in facilities that employ semi-automatic or fully automatic Clean In Place (CIP) systems, can represent a significant concern. For example, sanitary type pipework without ball valves should be used, since non-sanitary ball valves make the cleaning process more difficult. Such difficult to clean systems should be properly identified and validated, and it is important that operators performing these cleaning operations are aware of potential problems and are specially trained in cleaning these systems and valves. Furthermore, with systems that employ long transfer lines or pipework, clearly written procedures together with flow charts and pipework diagrams for the identification of valves should be in place. Pipework and valves should be tagged and easily identifiable by the operator performing the cleaning function. Sometimes, inadequately identified valves, both on diagrams and physically, have led to incorrect cleaning practices. Equipment in CIP systems should be disassembled during cleaning validation where practical to facilitate inspection and sampling of inner product surfaces for residues or contamination, even though the equipment is not normally disassembled during routine use.

Test until clean

Some manufacturers are known to test, re-sample and re-test equipment or systems until an 'acceptable' residue level is attained. For the system or equipment with a validated cleaning process, this practice of re-sampling should not be utilized and is only acceptable in rare cases. Constant re-testing and re-sampling can show that the cleaning process is not validated, since these re-tests actually document the presence of unacceptable residue and contaminants from an ineffective cleaning process. The level of testing and the re-test results should, therefore, be routinely evaluated.

Detergent
The manufacturer must consider and determine the difficulty that may arise when attempting to test for residues if a detergent or soap is used for cleaning. A common problem associated with detergent use is its composition — many detergent suppliers will not provide specific composition, making it difficult for the user to evaluate residues. As with product residues, it is important that the manufacturer evaluate the efficiency of the cleaning process for the removal of residues from the detergents. However, unlike product residues, it is expected that no (or for ultra sensitive analytical test methods — very low) detergent remains after cleaning. Detergents are not part of the manufacturing process and are only added to facilitate cleaning during the cleaning process, so they should be easily removable or a different detergent should be selected.

11.6 Process utilities and services

11.6.1 Water systems
Water is a very important component of bio-pharmaceutical processes. Water of suitable quality is required depending on the culture system used, the phase of manufacture and the intended use of the product. Tighter chemical and microbiological quality specifications are required during certain process steps such as cell culture, final crystallization and isolation, and during early process steps if impurities that affect product quality are present in the water and cannot be removed later. Where water is treated to achieve an established quality, the treatment process and associated distribution systems should be qualified, validated, maintained and routinely tested following established procedures to ensure water of the desired quality. The water used should meet the standards for potable water as a minimum for the production of bio-pharmaceuticals.

The potable water supply, regardless of source, should be assessed for chemicals that may affect the process, and information should be periodically sought from local authorities about potential contamination by pesticides or other hazardous chemicals. For example, if water is used for a final wash of a filter cake, or if the bulk active substance is crystallized from an aqueous system, then the water should be suitably treated, such as by de-ionization, ultrafiltration, reverse osmosis or distillation, and tested to ensure routine compliance with appropriate chemical and microbiological specifications. If the water is used for final rinses during equipment cleaning, then the water should be of the same quality as that used in the manufacturing process. Water used in the final isolation and purification steps of non-sterile bulk actives intended for use in the preparation of parenteral products should be tested and controlled for bioburden and endotoxins.

The quality of water, therefore, depends on the intended use of the finished product. For example, only Water for Injection (WFI) quality water should be utilized as process water; this is because, even though water may not be a component of the final sterile product, water that comes in contact with the equipment or that enters into the bioreactor can be a source of impurities such as endotoxins. On the other hand, for in-vitro diagnostics purified water may suffice. For heat-sensitive products where processing such as formulation is carried out cold or at room temperature, only cold WFI will suffice, and the self-sanitization of a hot WFI system at 75° to 80°C is lost. As with other WFI systems, if cold WFI water is needed, point-of-use heat exchangers can be used; however, these cold systems are still prone to contamination, and should be fully validated and routinely monitored both for endotoxins and microorganisms.

Water treatment plants and distribution systems should be designed, constructed and maintained to ensure a reliable source of water of an appropriate quality. They should never be operated beyond their designed capacity. For economic reasons, some biotechnology companies manufacture WFI utilizing marginal systems, such as single pass reverse osmosis, rather than by distillation. Many such systems have been found to be contaminated, typically because they use plastic pipes and non-sealed storage tanks, which are difficult to sanitize. Although some of the systems employ a terminal sterilizing filter to minimize microbiological contamination, the primary concern is endotoxins which the terminal filter may merely serve to mask. Such systems are, therefore, totally unacceptable. Moreover, the limitations of relying on a 0.1 ml sample of WFI for endotoxins from a system should also be recognized.

New water quality requirements were brought into effect in 1996. These updated requirements provide major cost savings to those manufacturers who needed to produce and maintain pure water systems, and allowed for the continuous monitoring of water systems with a reliance on instrumentation rather than laboratory work, thereby reducing labour and operating costs.

Previous standards required a battery of expensive and labour intensive chemical, physical, and microbiological testing, many of which only provided qualitative information. Advances in technology and instrumentation mean that simple, cost effective replacements have become available. However, before changing to the new testing standards, manufacturers should evaluate their existing water system in terms of compliance with existing operations, reliability, maintenance and improved monitoring.

11.6.2 Medical air

Medical air is a natural or synthetic mixture of gases consisting largely of nitrogen and oxygen, containing no less than 19.5 percent and not more than 23.5 percent by volume of oxygen. Air supplied to a non-sterile preparation or formulation area, or for manufacturing solutions prior to sterilization, should be filtered at the point of use as necessary to control particulates. However, air supplied to product exposure areas, where sterile bio-pharmaceuticals are processed and handled, should be filtered under positive pressure through high efficiency particulate air (HEPA) filters. These HEPA filters should be certified and/or Dioctyl Phthalate tested. Tests for oil (none discernible by the mirror test), odour (no appreciable odour), carbon dioxide (not more than 0.05%), carbon monoxide (not more than 0.001%), nitric oxide and nitrogen dioxide (not more than 2.5 ppm), and for sulphur dioxide (not more than 5 ppm) should also be carried out. Medical air is packaged in cylinders or in a low pressure collecting tank. Containers used should not be treated with any active, sleep-inducing, or narcosis-producing compounds, and should not be treated with any compound that would be irritating to the respiratory tract. Where it is piped directly from the collecting tank to the point of use, each outlet should be labelled Medical Air.

11.6.3 Heating, ventilation and air conditioning (HVAC) systems

A bio-pharmaceutical facility should have proper ventilation, air filtration, air heating and cooling. Therefore, adequate ventilation should be provided where necessary, and equipment for the control and monitoring of air pressure, microorganisms, dust, humidity and temperature should be provided when appropriate. This is especially important in areas where the product is exposed to the environment or handled in the final state. Air filtration, dust collection and exhaust systems should be used when appropriate, and if the air is recirculated, appropriate measures should be taken to control contamination and cross-contamination. For example, air from pre-viral inactivation areas should not be recirculated to other areas used for the manufacture of the sterile bio-pharmaceuticals. Regulatory authorities require the following information to be available for pre-approval inspection:

- A general description of the HVAC system(s) including the number and segregation of the air handling units, whether air is once-through or recirculated, containment features, and information on the number of air changes per hour;
- Validation summary for the system with a narrative description of the validation process (or protocol), including the acceptance criteria; the certification that IQ, OQ, and certification of filters has been completed; the length of the validation period; validation data should include Performance Qualification data accumulated during actual processing; and an explanation of all excursions or failures, including deviation reports and results of investigations;
- A narrative description of the routine monitoring programme including the tests performed and frequencies of testing for viable and non-viable particulate monitoring parameters; viable and non-viable particulate action and alert limits for production operations for each manufacturing area; and a summary of corrective actions taken when limits are exceeded.

11.6.4 Decontamination techniques and waste recovery

Air and gaseous waste streams

Filtration
The primary method of decontaminating exhaust gases mixed with liquid broth is through the use of filters. Before filtration, the mixture may be passed through a condenser, a coalescing filter and a heat exchanger. Filtration is accomplished either through pairs of high efficiency particulate air (HEPA) filters, or membrane filters used in series to decontaminate vent or exhaust gases.

Incineration
Another method of decontaminating air and gaseous waste streams is thermal destruction or incineration. Incineration may be used independently, or as a supplement to filtration, and is generally used for small volume gas streams. Automatic safety devices should be used with incinerators to protect against problems resulting from power failures and overheating.

Irradiation

Irradiation involves exposing the waste materials to x-rays, ultraviolet rays or other ionizing radiation to decontaminate them.

Liquid wastes

Liquid wastes can be decontaminated through chemical or heat treatment. When liquid wastes are of limited volume, chemical treatment is often used, whilst for large volumes of liquid wastes, heat treatment is generally preferred. Also, since proteins present in liquid wastes can deactivate the sterilant used in chemical treatment, thermal sterilization may be more appropriate for wastes involving bioengineered microorganisms.

Solid wastes

Solid wastes such as microbial cultures, cell debris, glassware, and protective clothing, are generally decontaminated by autoclaving, followed by incineration if necessary. To decontaminate laboratory devices exposed to genetically engineered products, the most common practice is the use of pressurized steam that contains an appropriate chemical. For heat-sensitive equipment, such as electronic instruments, decontamination is generally achieved through chemical sterilization or irradiation. Gaseous sterilants are applied by a steam ejector that sprays down from overhead. If decontamination by steam, liquid, or gas sterilization is not possible, ionizing or ultraviolet radiation is used. However, since irradiation methods do not always inactivate all types of microbes, steam or gaseous chemical sterilization should be used for devices contaminated with genetically engineered organisms.

Index